The Origins and Possibilities of Genetics Illustrated

A Mathematical Exploration in a Cosmology of Light

Pravir Malik, Ph.D.

Illustrated by Narendra Joshi, Ph.D.

ISBN-13: 979-8-9896169-3-0

Published by *Possibilities Publishing*

If there is a *Cosmology of Light Shankha* then imagine this book as contributing to the rhythm and sound that reveals something of the immense, largely untouched quantum computational possibility, that has yet to be wielded by practical technology.

Author's Note and Introduction

The Cosmology of Light book series explores a mathematical structure of existence based on infinite possibility. The infinite possibility is suggested as being the result of light imagined existing in a native state in which it travels infinitely fast. Properties of light traveling infinitely fast are those of omnipresence, omniscience, omnipotence, and omninurturance, and are derived by imagining a large sphere with a light source at the center. This light, being able to travel infinitely fast will be instantaneously present – omnipresent - across the entire sphere. If anything other than the nature of light arises it will sooner or later be overpowered by light – it will be omnipotent. Since anything that arises or disappears will do so in the all-present fabric of light, light will have an instantaneous knowledge – omniscience – of all that is happening in the sphere. And since everything is fundamentally connected in the nature of light, it will have an implicit property of omninurturance or omniharmony. In other words, being omnipresent, omniscient, omnipotent, omniharmonious, all information, all potential information, and all possibility is contained within Light, and this implicit codification of information in Light itself is the *origin of Genetics*.

Note that the altering of the speed of light to different constant speeds may be viewed as a device to create different mathematical spaces. If these mathematical spaces can then offer greater insight into phenomena, then whether the speed of light can actually exist as suggested here, is irrelevant.

As light is imagined projecting itself at slower speeds down to c, 186,000 miles per second in vacuum, it creates different realities that exist simultaneously such that the realities created when light travels slower are contained

14

within the realities already created where light travels faster. While there could be an infinite number of projected layers of Light, through logic of quantization these layers resolve themselves into a finite and distinct set that each embodies something fundamental to the creation of material existence. This gives rise to a quaternary-based – aka omnipresence, omnipotence, omniscience, omniharmony - multi-layered mathematical structure in which the unity of light traveling infinitely fast separates in an increasing display of distinct functionality to create the reality of infinite diversity we are familiar with in this reality where light travels at c. This happens through a process of precipitation that will be adequately explored in this book. This process of precipitation can also be perceived in the Light-Matrix as a "subtle" backbone or strand not unlike the backbone or downward-strand of the double helix DNA structure that animates every living cell. In other words, the *very basis of genetic structure* is potentially tied to this process of precipitation.

Further, as light slows down the codification implicit in its native state begins to further materialize and generates "libraries" of information as it were, that are accessible to all subsequent layers created by "slower" light. These libraries are the *basis of genetic information* and *heredity*.

The second strand of this subtle double helix DNA structure can be perceived as arising in the Space-Matrix as a function of time. As will be discovered the dynamics of the Space-Matrix are intimately connected with the realities set up by the Light-Matrix. The infinite variety of material existence exists as numerous materialized "ecosystems" bounded by these two strands. The very structure of chromosomes that we are familiar with, and within these of genes, articulating information through four repeating nucleotide bases, may be seen as similar to the quaternary-base structure that prescribes the logic of the subtle-ecosystems just behind the reality so set up by light traveling at c.

Further, the process of *mutation*, involving also the subtle-ecosystems, will be seen to vary depending on which layer of light is invoked. Massively significant mutations such as took place on the SRY gene and led to the evolution of human from the chimpanzee species are architected by a mathematical space generated when a layer of light is imagined traveling faster than c. The very process by which such significant mutation can take place will be articulated by the quaternary-based mathematical structure explored in this book. By contrast degenerative mutation leading to dysfunction and disease are linked to the layer of reality so set up by light imagined existing at zero speed.

The progressive creation of the material universe itself will be seen to be dependent on these antecedent subtle-libraries of genetic-type information. It will be proposed that in the adventure of material existence some initially simple ecosystem of genetic-type logic, perhaps describing the functioning of a space-time-gravity-energy quadrumvirate, or of the electromagnetic spectrum, was projected into the layer defined by light traveling at c. As conditions within that relatively simple reality ripened, an organic interaction between the layers of light initiated by the layer at c was put in place, and thus began the increasingly bold adventure of materialization of light we now find ourselves in. In other words, all evolution is of the nature of genetic-type evolution, with the material possibilities being endless.

Of necessity any discussion on genetics will have to explore the reality of information, as it imagined existing in the mathematical spaces generated by layers of light antecedent to matter. The notion of matter itself will have to become more sophisticated to admit of the reality by which its possibilities and changes are arbitrated by a constant "quantum" computation. Such computation has to be quantum since it is the quantum that elaborates the interface between material and more subtle light-based layers of existence.

Hence this book will explore the notion of content in Light as the origin of genetics, the process of precipitation of Light and the subsequent and mathematical creation of subtle-DNA replete with infinite four-base logic-ecosystems as the origin of genetic structure and heredity, and quantum computation and its intimate relationship with the structure and evolution in matter as the origin of genetic mutation and all further genetic and post-genetic possibility. This book hence will propose that the origin and possibilities of genetics is based on Light, and as nothing other than a key mechanism by which the infinite information present in Light progressively materializes.

Here I also draw attention to the Shankha (conch shell) displayed on the cover of this book. The Shankha is deeply significant in several regards. Its story winds through ancient traditions and even modern-day applications. From the Hindu god Vishnu wielding his conch Panchajanya – believed to have control over all the elements of nature, to Triton, son of Poseidon, sounding its depths, the conch has echoed through diverse civilizations. When it is blown its sound is creative, just as the universal quantum computational process highlighted in this book, and generating genetic-type output, is creative. The conch itself, is created by a vast number of lines, curves, symbols, that can be viewed as representative of the different elements of the comprehensive cosmological mathematics, just as the Moti Shankha having thousands of markings on it is significant of deep multi-dimensional power. This enduring symbol, with its Fibonacci spirals, bridges ancient wisdom with modern scientific exploration, and is one amongst close to 150 illustrations that appear in this book. Many of these illustrations connect the ancient with the contemporary and the artist's thought process for each illustration is captured in a separate section toward the end of this book.

The quantum computational process with its genetic-type output also draws attention to a new category of quantum computation, orthogonal to the pursuits of the quantum computing industry. In fact each of the 20 books across the Cosmology of Light book series – from the first 'A Story of Light,' to this book on genetics, to the others – may be viewed as an outpouring of a *Shankha,* progressively revealing something of this immense, largely untouched possibility, that has yet to be wielded by practical technology.

Pravir Malik,
San Francisco

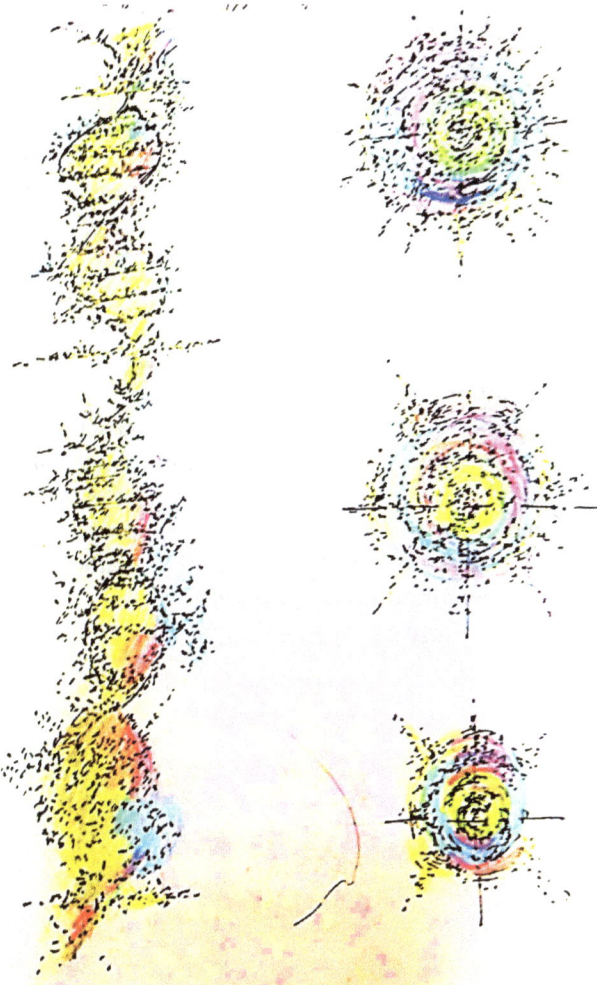

Chapter 1.1 explores the impact that light has on the experienced nature of reality. In particular the impact of light traveling at potentially different speeds will be examined for its ability to create realities based on its speed. Further the Big Bang and a pervasive seed-equation are posited as existing as a result of the slowing

down of Light. The information codified in Light will be positioned as being the origin of genetics.

Chapter 1.2 explores a theory for the creation of quanta related to the speed of light. Quanta are intimately associated with the seed-equation and acts as a doorway, as it were, into the multi-layered structure represented by the seed-equation. Further quanta and the to-be explored notion of quantum-computation are fundamental to genetics in this view and this chapter introduces some basic ideas to be leveraged later.

Chapter 1.3 summarizes key ideas about genetics already alluded to in Chapters 1.1 and 1.2 to lay out why light and genetics are intimately associated. These key ideas or concepts include the origins of genetics, the structure of subtle-DNA including the downward-strand and upward-strand, four-base logic-encoding ecosystems that exist within these strands, mutation of both a constructive and destructive nature, conceptualization of matter, and the process of evolution, amongst others.

Chapter 1.1: Light and the Realities it Creates

This chapter will look at the impact light has on creating practical realities.

Light@c, Quanta, and the Reality it Creates

Everything is made from light. But how can this be. How can the dense material objects that make up our world – the stones, cars, buildings – be made from light? How can our thoughts, our emotions, our cells and bodies be made from light? How can all the animals and plants and clouds be made from light? Here is an exploration on now this may be possible.

To begin this exploration think for a minute as to how fast light is traveling. If we turn on a light switch in a room it immediately gets filled with light. If we turn on a light torch on a dark night it immediately lights up a path through the darkness. And yet the light from the sun takes about eight minutes to reach the earth – this speed through a vacuum is referred to as the physical constant 'c'. So while light is moving incredibly fast it is still not moving at the infinite speed we may think it is moving at.

Now, it is because the light is not traveling at an infinite speed that quanta and subsequently the world of matter can appear. All matter is made from atoms. So let us enter into the world of a hypothetical atom in the process of creation to explore how this might be. Imagine that some form of light originates from what will become the nucleus within an atom. By virtue of its finite speed light would take some time, however small, to travel any minute distance within the forming atom. During that fraction of time there is a build up of energy and it is this build up of energy that forms a packet or quanta, as it were, that allows matter to form. In the absence of such a build-up potential-matter would simply disperse. Hence, it is this build up on energy that variously expresses itself as quarks and leptons and bosons and subsequently atoms – the very building blocks of everything that is.

So, in this view, it is because of the finite speed of light, c, that matter is created. That is not to say that matter will be created wherever light is traveling at c, but the possibility exists, so long as other conditions are also fulfilled. But let us leave aside the other conditions and the exact process of such a creation for now.

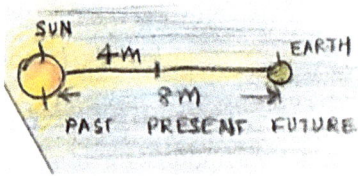

But what else is implicit or made possible because of the finite speed of light? Well, imagine traveling on a ray of light from the sun to the earth. Imagine that you are in minute 4 of the approximately 8 minute journey. As you look back you will see that 4 minutes in the past you were at the sun. 4 minutes in the future you will be at the earth. And in the present moment you are somewhere between the sun and the earth. So this limited speed of light already creates the concept of time and specifically of the past, the present, and the future.

So, four incredibly fundamental things are created because of the finite speed of light: quanta and therefore matter, the past, the present, and the future. It could be said, therefore, that this matter-based time-experienced universe is a result of the finite speed of light.

Looking at this in another way, it is known too that in a denser medium the speed of light further slows down. Hence, light travels at some fraction of c when moving through water for instance. So by reversing the logic of such a process this may show that if light slows down from an infinite speed to some other lesser speed whether a multiple or a fraction of c, then the material reality will have to alter, in a similar way as the material reality between vacuum and water is different.

Light @Infinity

But what if light can be experienced at other speeds? What if light can travel at other finite or an infinite speed beyond the c we experience it at, which by the way is an astronomical 186,000 miles per second in a vacuum?

Let us imagine for a minute that light can travel infinitely fast. Think about a big area or volume with a light source at the center. Now since the light travels infinitely fast it will fill up the entire volume instantly. This will be true no matter how large the volume is – it could be the entire solar system, an entire galaxy, or an entire universe. So that light is going to be instantaneously present everywhere – it is going to be omnipresent.

A further thought experiment may give insight into such omnipresence by considering the night sky. It is because of the finite speed of light that the night sky has only spots of light, however many, across it. But if light were traveling infinitely fast then the night sky would be a canvas or a sea of brilliant white since the light from every corner of the universe, wherever a light source or star existed, would immediately be present. It would appear, in such a thought experiment, that we existed in a sea of light.

Now, since the light is already present everywhere in whatever volume, that is, since that light has already filled up the entire volume, there is nothing else that can similarly arise there that is not of the nature of light. Even if something else were to arise, being surrounded by light it would eventually succumb to that light. So the light is all-powerful or omnipotent within that volume.

Further, since the light exists simultaneously everywhere in that space and has a complete knowledge of itself it therefore has a complete knowledge of that space or of anything that can arise in that space. So it is all-knowing or omniscient in that space.

Finally, the light connects everything together instantaneously and holds these connections and the things connected in its nature, so it is all-nurturing or omninurturing within that space.

So, as a result of the infinite speed of light, light appears to have properties of omnipresence, omnipotence, omniscience, and omninurturing.

The Big Bang, The Seed Equation, & Genetics

We hear about the Big Bang as the start of the universe. In this Big Bang matter is created. But in the examinations just presented the creation of matter is nothing other than the result of light traveling at the finite speed, and in this universe at 186,000 miles per second in vacuum. So we can say that the Big Bang, the apparent start of the

24

universe, is the result of a slowing down of light from an infinite speed to some finite speed.

When light slows down then energy accumulates in packets or quanta and this results in what was inexpressible being able to express itself as matter. This can also be thought of in terms of an incredibly rapidly moving stream of water. If the water is traveling so fast then no boundary will be able to contain it and the energy will be continuous over the length of the stream. If it is traveling slower though, then the water will be able to be held by boundaries along the length of the stream. The energy in this case will be discontinuous and will appear in "packets". In other words there is a process by which information implicit in light traveling infinitely fast is rearranged into a more and more material basis as light travels slower. This rearrangement or elaboration of implicit information is also the origin of genetics.

In this process of slowing down the implicit omnipresent-omnipotent-omniscient-omninurturing nature of light becomes or transforms into an explicit or emergent matter-past-present-future nature of light. So there is implicit in this transformation also a high degree of 'entanglement' as it were. That is information exists in a different and highly connected manner. Ab initio,

everything that appears in this universe is highly entangled. In other words every fundamental particle, whether it be a boson or lepton, or any subsequent emergence of matter is already deeply entangled by virtue of having emanated from the same single starting-point conceptualized as the Big Bang. This process of transformation creates a seed-equation that is present and synonymous with the Big Bang. The dynamics of a resulting universe are contained in the dynamics of such a seed-equation. The seed-equation provides insight into the subtle-structure which houses genetic information, and entanglement provides insight into the availability of such antecedent or subtle genetic information. The seed-equation will be explored in more detail in the section on mathematical foundations.

We have arisen in the field of light that has a finite speed. It is difficult therefore to feel the reality of the omnipresent-omnipotent-omniscient-omninurturing light. We are attuned to perceiving in the material world that has resulted due to the finite speed of light. But if we could step back into the fullness of light in its pure state at infinity from there we would likely see that there can be different universes created as a result of light selectively slowing down to

different levels. The slowing down to a different level will create a particular kind of universe.

Note that there have been experiments to slow down light so that it practically moves at a snail's pace. This slowing down has to be put into context. Even if light were to slow down to 1 mile per second, say, from 186,000 miles per second, keep in mind that it is possible that this change in speed is likely only a miniscule fraction of the change in speed from light traveling infinitely fast to light traveling at c, that is, 186,000 miles per second. So in effect what could be said is that when light is made to slow down in experiments, the range or band that it slows down to deviates only slightly from its relative-to-infinity slower speed in vacuum.

Necessity of Constant Speed of Light and Variable Time and Space

The constant speed of light, at a speed less than an infinite speed, allows a build up of energy at quantum levels, that in turn allows any properties or possibility in light to express itself materially. When light travels at the speed of c – 186,000 miles per second – then the result of that is the material universe as we see and experience it, also with its division of time into a past, a present, and a future. In fact, we could say that for the known universe, as we experience it, light had to travel at c for it to arise. The speed of light had to be constant else there would be a variable, fluid reality to matter, likely displaying barely forming islands of matter subject to sudden disappearance, reappearance, continuing ad infinitum.

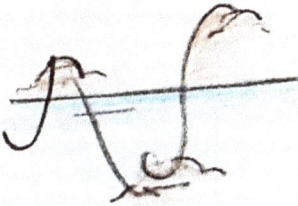

But also since the distance traveled by an object or a ray is the result of the speed it is traveling at for the time involved, this gives us an interesting insight that Einstein based his Theory of Relativity on (Einstein, 1995). Since the speed of light is constant, and has to be for the universe to be in its observable stable condition, this means time and distance (or space) potentially can vary. So if an object is traveling very fast, then time and space are going to be experienced differently by it, as compared with an object that is traveling considerably slower: as the speed of an object approaches c, time slows down and distance contracts.

So an object that manages to travel at a speed of c will in some sense partake of the reality as experienced by c. It will transcend the conventions of time and space that are set up because of c and experience these differently.

Light at Speed Close to Zero

On the flip-side imagine light traveling at a speed close to or approaching zero. In this case the experienced reality is going to be the opposite of reality as experienced when light is traveling at an infinite speed.

First of all, light emanating from any point will remain fundamentally isolated and become the basis of extreme fragmentation. This has to be since regardless of the amount of elapsed time the light will still be only at its point of origin or source. So instead of a Presence of Light in any considered volume, there will be an Absence of Light.

Further, in imagining an area or volume, such light will have no knowledge of what is going on in the volume, and hence be completely ignorant, will further, have no power to effect any circumstances and hence be completely powerless, and finally, will not be able to coordinate or be present in any relationships. Hence it will be detrimental to or agnostic of any relationships instead of harmonious or nurturing to relationships.

Hence, reality in such a scenario would be stark, dark, desolate, fruitless, completely fragmented, and even perhaps pointless.

Such a reality may represent a negative-infinity and may be thought of as an extreme limit of what may be possible in worlds in which Light projects itself.

Seen from this point of view a Big Bang would be a fortuitous event since it will allow a fruitful reintegration of different worlds, each created by light traveling or projected at a different speed, to perhaps begin to take place as will be discussed further.

Impact of Superposition on Genetics in a Light-Based Model

Light traveling at different speeds can, as illustrated by the cases of an infinite speed, c, and 0, create different realities. But if the infinite speed case is the native state, and if every other case is a projected state, then it is possible that these realities all exist simultaneously. In other words, phenomena resident in the reality created by the native state and by multiple projected states can be superposed. This also implies that categorization of genetic-type information and interrelation of genetic-type information will be superposed.

Further, as will be explored in detail in subsequent chapters such states of superposition must be considered as superposition because they are inherently related to each other. A phenomenon that occurs in the reality where light travels at c, has its origin in phenomena or functionality existing in reality where light travels at a speed greater than c, and is influenced by phenomena existing in a reality where light travels at 0 miles per second.

Chapter 1.2: Light & Quanta

This chapter explores quanta, its relation to light in more detail, and other emergent properties that must also be quantized because of multiple speeds of light.

Light, Quanta, & Relationship to Genetics

When we think of light traveling at the speed of c, there are properties in the reality that emerge as a result of this. These properties as we have examined are of the nature of matter, the past, the present, and the future. These properties are the result of light traveling at c and so can be thought of as emerging from light.

But what do these properties actually mean? And further, how do these properties relate to the apparently quite different properties of omnipresence, omnipotent, omniscience, omninurturing that can appear in a reality where the speed of light is infinite?

Let us consider first the properties of light that emerge when light is traveling at c: the past, the present, the future, and matter.

What is the past? It is the perceivable result of all the work and effort that has taken place so far. It is the foundation upon which the present and future will be built. It represents a status quo, a stability, and even a rigidity, and given that it is the result of the long play of time, it will not easily be persuaded to become another thing. It can be thought of as that which the eye can see when it looks around it. There is "physicality" to what the eye can see and so the essence of the past is a kind of physical-ness. So, ingrained in light, is this ability to project or create physical-ness.

What is the present? It is the tremendous play of forces of all kinds to express themselves here and

now. There is "vitality" that is present in this play and often it is the most energetic or forceful of the forces that will win out, as opposed to the most insightful or thoughtful. All the tremendous possibility of the future is seeking for expression now and so this essential vitality can also be thought of as a projection or possibility implicit in light.

What is the future? It is the inevitability of what will manifest. The great thoughts, the great ideas, the purpose, the possibilities will sooner or later express themselves in what we call the future. And the essence of this is thoughtfulness or a curiosity or a purpose that we can summarize as an essential "mentality". So embedded in light is this ability to project mentality.

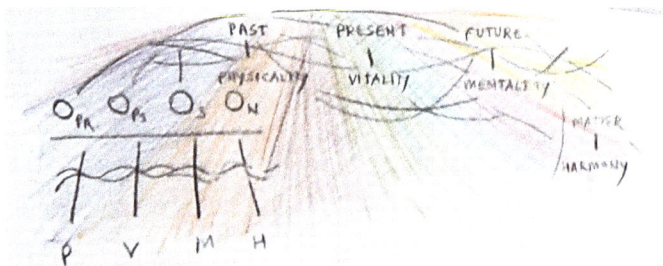

And what is matter? It is the myriad crystallization into apparent diversity of the one essential reality of light, to allow for a play between these different sides or possibilities in an increasingly harmonious interaction. So its essence is "harmony", and this too can be thought of as an essential property projected from or implicit in light.

But what about when light is considered to move at an infinite speed? Then the properties that become apparent, as already explored, are those of omnipresence, omnipotence, omniscience, and omninurturing. But omnipresence has physical-ness to it, omnipotence has vitality to it, omniscience a mentality to it, and omninurturing a harmony to it. It may be said that physical-ness comes from omnipresence, vitality comes from omnipotence, mentality comes from omniscience, and matter comes from omninurturing.

So whether light is traveling at an infinite speed or the speed we know as c, there is something about the properties it projects, that in essence is the same. So let us refer to these essential properties made apparent through the worlds that are created, as Presence (from omnipresence), Power (from omnipotence),
Knowledge (from omniscience), and Harmony (from omninurturing).

Further, quanta can be thought of as existing on that very border of the world or reality created due to Light traveling at c, and realities created as Light travels faster, which at its limit is an infinite speed. So quanta are a doorway or an interface into worlds of Light, and a doorway by which possibilities in deeper worlds of Light can express themselves here in this material realm. Being so, phenomenon such as superposition and entanglement are natural to quanta. But further, any process of genetics, which by definition is associated with dynamics of superposition and entanglement, will also intimately be associated with quanta. This has to be since quanta are the interface between different layers of light. In fact a process of quantum computation will be proposed and elaborated as the means by which genetic mutation takes place and any future genetic-type possibilities unveiled.

Structured Time, Structured Space, and Their Relation to Quanta

In considering the worlds that are created due to the way light moves or the play of light, we see properties that are projected because of it. In the finite world, the world that results from light traveling at the speed c, there is, relatively, smallness we can grasp and on which we can build. In the infinite world that results from light traveling at an infinite speed there is a vastness and fullness that is difficult to grasp.

33

The notions of time and space are something entirely different in both worlds, and it could be said that Space allows the full play of everything meant by Power, Knowledge, Harmony, and Presence to be seeded in it, and that Time allows that seeding to flower into fuller forms with its passage.

In other words both space and time are not just abstract concepts but are essentially highly structured to allow physical-ness, vitality, mentality, and matter to become Presence, Power, Knowledge, and Harmony.

Space allows all the possibilities present in Presence, Power, Knowledge, and Harmony and seeded in vast diversity, to evolve into more fullness through the time stages of physical-ness, vitality, mentality, and evolving matter.

Such a creation of space and time is synonymous with the creation of quanta. Quanta become the means for the possibilities inherent in the anterior worlds of Presence, Power, Knowledge, and Harmony to express themselves in a structured space and time. Quanta are therefore a passage into deeper worlds of Light, and a means for possibilities in these deeper realms to express themselves materially. A process of quantum computation involving anterior-layer codified possibilities, or "pre-genetic" information, and a range of shifting forces arbitrates the expressions that manifest in space and time. This will be described in detail in Section 3 that focuses on the mathematical foundations for the four-base logic-encoding ecosystems.

But further, in this view it may also be proposed that space and time being so structured by the four properties of Light, need also to be quantized. That is, space and time must be experienced as quanta as well.

Chapter 1.1 proposed that quanta is the result of the slowing down of light, and as such becomes the basis for material expression. In other words, for matter to express itself requires quanta. But further it was also just proposed that space, consisting of seeds of the properties of light, would also require quanta to express itself. Philosophically this has to be in a world where light is traveling at a fraction of its possible speed. Further, it was also proposed that time is highly structured, expressing the growth of the seeds in space through definite phases. As such, growth through such phases can also be thought of as happening due to quantization that so allows phases to express themselves.

But if we take a deeper look at space, time, and matter in light of the properties of Light, it can be deduced that space, consisting of a vast array of seeds derived from the properties of Light is itself an expression of Light's property of Knowledge.

Time, bringing forth the meaning contained in the seeds, regardless of circumstance, and even being opposed by circumstance, can be thought of as Light's property of Power.

Matter itself, being a container in which space and time can allow deeper properties of Light to become materially tangible, must be an expression of Light's property of Presence.

But it is also known from Einstein's General Theory of Relativity that gravity is associated with mass and space, in that, as the physicist John Wheeler has put it, it is none other than a mass's instruction telling space how to curve, and again is nothing else that space's instruction telling mass how to move through it (Wheeler, 2000). As such, where mass and space exist, there gravity has to exist as well. Hence it may be inferred that gravity is none other than an expression of Light's property of Harmony, which fixes the collective relationship between object and object.

But if as proposed, matter, space, and time all need to be quantized in order to express themselves, then this must also be true of gravity.

But also, the emergence of space, time, matter, and gravity can be seen as a quantum computational outcome of light precipitating from the reality where it travels at an infinite speed to the reality where it travels at c.

Subsequent chapters will explore these claims in greater detail.

Chapter 1.3: Some Concepts of Genetics in a Cosmology of Light

This chapter briefly summarizes some concepts of genetics in a Cosmology of Light, already alluded to in the discussion in Chapters 1.1 and 1.2 and introduces additional concepts to be further explored in this book.

The Origins of Genetics

The infinite information codified in Light is the origin of genetics. As discussed previously Light in its native state, possessing characteristics of omnipresence, omnipotence, omniscience, and omninurturance, contains all-possibility within it. This all-possibility can be thought of as an infinite amount of information, and its structure related to the four characteristics, as the basis of genetics.

The Light-Matrix or Downward-Strand

All-possibility that exists in the reality of light traveling infinitely fast is progressively materialized through a mathematical arrangement by which the subtle-infinite becomes the astounding material-diversity experienced when light travels at c. The mathematical process, by which this transformation takes place, creates the Light-Matrix or downward-strand. This will be explored in detail in Section 2, on the mathematical structure of subtle-DNA.

Subtle-Libraries of Pre-Genetic Information

The progressive materialization of light can be modeled by a series of mathematical transformations. These transformations take the infinite amount of information existing in Light's native state, to effectively create a series of precipitated subtle-libraries also of infinite pre-genetic information. This series of subtle-libraries subsequently allows infinite material diversity to come into being.

The Space-Matrix or Upward-Strand

The possibilities seeded in the structure of space arise or mature through the passage of time, and this process is summarized by the Space-Matrix or upward-strand. The fundamental structure of the upward-strand mirrors the downward-strand and its possibilities are intimately tied to the levels that exist in the downward-strand. This will be explored in detail in Section 2, on the mathematical structure of subtle-DNA.

The Essential Structure of Subtle-DNA

The essential structure of subtle-DNA comprises of a largely pre-existent Light-Matrix or downward-strand and a resulting Space-Matrix or upward-strand. Libraries of subtle pre-genetic information exist in the downward-strand. Genetic-type information expresses itself through the upward-strand as it were, and is due to the interaction between the possibilities embodied by the downward and upward strands.

Four-Base Logic-Encoding Ecosystems

Four-base logic-encoding ecosystems contain logic in the quantum-layer antecedent to the material layer relating to constructs that exist in the material layer. These four-base logic-encoding ecosystems are subject to change in the interplay with the material layer. In a sense the call from 'below' is envisioned to potentially meet with some precipitation from 'above' of one of an infinite number of functions already created in the subtle-libraries.

Material-Fabric

Just as genetic information is housed in DNA in living cells, there has to exist some structure to house the proposed pre-genetic information that exists at a pre-cellular stage. It is proposed that such pre-genetic information is housed in a structure termed 'Material-Fabric'. This material-fabric exists at the interface or could be the interface between the antecedent quantum-layer and Space.

Space, Time, Energy, Gravity Quantization & the Material-Fabric

The structures of Space, Time, Energy, and Gravity as introduced in the previous chapter, and as discussed further in Chapter 3.5, are critical quantization mechanisms that emerge in the downward-strand where Light travels at speed c. Their initial appearance imbues the material-fabric with reality when the fourfold space-time-energy-gravity quantization is itself generated as pre-genetic code. This code must exist in every iota of the material-fabric and the material-fabric can be thought of as an emergent property of such fourfold quantization. Further, all subsequent materialization of information must involve a composite space-time-energy-gravity quantization.

Superposition in Genetics

In the process of quantum-computation, by which four-base logic-encoding ecosystems, the pre-genetic and the genetic libraries, can be altered, the different dynamics representative of realities created by light traveling at different speeds are always present. This presence exists in superposed fashion and the real-time quantum-computation determines which superposed possibility will manifest materially.

Entanglement in Genetics

The information inherent to a particular layer, created through light traveling at a different speed, generates libraries of possibility through a mathematical process. Due to a different dynamics of space and time representative of the layer created by light at a particular speed, these libraries exist differently in an entangled state, therefore being subtly present or influencing layers of light traveling at a slower speed relative to that layer. Common DNA existing in every cell at the material layer is a logical outcome of this process of antecedent-entanglement.

Heredity

The presiding or generally accessible four-base logic-encoding ecosystems are the bases of heredity.

Constructive-Mutation

Constructive-mutation occurs when patterns of a largely obstinate nature at the material level are broken as a result of which the possibilities existing by the nature of the upward-strand are allowed to manifest. Of necessity this means that an inherent process of integration is taking place since light is unifying with its deeper nature. Constructive-mutation is intimately tied to this notion of integration.

Destructive-Mutation

When obstinate or disintegrating patterns persist or are chosen destructive-mutation will result. In its essence this means that light is moving away from its essential unified reality more towards the reality typified by light existing at zero-speed. Further the process of destructive-mutation while similar to constructive-mutation will be seen to have different boundary conditions and a different range of inputs of a different nature than those that take part in constructive-mutation.

Mutational Sequence

In general there is a mutational sequence dictated as it were by the levels in the downward-strand. Over time, and as expressed in the upward-strand, the nature of mutation will be such that it will be primarily constructive, and further, will be an outcome of the subtle-libraries existing at layers of reality set up by faster and faster speeds of light.

Interpretation of Matter

Matter is the result of a constant computation involving the existing material layer, the material-fabric, four-base logic-encoding ecosystems, and the antecedent light layers. This means that matter can and will change as the interplay between these layers changes.

Evolution and the Possibilities of Genetics

Evolution is a process by which pre-genetic possibilities materialize due to the existence of the right material conditions. This may also mean that the pre-genetic material existing in the four-base logic-encoding ecosystems will change.

Post-Genetic Code

In its possibilities of materialization information in Light may house itself in subtle-libraries generated at various layers of Light, four-base logic-encoding ecosystems at the quantum-level, the material-fabric, or in genetic code in living cells. But it is also possible through the process of evolution that the "structure" housing further possibilities in Light may take on a hybrid form that will be referred to as Post-Genetic Code.

SECTION 2: THE MATHEMATICAL STRUCTURE OF SUBTLE-DNA

This section explores the mathematical structure of subtle-DNA. In this structure there is a downward-strand and an upward-strand. These strands form the backbone of all possibility. All possibility originates by the structure and implicit functionality of these strands. There are five representative levels that are illustrated in this section, that are themselves determined by the play of four ubiquitous bases.

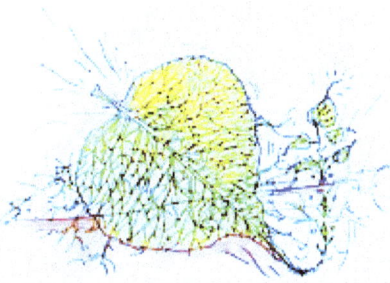

While the number of levels may potentially be infinite, the five levels seem to represent important and may be even fundamental quantum-based shifts, as light projects itself at slower and slower speeds to progressively reveal more of the vast information implicit in it.

Chapter 2.1 explores the Light-Matrix or downward-strand, positioned as being the primary and originating strand. Levels in this strand describe an originating or creative structure that gives rise to the first apparent layer of the "ubiquitous-point-instant" and to subsequent layers culminating in the layer that results in the vast diversity of life. Each fundamental layer in the matrix and the subtle-libraries of information they automatically spawn, is then explored in more detail in subsequent chapters through the ubiquitous-point-instant (Chapter 2.2), the architectural forces (Chapter 2.3), and uniqueness of organizations (Chapter 2.4). The focus then shifts to a possible mutational-sequence (Chapter 2.5) and systematization of that (Chapter 2.6) to set up the basis for the upward-strand of the ubiquitous subtle-

DNA.   The upward-strand is envisioned as being organic, depending on the active play of influences, while at the same time eventually prescribing the sequence discussed. The sequence itself is determined by the layers described in the downward-strand.

The mathematical foundation of subtle-DNA explored in Section 2 will set up the basis for the creation and dynamics of chains of four-base ecosystems that encode logic, as explored in Section 3.

Chapter 2.1: Light-Matrix & the Ubiquitous Downward-Strand

This chapter will explore the Light-Matrix, a mathematical expression of the codification in Light formed at the time of the Big Bang (Malik et al., 2018). Further a case will be made equating the Light-Matrix as the primary or downward-strand in a ubiquitous subtle-DNA.

This Light-Matrix can be thought of as a seed-equation that provides insight into dynamics of the universe. The Light-Matrix codifies dynamics that may be experienced at the quantum borders of realities created through light traveling at different speeds, and further in the worlds that the quantum veil or window provide access to. Weaving together possible realities created through light traveling at different speeds, the Light-Matrix also provides a model of superposition and entanglement.

Light slowing down or projecting itself at a particular speed is envisioned to create a particular type of universe that will exist subtly in all universes where light is traveling at any faster projected speed. Hence, all universes will exist in that created where light is in its native state traveling at an infinite speed. This implies that all properties true of the native state will also be present in any sub-universe. It is the presence of such properties and dynamics true of supra-universes that that gives rise to the phenomenon of superposition and potential entanglement in any sub-universe.

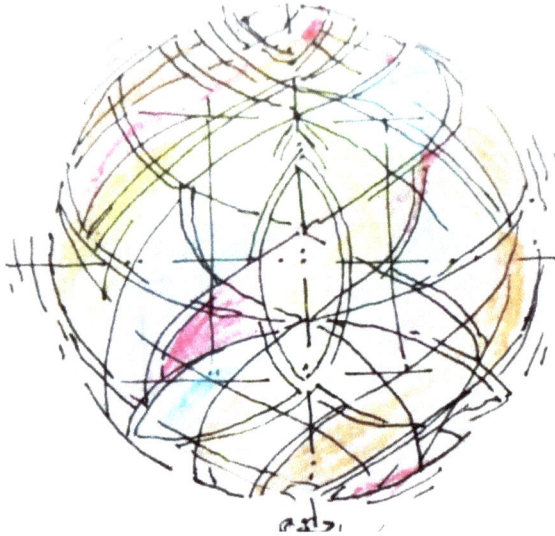

But further, such superposition also creates a "strand" as it were, that intimately connects together these universes and the realities represented by them. Entanglement creates the dynamic of influence. So it is the combined action of superposition and entanglement that also creates the notion of ubiquity of the structure of "subtle-DNA".

It is interesting to note that Stanford anthropologist, Jeremy Narby, in his book 'The Cosmic Serpent: DNA and the Origins of Knowledge' (Narby 1998) points first to a shift in vision experienced by shamans around the world after the ingestion of ayuhuasca, and second to the commonality of the ensuing vision marked primarily by two entwining serpents. His interpretation is that these shamans enter into the molecular realm and actually perceive DNA and accompanying structures at the cellular levels. This book also requires a shift in vision, one bought about by perceiving Light in a precipitating series of layers caused by different speeds of light. The result of such a perception is the view of a ubiquitous subtle-DNA and a vast subtle structure continually arbitrating genetic-type information to create the realities we exist in materially. This book aims to bring such an enterprise into view.

Nature of Light at U

As laid out in Section 1, it is perhaps fair to say that the speed of light has significant implications on the experienced nature of reality. The finiteness, c, at 186,000 miles per second in a vacuum creates an upper bound to the speed with which any object may travel also implying, and as discussed in Section 1, that objective reality will be experienced as a past, a present, a future from the point of view of that object (Einstein, 1995). These characteristics – a past, a present, a future – are implicit in the nature of light and become part of objective reality because of the speed of light. So the way we experience time seems to be determined by c.

Further, c also creates a lower bound when inverted $(1/c)$ being proportional, or arguably even determining Planck's constant, h, that pegs the minimum amount of energy or quanta required for expression at the sub-atomic level. Planck's constant, h, pegging the amount of energy required for expression, therefore may allow matter to form as suggested by the physicist Lorentz a century ago (Lorentz, 1925). Note that Einstein postulated quanta as a fundamental property of light itself, rather than as something that arose in the interaction of light with matter as suggested by Max Planck (Isaacson, 2008). Hence, c contributes to or even establishes a reality of nature with a past, present, and future, to also be experienced as a phenomena of connection between seemingly independent islands of matter. This characteristic of 'connection' is therefore also proposed to be implicit in the nature of light and becomes part of objective reality because of the speed of light.

But as explored in Chapter 1.2, a 'past' can also be viewed as established reality as defined by what the eye or other lenses of perception can see. Hence, 'It is the perceivable result of all the work and effort that has taken place so far. It is the foundation upon which the present and future will be built. It represents a status quo, stability, and even rigidity, and given that it is the result of the long play of time, it will not easily be persuaded to become another thing. It can be thought of as that which the eye can see when it looks around it. There is "physicality" to what the eye can see and so the essence of the past is a kind of physical-ness. So, ingrained in light, is this ability to project or create physical-ness.'

Also as explored previously, the 'present' is 'the tremendous play of forces of all kinds to express themselves here and now. There is "vitality" that is present in this play and often it is the most energetic or forceful of the forces that will win out, as opposed to the most insightful or thoughtful. All the tremendous possibility of the future is seeking for expression now and so this essential vitality can also be thought of as a projection or possibility implicit in light.'

The 'future' is 'the inevitability of what will manifest. The great thoughts, the great ideas, the purpose, the possibilities will sooner or later express themselves in what we call the future. And the essence of this is thoughtfulness or a curiosity or a purpose that we can summarize as an essential "mentality". So embedded in light is this ability to project mentality.'

These implicit characteristics of the nature of light as experienced at the layer of reality so set up by a finite speed of light may hence be summarized by Equation 2.1.1, where c_U refers to the speed of light of 186,000 miles per second, that has created the perceived nature of reality, U, as expressed in Equation 2.1.1:

c_U: [*Physical, Vital, Mental, Connection*]

Eq. 2.1.1: Implicit Characteristics of the Nature of Light at U

Nature of Light at ∞

Exploring further, it is thought however that at quantum levels the nature of reality is characterized by wave-particle duality.   Light itself and matter may be experienced as both particles and waves. But for matter to be experienced as waves implies that 'h' must have become a fraction of itself, $h_{fraction}$, to allow the concentration or possibility of quanta to have dispersed into wave-form.   This further implies that c must have become greater than itself, c_N, such that the inequality specified by Equation 2.1.2 holds:

$c_N > c_U$

Eq. 2.1.2: Layer N, Layer U Inequality of Speed of Light

Note that what is implied here is that just as there is a nature of reality specified by U that is the result of the speed of light being 186,000 miles per second, so too there is another nature of reality specified by N that is the result of a speed of light greater than 186,000 miles per second.

This is akin to recent developments in physics with the notion of property spaces being separate from but influencing physical space as explored by Nobel Physicist Frank Wilczek in his book 'A Beautiful Question" (Wilczek, 2016). But further in "Slow Light" Perkowitz's recent treatment of today's breakthroughs in the science of light (Perkowitz, 2011) he states: "Although relativity implies that it's impossible to accelerate an object to the speed of light, the theory may not disallow particles already moving at speed c or greater."

So light traveling at c_N may be possible. Current instrumentation, experience, and normal modes of thinking though, having developed as a bi-product of the characteristics so created in the layer of reality U may be inadequate to access N without appropriate modification. The notion of wave-particle duality already challenges the notion of normal thinking perhaps because wave-like phenomena could be a function of faster than c motion, and particle-like phenomena a function of equal to c motion. That these may be happening simultaneously is reinforced by principles such as complementarity in which experimental observation may allow measurement of one or another but not of both as pointed out by Whitaker (Whitaker, 2006). But further the very notion of the Pilot-Wave Interpretation of Quantum Mechanics, as modeled by the physicists Bohm and DeBroglie (Holland, 1995), is that constructive interference of waves causes particles to move toward these areas. Hence in this interpretation both particle and wave always exist simultaneously.

But then taking this trend of a possible increase in the speed of light to its limit, this will result in a speed of light of infinite miles per second. The question is, what is the nature of reality when light is traveling at infinite miles per second? As explored in Section 1, in any continuum light originating at any point will instantaneously have arrived at every other point. Hence light will have a full and immediate *presence* in that continuum. Further, that light will *know* everything that is happening in that continuum completely and instantaneously – that is know what is emerging, what is changing, what is diminishing, what may be connected to what, and so on - or have a quality of *knowledge*. It will connect every object in that continuum completely and therefore have a quality of connection or *harmony*. Finally nothing will be able to resist it or set up a separate reality that excludes it and hence it will have a quality of *power*.

These implicit characteristics of the nature of light as experienced at the layer of reality so set up by an infinite speed of light may hence be summarized by Equation 2.1.3, where c_∞ refers to the speed of light of ∞ miles per second, that has created the perceived nature of reality, ∞:

$$c_\infty: [Presence, Power, Knowledge, Harmony]$$

Eq. 2.1.3: Implicit Characteristics of Nature of Light at ∞ Speed

Transformation from ∞ to U

But by (2.1.3) it can also be noticed that 'physical' is related to Presence, 'vital' is related to Power, 'mental' is related to Knowledge, and 'connection' is related to Harmony.

The question then, is how do these apparent qualities at ∞ precipitate or become the physical-vital-mental-connection based diversity experienced at U? This may be achieved through the intervention or action of a couple of mathematical transformations acting on the implicit characteristics of nature of light at ∞ speed as summarized by (2.1.3).

First, the essential characteristics of Presence, Power, Knowledge, Harmony that it is posited exists at every point-instant by virtue of the ubiquity of light at ∞ will need to be expressed as sets with up to infinite elements. Second, elements in these sets will need to combine together in potentially infinite ways to create a myriad of seeds or signatures that then become the source of the immense diversity experienced at U. Note that these mathematical transformations suggest that light may gather itself in such a way so as to first release, as it were, from its essential nature such sets with infinite variation around the essential characteristics, and second, an infinite combination of such vast variation. This suggests

51

that all that is seen and experienced at U may be nothing other than 'information' or 'content' of light and as such that there are fundamental mathematical symmetries at play where everything at U is essentially the same thing that exists at ∞. This then is the primordial basis of genetics. Further, the heart of creation, in this view, is likely a constant quantum-computation that interrelates various realities so set up by light traveling at different speeds. The essential "nodes" of such quantum-computation take place at distinct levels or realities created at an interface of light traveling at different speeds, and these nodes distinguish the precipitating structure of a downward-strand of ubiquitous subtle-DNA.

Assuming that the first transformation occurs at a layer of reality K where the speed of light is c_K, such that $c_U < c_K < c_\infty$, this may be expressed by Equation 2.1.4:

$$c_K : [S_{Pr}, S_{Po}, S_K, S_H]$$

Eq. 2.1.4: The First Transformation at Layer K

S_{Pr} signifies 'Set of Presence', S_{Po} signifies 'Set of Power', S_K signifies 'Set of Knowledge', S_H signifies 'Set of Harmony/Nurturing' (note: Harmony and Nurturing will be used interchangeably through this book).

Assuming that the second transformation occurs at a layer of reality N where the speed of light is c_N, such that $c_U < c_N < c_K < c_\infty$, this may be expressed by Equation 2.1.5:

$$c_N : f(S_{Pr} \times S_{Po} \times S_K \times S_H)$$

Eq. 2.1.5: The Second Transformation at Layer N

The unique seeds are therefore a function, f, of some unique combination of the elements in the four sets S_{Pr}, S_{Po}, S_K, S_H.

The relationship between the layers of light may be modeled by the following matrix in Equation 2.1.6:

$$Light_{Matrix} = \begin{vmatrix} c_\infty: [Pr, Po, K, H] \\ (\downarrow R_{C_K} = f(R_{C_\infty})) \\ c_K: [S_{Pr}, S_{Po}, S_K, S_H] \\ (\downarrow R_{C_N} = f(R_{C_K})) \\ c_N: f(S_{Pr} \; x \; S_{Po} \; x \; S_K \; x \; S_H) \\ (\downarrow R_{C_U} = f(R_{C_N})) \\ c_U: [P, V, M, C] \end{vmatrix}$$

Eq. 2.1.6: Light-Matrix & Downward-strand of Subtle-DNA

The matrix should be read from the top row down to the bottom row as indicated by the $\downarrow$ between rows, and suggests a series of transformations leading from the ubiquitous nature of light implicit in a point – presence, power, knowledge, harmony - to the seeming diversity of matter observed at the layer of reality U which is fundamentally the same presence, power, knowledge, and harmony projected into another form of itself.

The first transformation is summarized by Equation 2.1.7:

$$R_{C_K} = f(R_{C_\infty})$$

Eq. 2.1.7: Light-Matrix First Transformation

This is suggesting that the reality at the layer specified by the speed of light c_K, R_{C_K}, is a function of the reality at the layer specified by the speed of light c_∞. The function itself is a quantization-function that allows some essential characteristics in the point-nature of light to express itself more "materially", relatively speaking, in sets described by (2.1.4), by light traveling at a relatively slower speed. This quantization-function can be thought of as a node in which a higher-level quantum-computation occurs by which infinite possibility begins to disaggregate itself.

The second transformation is summarized by Equation 2.1.8:

$$R_{C_N} = f(R_{C_K})$$

Eq. 2.1.8: Light-Matrix Second Transformation

This is suggesting that the reality at the layer specified by the speed of light c_N, R_{C_N}, is a function of the reality at the layer specified by the speed of light c_K. This transformation combines elements of the sets into unique seeds as suggested by Equation 2.1.5, and is envisioned as also being a quantization-function that allows what is expressed by the sets to gather into unique seeds.

The third transformation is summarized by Equation 2.1.9:

$$R_{C_U} = f(R_{C_N})$$

Eq. 2.1.9: Light-Matrix Third Transformation

This is suggesting that the reality at the layer specified by the speed of light c_U, R_{C_U}, is a function of the reality at the layer specified by the speed of light c_N. This transformation builds on the unique seeds suggested by (2.1.5) to create the diversity of U as specified by (2.1.1). This too must be a quantization-function that further allows all the combination of possibility to express itself in more material form.

In this framework the notion of wave-particle duality hence may become complementary block-field-wave-particle "quadrality" where block refers to phenomenon resident to ∞, field to phenomenon resident to N, wave to phenomenon resident to K, and particle to phenomenon resident to U. The block is all the reality always present behind the surface and is captured by (2.1.3) or the top line in the Light-Matrix as expressed in (2.1.6). The field is captured by (2.1.4) or the second major (non-parenthetical) line from the top in (2.1.6) and can be thought of as layers of possibility existing in each of the sets. The wave is captured by the creation of seeds represented by (2.1.5) or by the third major (non-parenthetical) line in (2.1.6). The particle that is apparently disconnected from the whole is captured by (2.1.1) or the bottom line in (2.1.6).

So what we arrive at is a fundamental Light-Matrix or the downward-strand of a ubiquitous subtle-DNA that suggests key dynamics for different layers of Light.

Layer Zero (0)

For the sake of completeness the thought-experiment to do with light traveling at a speed of zero must now be brought up. In Chapter 1.1 a possible reality that would result if light were to exist at this speed was described as opposite to the reality were light to travel at an infinite speed.

Hence, ubiquitous Presence of Light would become the ubiquitous Absence of Light, or Darkness. The aspect of Power would become utter Weakness. The aspect of Knowledge would become complete Ignorance. The aspect of Harmony would become total Chaos.

And all this because Light would be unable to travel from where it is, in complete opposition to its known nature of traveling at a speed of c, and perhaps its true nature when traveling at a speed of ∞. Light, hence, would be hidden in itself, so as to speak, in some sort of a negative infinity. An Equation, 2.1.10, Nature of Light at Speed Zero (0), would hence be:

c_0: $[Darkness, Weakness, Ignorance, Chaos]$

Eq. 2.1.10: Nature of Light at Speed Zero (0)

Further, (2.1.6), the Light-Matrix, would be modified to become Equation 2.1.11, Light-Matrix with Zero-Limit, where c_0 implies light at zero-speed, and D, W, I, and C imply Darkness, Weakness, Ignorance, and Chaos respectively:

56

$$Light_{Matrix}(0_{limit}) = \begin{vmatrix} c_\infty: [Pr, Po, K, H] \\ (\downarrow R_{C_K} = f(R_{C_\infty})) \\ c_K: [S_{Pr}, S_{Po}, S_K, S_H] \\ (\downarrow R_{C_N} = f(R_{C_K})) \\ c_N: f(S_{Pr} \times S_{Po} \times S_K \times S_H) \\ (\downarrow R_{C_U} = f(R_{C_N})) \\ c_U: [P, V, M, C] \\ \Uparrow \\ c_{0:[D,W,I,C]} \end{vmatrix}$$

Eq. 2.1.11: Light-Matrix with Zero-Limit

The implication of the bottom zero-limit line is that just as there is proposed to be an influence from the upper layers of light on U, as will be explored in subsequent chapters, so too there is a subtle influence from this lower layer of "light" that perhaps is responsible for the obstinacy of the untransformed (U) nature of practical reality. Such obstinacy will be further explored in Chapters 2.5, 2.6, and 3.7 and reinforces the need for the notion of 'transformation' in the first place. Further, such obstinacy also has a bearing on the phenomena of genetic mutation that will also be discussed in greater detail subsequently.

Superposition, Entanglement, Quantum Computation and their Relationship to Genetic Mutation

As may be apparent in the Light-Matrix (2.1.6) and the complete form (2.1.11), circumstance at U is the outcome of a number of influences from layers ∞, K, N, U itself, and Zero (0). But further since each of these layers is itself a reality caused by a different speed of light, the interface between these layers occurs at the quantum-level. Hence the quantum-level is replete with superposition emanating from layers ∞, K, N, U itself, and Zero (0). There is simultaneity and multiple possibilities that will determine what manifests at U. But as will be explored in more detail in subsequent chapters there is a logic or process to what manifests. This logic and process provides insight into the reality of genetic mutation.

The notion that a quantum-object – an entity at the quantum-level – can be in an infinite number of superposed states as is presumed in several leading interpretations of quantum mechanics, is called into question in the interpretation offered in this book. There may in fact be infinite number of possibilities vying for manifestation, but there is likely a more precise process by which this manifestation takes place. In other words, genetic mutation is not random, but follows process at the subtle-levels.

Further, the process of entanglement by which quantum-objects are made to relate to each other, may be otiose since in this interpretation offered in this book entanglement exists ab initio at the layer ∞. This entanglement will be referred to as ∞-entanglement and will be further explored in Chapter 2.2. Further, there are additional forms of entanglement that occur even before a quantum-object becomes perceivable at layer U. There is entanglement at layer K by virtue of field-type action of architectural sets. This will be referred to as K-entanglement and will be further explored in Chapter 2.3. There is an entanglement at layer N by virtue of the wave-type action of seeds. This will be referred to as N-entanglement and will be further explored in Chapter 2.4. Any process of entanglement is relevant from the point of view of genetic mutation as it can be thought of as a source of mutation.

Quantum computation is therefore a reality in the manifestation of the minutest of circumstances at U and perhaps can also be thought of as integral to the process of genetic mutation.

Chapter 2.2: The Ubiquitous-Point-Instant Pre-Genetic ∞-Entanglement Library

The 'ubiquitous-point-instant' captures the inherent nature that appears to exist in the system and is represented by the top-line in Equation (2.1.6) and (2.1.11), different forms of the Light-Matrix.

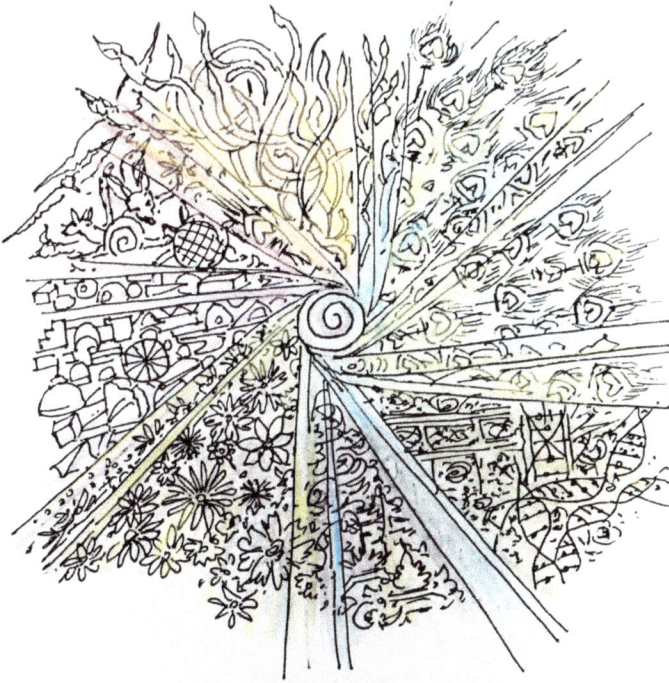

The ubiquitous-point-instant is a function of the dynamics of that reality where light travels at ∞ speed and is envisioned as being infinitely entangled by virtue of light being omnipresent-omnipotent-omniscient-omninurturing in that realm. In other words this nature is a function of the properties of light derived in Equation (2.1.3) - Presence, Power, Knowledge, and Harmony – reproduced here for convenience:

c_∞: [$Presence, Power, Knowledge, Harmony$]

The 'point' aspect of the ubiquitous-point-instant suggests the space-dimension and that space is seeded with the possibilities inherent in the properties of Presence, Power, Knowledge, and Harmony. The 'instant' aspect suggests the time-dimension and gives insight into the process of emergence that the possibilities in space progressively surface as. Therefore, the 'point' or space-aspect provides insight into structural aspects of genetic information. The 'instant' or time-aspect provides insight into the mutational aspects of genetics. Note though that such separation into point and instant aspects is just a way of trying to unpack some of the dynamics in that realm where everything, even space and time, are all one thing.

In its ubiquitous-point-instant, Presence-Power-Knowledge-Harmony wholeness, Presence allows emergences to continue to develop as per the possibilities implicit in the past-present-future or physical-vital-mental pathway.

Beginning to translate this into an equation, the notation $System_{Pr}$ is given to system-presence. Note that the derivation of this and many subsequent equations was part of my doctoral work at University of Pretoria (Malik, 2017a). This system-presence is true across any considered Time-Space continuum starting from a time-space boundary '0' to a time-space boundary 'N'. This notion is characterized by the notation $TS_{0 \to N}$. Within that boundary from 0 to 'N', the 'presence' is such that it will always seize an opportunity to cause a shift from the physical-leading to the vital-leading, and from the vital-leading to the mental-leading. Research shows (Malik, 2009) that greater degrees of freedom is afforded by such a shift.

The notion that the 'presence' seizes on 'opportunity' is characterized by the notation:

Presence
↓
Opportunity

The shift from physical-leading (P_L) to vital-leading (V_L) and vital-leading (V_L) to mental-leading (M_L) is characterized by:

$$P_L \rightarrow V_L$$
$$V_L \rightarrow M_L$$

Hence in this approach it is suggested that:

$$System_{Pr} \equiv TS_{0 \rightarrow N} \left[\begin{matrix} Presence \\ \downarrow \\ Opportunity \end{matrix} \begin{bmatrix} P_L & \rightarrow & V_L \\ V_L & \rightarrow & M_L \end{bmatrix} \right]$$

But there is something else about this Presence as well. All other developments take place in it. That is, it provides a container of sorts in which the plays of system-power, system-knowledge, and system-harmony/system-nurturing can take place. This notion is summarized by the notation:

$$Container \begin{bmatrix} System_P \\ System_K \\ System_N \end{bmatrix}$$

Hence, combining these various components, an equation for 'system-presence', Equation 2.2.1, arises:

$$System_{Pr}$$
$$\equiv TS_{0 \rightarrow N} \left[\begin{matrix} Presence \\ \downarrow \\ Opportunity \end{matrix} \begin{bmatrix} P_L & \rightarrow & V_L \\ V_L & \rightarrow & M_L \end{bmatrix} \& Container \begin{bmatrix} System_P \\ System_K \\ System_N \end{bmatrix} \right]$$

Eq 2.2.1: System Presence

In its ubiquitous-point-instant, Presence-Power-Knowledge-Harmony wholeness, Power allows emergences to continue to happen in spite of tremendous oppositions of all kinds; this too, regardless of field or area.

Constructing an equation for system-power, the notation $System_P$ is used to represent system-power. Any endeavor will always be met with resistances of various kinds. The resistances that arise along the physical dimension are referred to as P_R. The resistances that arise along the vital dimension are referred to as V_R. The resistances that arise along the mental dimension are referred to as M_R. In the fruition of any endeavor one or all of these types of resistances may arise. Further, resistance of one kind often feeds on resistance of another kind, and to generalize the resistances encountered in an endeavor, these may be characterized as the product of the three types of resistance:

$$P_R \ * \ V_R \ * \ M_R$$

These resistances arise across any considered Time-Space boundary from 0 to 'N', and therefore it may be said that the power of the system is such that:

$$power > \sum_{TS=0}^{N} P_R \ * \ V_R \ * \ M_R$$

An equation for 'system-power', Equation 2.2.2, hence, is the following:

$$System_P \equiv power > \sum_{TS=0}^{N} P_R \ * \ V_R \ * \ M_R$$

Eq 2.2.2: System Power

In its ubiquitous-point-instant, Presence-Power-Knowledge-Harmony wholeness, Knowledge orchestrates emergences to continue to happen by leveraging the right instruments and circumstances.

Translating this into an equation, the notation, $System_K$, is used for system-knowledge. This $System_K$ is such that it leverages the right instrumentation and circumstance to bring about the progress that is possible. This concept of 'instrumentation' is denoted by the subscript 'I'. The concept of 'circumstance' is denoted by the subscript 'C'. Both instrumentation and circumstance can be of a physical, vital, or mental type and this possibility is denoted by:

$$\begin{bmatrix} P_{I,C} \\ V_{I,C} \\ M_{I,C} \end{bmatrix}$$

Further, the notion that the 'knowledge' is such that it 'leverages' the right instrumentation and circumstance is depicted by:

$$Knowledge$$
$$\downarrow$$
$$Leverage$$

This act of leveraging results in a fundamental shift so that the physical-leading yields to the vital-leading, and the vital-leading yields to the mental-leading. Hence:

$$\begin{matrix} Knowledge \\ \downarrow \\ Leverage \end{matrix} \begin{bmatrix} P_{I,C} \\ V_{I,C} \\ M_{I,C} \end{bmatrix} \rightarrow \begin{bmatrix} P_L & \rightarrow & V_L \\ V_L & \rightarrow & M_L \end{bmatrix}$$

Since this behavior may exist across any Time-Space continuum an equation for system-knowledge, Equation 2.2.3, is suggested:

$$System_K \equiv TS_{0 \to N} \begin{bmatrix} Knowledge & \begin{bmatrix} P_{I,C} \\ \downarrow & V_{I,C} \\ Leverage & M_{I,C} \end{bmatrix} & \to & \begin{bmatrix} P_L & \to & V_L \\ V_L & \to & M_L \end{bmatrix} \end{bmatrix}$$

Eq 2.2.3: System Knowledge

In its ubiquitous-point-instant, Presence-Power-Knowledge-Harmony wholeness, Harmony or nurturing allows emergences to continue to happen with more and more degrees of freedom coming to the surface.

The characteristic of this implicit-nurturing may be referred to as 'system-nurturing'. Like the other characteristics it is suggested to exist across a Time-Space continuum. This is depicted by:

$$TS_{0 \to N}$$

There is an action of nurturing such that any state is always advanced to a higher level. This is depicted by:

$$\coprod_{Nurturing} \begin{pmatrix} P_- & M_+ \\ V_- & V_+ \\ M_- & P_+ \end{pmatrix}$$

Hence, there is a 'union', depicted by 'U' that 'nurtures' the negatives towards their positives.

Further, there is an increasing action of nurturing such that the possibility of integration is always increased to form a larger and larger basis. This increasing basis is depicted as being modulated by the polar coordinates 'r' and 'θ', where r is the radius which increases from an initial value of '0', and 'θ' is an angle from '0' to '360'.

Hence, the equation of system-nurturing, Equation 2.2.4, is depicted as:

64

$$System_N \equiv TS_{0 \to N} \left(\coprod_{Nurturing} \begin{pmatrix} P_- & M_+ \\ V_- & V_+ \\ M_- & P_+ \end{pmatrix} mod\ (r, \theta) \right)$$

Eq 2.2.4: System Nurturing

It is suggested that these four characteristics exist across any system, and to denote this it is generalized that every point in any system is embedded with this four-fold intelligence. In other words there is in effect a library of information represented by Equations 2.2.1 – 2.2.4 whose operation is governed by ∞-entanglement. Further, there is a category of constructive genetic mutation, ∞ - entanglement mutation that may be activated to bring about change at the material level.

Libraries created at subsequent levels of the downward-strand, and the very process of genetic mutation involving the material level, are influenced by this subtlest level ∞-entanglement library.

Chapter 2.3: - The Architectural Forces K-Entanglement Pre-Genetic Library

The characteristics embedded in a ubiquitous-point-instant as in (2.1.3) suggest a possibility that is hard to fathom. One can only glimpse this extraordinary nature. And yet it can be suggested that this extraordinary nature is barely visible unless the right analytical lens of the sort being suggested in this book is first set up. Further, it is suggested that this extraordinary nature is responsible for a broader set of architectural forces that exist behind the visible face of things. Such architectural forces can be envisioned as emanating from a pre-genetic library existing in a reality where light travels at the speed c_K (above c, and closer to ∞, as defined in Chapter 2.1). Due to the process of K-entanglement this pre-genetic library can be thought of influencing genetic mutation.

Hence, system-presence, system-power, system-knowledge, and system-nurturing that define the nature of every point in our system, become more tangible as a broader set of architectural forces that emanate from each of them.

Considering system-presence, here is a characteristic that appears to be everywhere (Malik, 2015) at the service of all the constructs that develop within it. There is a diligence and perseverance by which any opportunity for progress is seized. Further, if one considers the extraordinary detail that appears in any construct, whether an atom, a body, a planet, or a galaxy, one is struck by the high degree of perfection that surfaces in this presence.

So if one contemplates the nature of this system-presence there is a set of forces that surface. Depicting such a set as $S_{System_{Pr}}$, one can arrive at elements such as Service, Perfection, Diligence, Perseverance, amongst others, that are part of this set. Hence, the set can be described by Equation 2.3.1:

$$S_{System_{Pr}} \ni [Service, Perfection, Diligence, Perseverance, ...]$$

Eq 2.3.1: Set of System Presence

But from the point of view of the speed of light, K is a reality set up by light traveling at c_K, where $c_U < c_N < c_K < c_\infty$. Conversely assuming an inverse proportionality with 'h', as c_K is closer to ∞, h will be closer to zero, and 'matter' or form will be highly dispersed. This dispersal is presumed to be field-like so that in effect any element of the set described in (2.3.1) or in the subsequent sets (2.3.2-4) will have a field-like reality and express K-entanglement such that all emergences emanating or comprised of that element will automatically partake in an entanglement with every other unique emergence having that element as part of its foundation. K-entanglement is therefore different from ∞-entanglement and every emergence will at least have both these types of entanglements subtly coordinating or influencing its action.

Similarly, considering the characteristic of system-power, one can hypothesize that there is a family of forces that emanates from it. The kinds of forces may be thought of as Power, Courage, Adventure, Justice, amongst others. The set for system-power can hence be depicted by Equation 2.3.2:

$$S_{System_P} \ni [Power, Courage, Adventure, Justice, ...]$$

Eq 2.3.2: Set of System Power

Similarly, considering the system-knowledge as the root of various powers that emanate from it, one may characterize the set for system-knowledge by Equation 2.3.3:

$S_{System_K} \ni [Wisdom, Law\ Making, Spread\ of\ Knowledge \ldots]$

Eq 2.3.3: Set of System Knowledge

The set for system-nurturing is depicted by Equation 2.3.4:

$S_{System_N} \ni [Love, Compassion, Harmony, Relationship \ \ldots]$

Eq 2.3.4: Set of System Nurturing

Equations 2.3.1 – 4 describe potentially infinite-element sets that comprise the Architectural Force K-Entanglement Pre-Genetic Library. Further, there is a category of constructive genetic mutation, K-entanglement mutation that may be activated to bring about change at the material level.

Chapter 2.4: The Organizational-Uniqueness N-Entanglement Pre-Genetic Library

Philosophically, even in the absence of the mathematical model that leads from unity to unique diversity as is being discussed in this book, one would have to arrive at a thesis of vast and unique diversity just from observation of phenomena.

The hypothesis based on such observation, therefore, is that every organization, whether an atom, cell, person, team, corporation, market, or country is unique and that this uniqueness can be specified in terms of elements of the derived sets for power, knowledge, presence, and nurturing.

In other words there is a unique logic that defines the functioning of any organization at the deeper subtle level, and this is encapsulated by logical-ecosystems spawned by subtle-DNA. While we are currently exploring the structure of the downward-strand of subtle-DNA in nodes that also define different schemes of entanglement, we will in subsequent chapters also explore the upward-strand, and the manner in which specific four-base logical-ecosystems are spawned. This hypothesis for uniqueness stems from observations at multiple levels.

At the sub-atomic level Nobel Laureate Wolfgang Pauli's 'Pauli Exclusion Principle' states that no two similar fermions, which include fundamental particles with half-integer spin such as protons, neutrons, and electrons, can occupy the same quantum states simultaneously (Pauli, 1964). Spin has to do with the angle that the particle has to rotate through before being symmetrical with its original state. Half-integer spin particles need to rotate through 720 degrees before being symmetrical with their original state. The implication of the Pauli Exclusion Principle is that fundamental structure and consequently stability comes into being at the atomic level, which as is evident in the Periodic Table allows the separation of function related to form. This stability related to the underlying structure of atoms implies the basis of uniqueness and diversity. In the absence of the Exclusion Principle matter would just be a dense soup (Hawking, 1988) with particles occupying overlapping space.

At the observable level uniqueness is evident from the immense diversity of distinct species on earth (Mora, 2011) estimated to be over 2 million, and further the uniqueness of every member of each species. This member-level uniqueness is suggested by the difference in non-coding regions of the DNA that may vary in their sequence by about 1 to 4 percent, which in turn result in unique protein binding sequences of each human (Snyder, 2010), as an example, which in turn results in unique observable qualities.

At the astronomical level Einstein's Special Theory of Relativity (Einstein, 1995) suggests that every coordinate system potentially has its own space-time rendering as opposed to there being one absolute space and time. This implies the notion of uniqueness as an implicit property of space.

The four properties explored in Chapter 2.1 and 2.2 define the source of that uniqueness. From this source emanate 4 sets of forces that suggest the boundaries of that uniqueness as explored in Chapter 2.3.

Assuming then that the fount of uniqueness is system-presence, a general equation for organizations that belong to the family of system-presence can be derived. Such uniqueness can be depicted as Sig_x where the subscript 'x' refers to the source family, and 'Sig' or signature to 'uniqueness'. Hence the uniqueness of an organization in the family of system-presence would be notated by $Sig_{System_{Pr}}$.

In line with the development of properties of a point and the precipitating architectural forces as discussed in Chapter 2.1 and 2.2 respectively, an approach to constructing such uniqueness is to assume a primary factor X that drives the uniqueness that belongs to the set $S_{System_{Pr}}$. Further, assume that the uniqueness is qualified by a number of secondary factors Y that may belong to any of the 4 sets - $S_{System_{Pr}}, S_{System_P}, S_{System_K}, S_{System_N}$. The primary factor X would have a greater weightage than any of the secondary factors Y. The weightage of X hence could be depicted by the number 'a', and the weightage of Y a number 'b_{0-n}', such that a > b. Further, the secondary element can repeat from '0 – n' times, and is hence depicted as $\overline{Yb_{0-n}}$.

The equation, Equation 2.4.1, hence for a unique organization derived from the family of system-presence is:

$$Sig_{Pr} = Xa +$$

$$\overline{Yb_{0-n}} \ where \ \begin{bmatrix} X \in [S_{System_{Pr}}] \\ Y \in [S_{System_{Pr}}, S_{System_P}, S_{System_K}, S_{System_N}] \\ a, b \ are \ integers; a > b \end{bmatrix}$$

Eq 2.4.1: System Presence Based Unique Organization

But from the point of view of the speed of light, N is a reality set up by light traveling at c_N, where $c_U < c_N < c_K < c_\infty$. Conversely assuming an inverse proportionality with 'h', as c_N is closer to but greater than c_U, h_U will be less than h, and 'matter' or form will be unable to accumulate as it does at U, instead tending to disperse like a wave. The wave-like nature implies an entanglement, N-entanglement, such that any emergence will always be unique. Uniqueness could not be unless there was a dynamic such as N-entanglement in place.

Similarly, an equation, Equation 2.4.2, for a unique organization derived from the family of system-power is:

$$Sig_P = Xa + \overline{Yb_{0-n}} \quad where \quad \begin{bmatrix} X \in [S_{System_P}] \\ Y \in [S_{System_{Pr}}, S_{System_P}, S_{System_K}, S_{System_N}] \\ a, b \text{ are integers}; a > b \end{bmatrix}$$

Eq 2.4.2: System Power Based Unique Organization

An equation, Equation 2.4.3, for a unique organization derived from the family of system-knowledge is:

$$Sig_K = Xa + \overline{Yb_{0-n}} \quad where \quad \begin{bmatrix} X \in [S_{System_K}] \\ Y \in [S_{System_{Pr}}, S_{System_P}, S_{System_K}, S_{System_N}] \\ a, b \text{ are integers}; a > b \end{bmatrix}$$

Eq 2.4.3: System Knowledge Based Unique Organization

An equation, Equation 2.4.4, for a unique organization derived from the family of system-nurturing is:

$$Sig_N = Xa + \overline{Yb_{0-n}} \quad where \quad \begin{bmatrix} X \in [S_{System_N}] \\ Y \in [S_{System_{Pr}}, S_{System_P}, S_{System_K}, S_{System_N}] \\ a, b \text{ are integers}; a > b \end{bmatrix}$$

Eq 2.4.4: System Nurturing Based Unique Organization

The four preceding equations can be generalized by Equation 2.4.5:

$$Sig = Xa + \overline{Yb_{0-n}} \quad where \quad \begin{bmatrix} X \in [S_{System_{Pr}}, S_{System_P}, S_{System_K}, S_{System_N}] \\ Y \in [S_{System_{Pr}}, S_{System_P}, S_{System_K}, S_{System_N}] \\ a, b \text{ are integers}; a > b \end{bmatrix}$$

Eq 2.4.5: Generalized Equation for Unique Organization

Equations 2.4.1 – 5 describe potentially infinite-element sets that comprise the Organizational-Uniqueness N-Entanglement Pre-Genetic Library.

The N-entanglement level, like the ∞-entanglement and K-entanglement levels is also hypothesized as being that part of the backbone of subtle-DNA, where distinct elements of the deeper fourfold-based sets combines, to create unique seeds. These seeds are the primary organizing force at the center of any and every organization. Hence the subtle-library behind any emergence of matter and life derives primarily from the action on the N-entanglement level, and further, there is a category of constructive genetic mutation, N-entanglement mutation, that may be activated to bring about change at the material level.

The next two chapters turn away from the structural or space-aspect or structure of genetics, and focus on the emergent, mutational, or time-aspect of genetics. Hence Chapter 2.5 will focus on the possibilities that can come about through mutation. Essentially the deep structural basis of genetics hinted at by the ∞-entanglement, K-entanglement, and N-entanglement levels, also becomes the basis for mutational-emergence. In other words, the possibilities inherent in these levels can change life at the genetic-level through mutation.

Chapter 2.6 will focus on the systematization of the mutational-sequence discussed in Chapter 2.5 into a mathematical framework.

Chapter 2.5: A Possible Mutational-Sequence and the Upward-Strand

While the uniqueness of organizations as represented by the Signature is a seed, like any seed there is a process for its emergence (Kaufmann, 1995; Portugali, 2012; Yates, 2012), and the uniqueness will often be hidden or very much behind the scene until certain conditions are fulfilled (Malik, 2009). What this implies is that possibility may emerge through mutation, and further some of that mutation may either tap into the subtle-libraries created through the structures already set up by the ∞ -entanglement, K-entanglement, and N-entanglement levels, or by a fresh interaction with the layer represented by these nodes, as will be further explored in Chapter 2.6.

Note that the libraries so set up by the ∞-entanglement, K-entanglement, and N-entanglement levels would be quite different from the Library of Babel, referred to in Beinhocker's Origin of Wealth (Beinhocker, 2006). The Library of Babel is imagined to contain all the possible 500-page books in the English language and would be vastly larger than the universe. A vast majority of these books would be gibberish with characters strung together in random fashion. By contrast the subtle-libraries referred to in this book are a play on qualities or functions related to the four-foldness implicit in Light. These libraries potentially contain an infinite number of positive and useful functions that will be subtly available to inform any play involving genetic-type information.

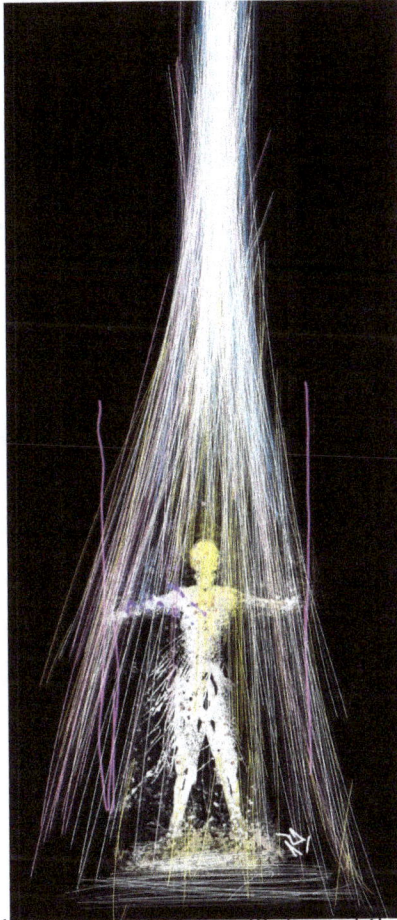

The implicit nature of Time and Space suggest a universal developmental model that provides a cue as to the process for emergence. In this model the four sets of architectural forces and the combination of their elements form a pool in space, as already discussed, from which possibility arises. Possibility itself is unique from point to point and is governed by the Equation for Uniqueness (Equations 2.4.1 through 2.4.5) described in the previous chapter.

In other words, Space contains seeds, as suggested in Section 1, and as just discussed in the previous chapters in Section 2, seeds can be thought of as the result of the superposition of ever-present ∞-entanglement, K-entanglement, and N-entanglement. Thus it could be said that Space is filled with multiple levels of superposed entanglements in static form.

Time on the other hand appears to be the working out of the possibilities implicit in these superposed entanglements. The inevitable trajectory implicit in these superposed entanglements will typically follow a mutational-sequence to be described in this chapter. What is variable is the amount of time for the mutational-sequence to express itself, as this will depend on the strength of the different forces or influences from each of the different layers so set up by light, that are active.

In equation form Space could be described by Equation 2.5.1:

$Space = STATIC(superposition$

$(\infty - entanglement, K - entanglement, N - entanglement)$

Eq 2.5.1: Space

In equation form Time could be described by Equation 2.5.2:

$Time = DYNAMIC(superposition$

$(\infty - entanglement, K - entanglement, N - entanglement)$

Eq 2.5.2: Time

Hence it is observed that initially the mutational-sequence takes a 'physical' form, moving on to a 'vital' form, and then onto a 'mental' form. Relating these forms to the equation-segment in (2.1.11), the shaded lines suggest which set of dynamics tend to be more active:

$$c_\infty : [Pr, Po, K, H]$$
$$(\downarrow R_{C_K} = f(R_{C_\infty}))$$
$$c_K : [S_{Pr}, S_{Po}, S_K, S_H]$$
$$(\downarrow R_{C_N} = f(R_{C_K}))$$
$$c_N : f(S_{Pr} \times S_{Po} \times S_K \times S_H)$$
$$(\downarrow R_{C_U} = f(R_{C_N}))$$
$$c_U : [P, V, M, C]$$
$$\Uparrow$$
$$c_{0:[D,W,I,C]}$$

Once the characteristics implicit in each of these phases are assimilated, then the mutational-sequence takes on an 'integral' form. In this case the more active sets of dynamics would be:

$$c_\infty : [Pr, Po, K, H]$$
$$(\downarrow R_{C_K} = f(R_{C_\infty}))$$
$$c_K : [S_{Pr}, S_{Po}, S_K, S_H]$$
$$(\downarrow R_{C_N} = f(R_{C_K}))$$
$$c_N : f(S_{Pr} \times S_{Po} \times S_K \times S_H)$$
$$(\downarrow R_{C_U} = f(R_{C_N}))$$
$$c_U : [P, V, M, C]$$
$$\Uparrow$$
$$c_{0:[D,W,I,C]}$$

The integral form is a threshold phase, and allows the uniqueness suggested by the Signature to emerge in fuller force or in its 'force' form.

The final phase is the 'contextual form' that allows the signature to act with impunity within a considered context.

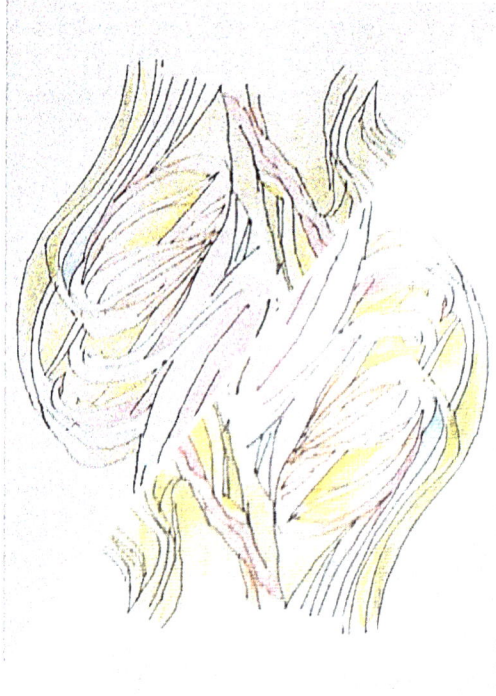

$$c_\infty : [Pr, Po, K, H]$$
$$(\downarrow R_{C_K} = f(R_{C_\infty}))$$
$$c_K : [S_{Pr}, S_{Po}, S_K, S_H]$$
$$(\downarrow R_{C_N} = f(R_{C_K}))$$
$$c_N : f(S_{Pr} \: x \: S_{Po} \: x \: S_K \: x \: S_H)$$
$$(\downarrow R_{C_U} = f(R_{C_N}))$$
$$c_U : [P, V, M, C]$$
$$\Uparrow$$
$$c_{0:[D,W,I,C]}$$

Mathematically, if an organization exists at the physical phase, it may be suggested that its signature or uniqueness is modulated by the constant 'π'. π is the seed of a circle or sphere and can be thought of as defining behavior that is tightly bound. Within such a tightly bound volume it will likely not even be apparent what the uniqueness of an organization necessarily is. Assuming the uniqueness to be defined by the derived equation *Sig*, the physical-level (P) behavior can be described by the following equation-segment where 'mod' signifies modulated-by:

$P: Sig * mod (\pi)$

If an organization exists at the vital level, it may be suggested that its uniqueness is modulated by the Euler-constant 'e'. e is at the root of exponential behavior. The vital, by definition, is about assertive and aggressive growth the symbol of which is 'e'. Hence vital-level (V) modulation (represented by 'mod') can be described by the following equation-segment:

$V: Sig * mod (e)$

If an organization exists at the mental level, it may be suggested that its uniqueness is modulated by the Gaussian Distribution 'G'. G summarizes rational behavior with a key direction followed by most, and directions more on the edge followed by outliers. Mental-level dynamics are arguably quite similar, and it can be suggested are best modeled by such a distribution (Salkind, 2007). Mental-level (M) modulation (mod) can hence be described by the following equation-segment:

$M: Sig * mod (G)$

The physical, the vital, and the mental levels are orientations in which patterns of perceiving, being, behaving are set in their ways. Each pattern has its purpose and its limitation and it can be argued that being able to learn from each orientation and yet being able to move beyond that, is the next logical step in any developmental model. The integral level hence, is about being able to leverage each of the patterns that naturally arise at the three preceding levels at will, and about further, being able to integrate these and arrive at new ways of perceiving and being.

Mathematically such behavior may be represented as being an integrative function ($\int x$) where 'x' is the ability to move between the patterns emanating from G, e, π, at will, represented by $\overline{G, e, \pi}$. Integral-level (I) modulation (mod) of uniqueness (*Sig*) can hence be represented by the following equation-segment:

$$I:\ Sig * mod \left(\int \overline{G, e, \pi} \right)$$

The condition of overcoming any fixed and limiting patterns is the prerequisite for the emergence of 'Force' or for entering into the force-level. At this level the uniqueness behind the particular development being considered can emerge in its purity and become a truly creative dynamic. This aspect of creativity that is in a sense not bound by circumstance may be represented by the constant 'c', the speed of light in a vacuum, which is an upper limit of the layer that systems practically operate in. This is also likely the level at which N-entanglement is overtly active. Force-level (F) modulation (mod) of uniqueness (*Sig*) can hence be represented by the following equation-segment:

$$F:\ Sig * mod\ (c)$$

Once the signature of an organization arises and continues to exercise itself in its purity, it achieves contextual-mastery (C) and is able to exercise itself as though the context it is acting in, that can vary in scale and complexity, were all of the same substance as itself. This is likely the level at which K-entanglement is active. This equality may be represented by the integrative function '$\int = 1$'. The equation-segment that notates this contextual-level (C) modulation (mod) applied to organizational uniqueness (*Sig*) is hence:

$$C: Sig * mod \left(\int = 1 \right)$$

Piecing all the equation-segments together the equation for the emergence of uniqueness (*Sig$_E$*), where 'X' can be any of the discussed modulations at the respective development-model levels (P, V, M, I, F, C), is hence summarized by Equation 2.5.3:

$$Sig_E = X \begin{vmatrix} C: Sig * mod \left(\int = 1 \right) \\ F: Sig \, mod \, (c) \\ I: Sig \, mod \left(\int \overline{G, e, \pi} \right) \\ M: Sig * mod \, (G) \\ V: Sig * mod \, (e) \\ P: Sig * mod \, (\pi) \end{vmatrix}$$

Eq 2.5.3: Mutational-Sequence or Emergence of Uniqueness

What is implied by (2.5.3) is that there is a bias in the phenomenon of mutation so that an implicit sequence is followed such that more and more of the uniqueness of an organization may emerge. Material change is therefore tightly tied in with implicit function. But this has to be if all is just a play of Light.

Further, the mutational-sequence also implies an upward-strand to the ubiquitous subtle-DNA.

Having thus explored a possible mutational-sequence, Chapter 2.6 will further systematize this. The P, V, M phases of the sequence will be found to belong to the untransformed or U-layer as will be further discussed.

Chapter 2.6: Systematization of Mutational-Sequence Along the Time-Based Upward-Strand

So far the inherent creativity in Light as summarized by four overarching properties in the nature of a point and its attendant '∞ - entanglement' has been considered. Further, how this deep fount of creativity is present everywhere and how sets with their attendant 'K-entanglement' that make more practical the range of creative forces available in each of the four components of a point have also been considered. These architectural forces further define the possibility inherent in any system. Leveraging these sets of forces by virtue of 'N-entanglement' an equation for the uniqueness of an organization, regardless of scale, was arrived at.

In some sense the precipitation of creativity from the barely perceptible nature of the ubiquitous point, to how this reveals a play of forces, to how organizations take their seed and grow from these forces, giving insight too into the dynamics of entanglement and superposition in Space and Time, has been traced. It is proposed that such creativity is what also spawns subtle-libraries that elaborate the structure of subtle-DNA, and this will be explored in more detail in later chapters.

In Chapter 2.5 the process of a possible mutational-sequence arising in time was explored. While such a sequence is intricately tied with the precipitating layers of light, and in fact can be thought of as a reversal of the downward precipitation, thereby reinforcing the notion a subtle-DNA since DNA at the material level is composed of two tightly related strands in opposite directions, this chapter will further explore a systematization of such a mutational-sequence.

Such an upward-strand is organic in nature suggesting likely paths or mutational-sequences determined by what sets of influences are active. Yet the possibilities are wholly determined by the downward-strand and the overarching organizational principles it puts in place in its precipitating scheme from an infinite to zero speed.

Hence, starting with the physical, which recall is suggested as being a projection of Light's property of Presence, an equation, Equation 2.6.1, is summarized as:

$$Physical = \begin{bmatrix} M_3 \rightarrow System_{Pr} \\ (\uparrow F \rightarrow I) \\ M_2 \rightarrow S_{System i_{Pr}} \\ (\uparrow Sig \rightarrow F) \\ M_1 \rightarrow Sig_P \\ (\uparrow > P_P) \\ U \rightarrow Physical_U \end{bmatrix} TC \rightarrow Physical_T$$

$$Where \begin{bmatrix} Physical_U \ni [inertia, lethargy, status\ quo, ...] \\ Physical_T \ni [adaptability, durability, strength, ...] \end{bmatrix}$$

Essentially this equation is laying out the conditions of moving from the untransformed or negative physical state represented by *Physical$_U$* to the transformed or positive physical state represented by *Physical$_T$*.

The first matrix should be read from the bottom to the top:

$$\begin{bmatrix} M_3 \rightarrow System_{Pr} \\ (\uparrow F \rightarrow I) \\ M_2 \rightarrow S_{System_{Pr}} \\ (\uparrow Sig \rightarrow F) \\ M_1 \rightarrow Sig_P \\ (\uparrow > P_P) \\ U \rightarrow Physical_U \end{bmatrix}$$

Hence, at the bottom is the starting point ' $U \rightarrow$ *Physical$_U$* ' which identifies the default or untransformed (U) level of the physical. The next row up, $(\uparrow > P_P)$, states that when the patterns of the untransformed physical (P_P) have been overcome (>), movement to the next level ($\uparrow$) is facilitated. Breaking through to the next level, $M_1 \rightarrow Sig_P$, allows its dynamics to become active. Hence, the signature or uniqueness of the physical (Sig_P) becomes active at meta-level 1 (M_1). As this signature becomes more like a Force ($Sig \rightarrow F$), the conditions for breakthrough ($\uparrow$) to the next level are achieved. This next level is referred to as meta-level 2

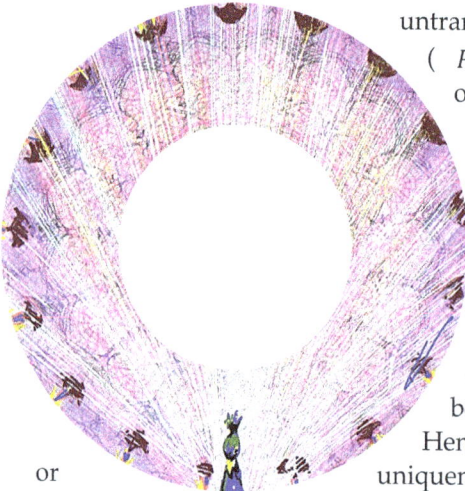

84

(M_2), and indicates that the architectural forces represented by the set of system-presence ($S_{System_{Pr}}$) have become more consciously active. When this Force becomes Integral ($F \rightarrow I$) then the conditions for breakthrough (↑) to the next level are achieved. The next level is notated as M_3 for meta-level 3, and the dynamics here indicate that the equation for system-presence becomes active. Becoming active basically means that the respective meta-level dynamic begins to act at the once 'untransformed' level (U) further modifying it. Modification or transformation began when M_1 became active. Transformation is accelerated when M_2 becomes active, and even further accelerated when M_3 becomes active.

The rate of the transformation can be better envisioned when considering action of the Transformation Circle, or TC. The TC can be thought of as 4 concentric circles, with M_3 at the center. M_3 is surrounded by M_2, which is surrounded by M_1. The outer circle is U. If TC is considered to be a clock, than at time 't = 0', the 'physical' can be thought of as being entirely in U. The clock starts ticking only when some initial patterns P_p are overcome (>P_p). From this point on as time proceeds the conditions for breakthrough become riper, and a sinusoidal wave begins to integrate more of the concentric circles together. The sinusoid wave (sin) is itself modulated by an euler function, e^x, where 'x' is determined by the strength to overcome patterns (↑) which will likely vary over time but will likely tend to be positive once the clock has started ticking because of the joy experienced with progressive movement. Being that the limit is the outer boundary of the concentric circles, there is further modulation by π until the 4 concentric circles have been integrated. TC, hence, may be represented by Equation 2.6.2:

$$TC \equiv (> P_p) \rightarrow \mathrm{mod}\,(\sin, e^x, \pi)$$

Hence, the initial nature of the physical that may be characterized by the set comprising of elements such as, lethargy, acceptance of the status quo, amongst other such elements, is represented by:

$$(Physical_U \ni [inertia, lethargy, status\ quo, ...])$$

This transforms into a physical more characterized by elements such as adaptability, durability, strength, and so on. That is:

$$(Physical_T \ni [adaptability, durability, strength, ...])$$

This transformation represents the inherent creativity-dynamic driving any mutation sequence within physical-type systems.

Such transformation as discussed in the previous chapters will implicitly involve the mechanisms of superposition and entanglement as emergence takes place.

Similarly, the equation for the 'Vital', Equation 2.6.3, which recall is suggested as being a projection of Light's property of Power, also shows the built-in transformation that represents the innovation-dynamic within the vital:

$$Vital = \begin{bmatrix} M_3 & \rightarrow & System_P \\ & (\uparrow F & \rightarrow I) \\ M_2 & \rightarrow & S_{System_P} \\ & (\uparrow Sig & \rightarrow F) \\ M_1 & \rightarrow & Sig_V \\ & (\uparrow > P_V) \\ U & \rightarrow & Vital_U \end{bmatrix} TC \rightarrow Vital_T ,$$

$$Where \begin{bmatrix} Vital_U \ni [aggression, self\ centeredness, exploitation, ...] \\ Vital_T \ni [energy, support, adventure, enthusiasm, ...] \end{bmatrix}$$

Eq 2.6.3: Possible Mutational-Sequence of Vital-Type Systems

The equation for the 'Mental', Equation 2.6.4, which recall is suggested as being a projection of Light's property of Knowledge, is similarly summarized as:

$$Mental = \begin{bmatrix} M_3 & \to & System_S \\ & (\uparrow F & \to I) \\ M_2 & \to & S_{System_S} \\ & (\uparrow Sig & \to F) \\ M_1 & \to & Sig_M \\ & (\uparrow > P_M) \\ U & \to & Mental_U \end{bmatrix} TC \to Mental_T$$

$$Where \begin{bmatrix} Mental_U \ni [fixation, fundamentalism, fragmentation, ...] \\ Mental_T \ni [understanding, imagination, inspiration, ...] \end{bmatrix}$$

Eq 2.6.4: Possible Mutational-Sequence of Mental-Type Systems

The equation for the 'Integral', Equation 2.6.5, suggested as being a projection of Light's property of Harmony, is similarly summarized as:

$$Integral = \begin{bmatrix} M_3 & \to & System_N \\ & (\uparrow F & \to I) \\ M_2 & \to & S_{System_N} \\ & (\uparrow Sig & \to F) \\ M_1 & \to & Sig_I \\ & (\uparrow > P_I) \\ U & \to & Integral_U \end{bmatrix} TC \to Integral_T$$

$$Where \begin{bmatrix} Integral_U \ni [possession, usurpation, hidden\ agendas, ...] \\ Integral_T \ni [appreciation, shift\ POV, MPV, synthesis, ...] \end{bmatrix}$$

Eq 2.6.5: Possible Mutational-Sequence of Integral-Type Systems

The preceding equations can be generalized by Equation 2.6.6:

$$Innovation_{orientation-x}$$

$$= \begin{bmatrix} M_3 \rightarrow System_X \\ (\uparrow F \rightarrow I) \\ M_2 \rightarrow S_{System_X} \\ (\uparrow Sig \rightarrow F) \\ M_1 \rightarrow Sig_x \\ (\uparrow > P_x) \\ U \rightarrow x_U \end{bmatrix} TC \rightarrow x_T, where \begin{bmatrix} x_U \ni [...] \\ x_T \ni [...] \end{bmatrix}$$

Eq 2.6.6: Generalized Mutational-Sequence Equation

In this generalized equation, $Innovation_{orientation-x}$, refers to the inherent innovation within a specific orientation. Orientation refers to the physical, the vital, the mental, or the integral.

Further, the notion of a core-matrix can be summarized by the following equation, Equation 2.6.7:

$$Core\_matrix = \begin{bmatrix} M_3 \rightarrow System_X \\ (\uparrow F \rightarrow I) \\ M_2 \rightarrow S_{System_X} \\ (\uparrow Sig \rightarrow F) \\ M_1 \rightarrow Sig_x \\ (\uparrow > P_x) \\ U \rightarrow x_U \end{bmatrix}$$

Eq 2.6.7: Core Matrix

SECTION 3: SOME ASPECTS OF THE MATHEMATICS RELEVANT TO GENETIC MUTATION

The previous section outlined the structure of the downward and upward strands of subtle-DNA. The downward-strand is caused by light slowing down in quantized-decelerations, as it were, progressively concretizing more of the information in light. The upward-strand is envisioned as prescribing a time-variable sequence wholly determined by the nature of the layers in the downward-strand. The time-variability is envisioned as being the result of the interplay of the active influences emanating from the layers of light in the downward-strand. But further, each of the layers of light spawn subtle-libraries of pre-genetic information. These subtle-libraries are effectively infinitely large encoding many different possibilities.

This section will illustrate some of the interplay involving these subtle-libraries and a first projection of their possibilities in a Space-Time-Energy-Gravity construct. The Space-Time-Energy-Gravity construct is positioned as being fundamental in altering genetic information at the material level. This will be elaborated in subsequent sections. Further, numerous other constructs that have their logic determined in 'four-base logic-encoding ecosystems', envisioned as existing in the quantum layer antecedent to the material layer, will also be explored in subsequent sections. It is such ecosystems that can change due to interplay with the material layer. In the interplay or call from 'below', as it were, there can be though to be a response from 'above' that may precipitate one of the infinite functions already created at subtle-libraries at the ∞-entanglement, K-entanglement, and N-entanglement levels.

As such, this section will reinterpret some basic equations and existing quantum mechanics interpretations based on the Cosmology of Light view adopted in this treatise. Further, it will focus on the derivation of the Light-Space-Time Emergence equation leveraged in subsequent quantization analyses. The Light-Space-Time Emergence equation models the basis for quantization suggesting emergent reality for all phenomena from Light. Schrodinger's wave equation and Heisenberg's uncertainty principle are also interpreted from the point of view of the Light-based Interpretation of quantum mechanics central to this treatise, to reinforce the notion that even when considered from these points of view the existence of multiple layers of light is feasible. The chapter on quantization of space, time, matter and gravity models how these phenomena are related to Light and subsequently also models how these fundamentals work together to potentially impact genetic-type information.

Chapters in this section will focus on:

- Equation for Light-Space-Time Emergence
- Speed of Light and Quanta
- Interpreting Schrodinger's Equation
- Interpreting Heisenberg's Uncertainty Principle
- A Deeper Look at Quantization of Space, Time, Matter, and Gravity
- Effect of Levels of Light on Genetic Mutation
- Application of Qualified Determinism to Genetic Mutation

Equation 2.6.6, the generalized equation for mutational-sequence, can be restated as an evolving form true for all time, as in Equation 3.1.1:

$$Innovation_{orientation-x}$$

$$= \left(\begin{bmatrix} M_3 \to System_X \\ (\uparrow F \to I) \\ M_2 \to S_{System_X} \\ (\uparrow Sig \to F) \\ M_1 \to Sig_x \\ (\uparrow > P_x) \\ U \to x_U \end{bmatrix} TC \to x_T \text{ , where } \begin{bmatrix} x_U \ni [...] \\ x_T \ni [...] \end{bmatrix} \right)_{\langle x_U | x_T \rangle}$$

Eq 3.1.1: Evolving Form of Generalized Equation of Mutational-Sequence

The added notation of $\langle x_U | x_T \rangle$ implies that the output of the previous iteration of the equation of innovation, x_T, where the subscript T implies relatively-transformed, now becomes the input, x_U, for the next iteration of the equation, where U implies relatively-untransformed. Hence through time there is greater and greater transformation that pushes experienced reality to greater and greater levels of functional-richness.

But further, given that quanta is proposed to be a doorway to deeper worlds of Light, that in fact allow aspects of those worlds or layers to become active at the surface layer U, the question is when have those aspects become active in manifest time. The following timeline based on generally accepted models of universal history (Particle Data Group, 2015) suggests when. Note too that the subsequent sections of exploring pre-genetic and genetic information at the levels of the electromagnetic spectrum, matter, and life, will explore in far greater detail some of the statements made in the following timeline:

- At time, $t \leq 0$ seconds, only M_3 is active, and then remains active for all $t < \infty$. Recall that M_3 represents the four-fold reality present in every ubiquitous-point-instant.
- At time, $0 \geq t > \infty$, M_2 the set of architectural forces continually gets added to, thereby increasing the size of the sets of forces.
- At time, $t \geq 0$, space, time, energy, gravity, the first clear expression of the four-fold order, emerges. This first expression is significant because it sets in motion the interplay between the antecedent quantum-layer and the layer where matter will materialize. Note that the antecedent quantum-layer likely houses the ever-enhanced four-base logic-encoding ecosystems critical to genetics and evolution.
- At time, $0 > t \geq 10^{-36}$ seconds, the equation of Innovation, $Innovation_{orientation-x}$, is such that M_1 also becomes active. The activation of M_1 begins to result in unique expressions or signatures of the set of architectural forces, and in this case in the reality of the essentially ubiquitous electromagnetic-spectrum (EM Spectrum) as a vehicle of the four-fold order that expressed itself in all that existed and in all that unfolded from that point in time on.
- At time, $t \sim 10^{-10}$ seconds, fundamental particles emerge as an essential material basis of the four architectural forces that frame all further development. As in the case of the EM Spectrum this implies the activity of M_1, and then also of U.
- At time, $t \sim 3 \times 10^5$ years light atoms emerge, and at time $t \sim 10^9$ years heavier atoms in the stars emerge. These also imply the continued activity of M_1 and U.

- At time, $t \sim 13.8 \times 10^9$ years, a further clear expression of the same fourfold order as the bases of an even more complex organization, that of cellular life and all that is founded on it comes into being. This time-point will be represented by the notation $t \sim E_{Cell}$, where 'E' stands for emergence. This too implies the activity of M_1. Note that the sets of architectural forces specified by M_2 continue to increase the number of elements they comprise of as the complex interaction between the layers continues.

- At time $t > 13.8 \times 10^9$ years, human-beings, and more complex social organizations emerge. Here TC acts with an implicit direction of operation from U to M_3. This time-point will be represented by $t \sim E_{Human}$.

Note that the emergence of space-time-energy-gravity, and subsequently of the electromagnetic spectrum, quantum particles, and atoms, implies that the pre-genetic information in their attendant four-base logic-encoding ecosystems must also be present in some form in genes as appear later with the advent of cellular life. As such, the logic of cosmic fundamentals and of the very basis of matter is deeply ingrained in all things, inanimate and animate.

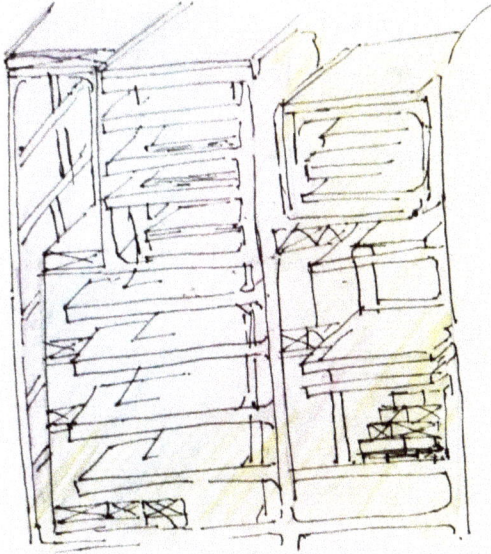

Based on the aforementioned timeline and description Equation 3.1.2 for Emergence true of any space-time scale may be generalized as the following:

$Emergence_{space-time}$

$$= \begin{vmatrix} \begin{bmatrix} M_3 \to System_X \\ (\uparrow F \to I) \\ M_2 \to S_{System_X} \\ (\uparrow Sig \to F) \\ M_1 \to Sig_x \\ (\uparrow > P_{x)} \\ U \to x_U \end{bmatrix}_{Space} \\ \begin{bmatrix} U \to \begin{matrix} M_3 : -\infty \leq t \leq \infty \\ \downarrow \\ M_2 : 0 \geq t > \infty \\ \downarrow \\ M_1 : 0 > t > \infty \\ \downarrow \\ t \leq E_{Cell}; \mathrm{TC}: M_3 \to \mathrm{U} \\ t \sim E_{Human}; \mathrm{TC}: U \to M_3 \end{matrix} \end{bmatrix}_{Time} \\ TC \to x_T, where \begin{bmatrix} x_U \ni [\dots] \\ x_T \ni [\dots] \end{bmatrix} \end{vmatrix}_{\langle x_U | x_T \rangle}$$

Eq 3.1.2: Space-Time Emergence

An implication of this equation, brought out more explicitly through the elaboration of the 'Time' component, is that the layers U, M_1, M_2, and M_3 exist simultaneously. Adding the Light-Matrix derived in Chapter 2.1 enhances Equation 3.1.2 to the Light-Space-Time Emergence form as represented by:

$Emergence_{light-space-time} =$

$$\left[\begin{bmatrix} c_\infty: [Pr, Po, K, H] \\ (\downarrow R_{C_K} = f(R_{C_\infty})) \\ c_K: [S_{Pr}, S_{Po}, S_K, S_H] \\ (\downarrow R_{C_N} = f(R_{C_K})) \\ c_N: f(S_{Pr} \times S_{Po} \times S_K \times S_H) \\ (\downarrow R_{C_U} = f(R_{C_N})) \\ c_U: [P, V, M, C] \\ \Uparrow \\ c_{0:[D,W,I,C]} \end{bmatrix}_{Light} \begin{bmatrix} M_3 \to System_X \\ (\uparrow F \to I) \\ M_2 \to S_{System_X} \\ (\uparrow Sig \to F) \\ M_1 \to Sig_X \\ (\uparrow > P_x) \\ U \to x_U \end{bmatrix}_{Space} \right.$$

$$\left. \begin{bmatrix} M_3 : -\infty \le t \le \infty \\ \downarrow \\ M_2 : 0 \ge t > \infty \\ \downarrow \\ M_1 : 0 > t > \infty \\ \downarrow \\ U \to \begin{array}{l} t \le E_{Cell}; TC: M_3 \to U \\ t \sim E_{Human}; TC: U \to M_3 \end{array} \end{bmatrix}_{Time} \quad TC \to x_T \right| \langle x_U | x_T \rangle$$

Eq 3.1.3: Light-Space-Time Emergence

In (3.1.3) there is a 1:1 mapping between the Light and Space matrices in that M_3 reflects the ever-present C_∞, M_2 reflects C_K, M_1 reflects C_N, and U reflects C_U. The Time matrix simply gives estimates at which time each of the layers became active.

Chapter 3.2: Speed of Light and Quanta

As discussed conceptually in Section 1 and also mathematically in Chapter 2.1, since c is finite and therefore there is past, present, and future implied by it, this implies that at U a point has to become quanta. This is implicit in the notion of finiteness. Since light takes a finite amount of time to get from A to B, a "unit" of light will require a finite time to traverse that. Quanta at the subatomic level can be thought of as related to this finite time and distance for a unit of light to be expressed.

Planck's discovery that energy at the subatomic level requires a minimum threshold 'quanta' to express itself therefore makes sense. It is to be noted though that Planck's treatment of quanta was more as a mathematical convenience that allowed the derivation of an equation that explained the curve of radiation wave-lengths at varying temperatures of a heated black-body (Isaacson, 2008). Einstein though postulated quanta as a fundamental property of light itself, rather than as something that arose in the interaction of light with matter as Planck thought. Einstein's theory produced a law of the photoelectric effect where the energy of emitted electrons would depend on the frequency of light. Einstein received the Nobel Prize for this discovery (Isaacson, 2008).

Summarizing, if c is the upper limit of the layer U, then it makes sense that the lower limit h (Planck's constant) should be inversely proportional to c. Hence:

$$h \propto \frac{1}{c}$$

This relationship is substantiated by combining two well-known equations: the first is the electromagnetic equation connecting speed of light with wavelength and frequency, and the second is Einstein's photoelectric equation connecting energy with frequency of light:

(1) $C = v\lambda$
(2) $E = hv$

Yields:

$$h = \frac{E\lambda}{C}$$

About h, H.A. Lorentz the Dutch scientist has commented in The Science of Nature (Lorentz, 1925): "We have now advanced so far that this constant not only furnishes the basis for explaining the intensity of radiation and the wavelength for which it represents a maximum, but also for interpreting the quantitative relations existing in several other cases among the many physical quantities it determines. I shall mention a few only, namely the specific heat of solids, the photo-chemical effects of light, the orbits of electrons in the atom, the wavelengths of the lines of the spectrum, the frequency of the Roentgen rays which are produced by the impact of electrons of given velocity, the velocity with which gas molecules can rotate, and also the distances between the particles which make up a crystal. It is no exaggeration to say that in our picture of nature nowadays it is the quantum conditions that hold matter together and prevent it from completely losing its energy by radiation."

So just as c sets up the past-present-future experience and reality of U, h suggests that this experience will take place in shells of matter. In the absence of the limit h, as pointed out by Lorentz, only radiation, and no matter would exist. This 'past-present-future-matter' dynamic reinforces the notion of the four-foldness implicit in the nature of Light as already discussed in Sections 1 and 2.

The suggested variance in the speed of light by meta-layer may also throw some further light on the quantum realm. First, summarizing:

1. At U the speed of light in a vacuum, c_U, is finite at 186,000 miles/sec. This finiteness creates the reality and experience of past-present-future, and further a sense of fragmentation and separation. Further, assuming that c_U is a fundamental upper-limit at U, the inverse of it, $\frac{1}{c_U}$, must define some fundamental lower limit at U. This is indeed the case as Planck's constant, h, is proportional to this. 'h' allows for matter to be sustained, as it fundamentally limits the dispersion of energy as suggested by Lorentz.

2. At M_3, the speed of light, c_{M_3}, is suggested as being ∞ miles/sec. This allows a reality of 'oneness' and the possibility of a suggested fourfold-intelligence existing in every ubiquitous-point-instant as already suggested.

3. As also already suggested the quantum world, here designated by Q, because it is at boundary of U, accesses and interrelates with the meta-levels. As such, the speed of light, c_Q, will appear as a hybrid as in the following figure. Note though that it is really the speed of light at the native or resident layer that becomes active, and that this is simply being represented as c_Q for convenience:

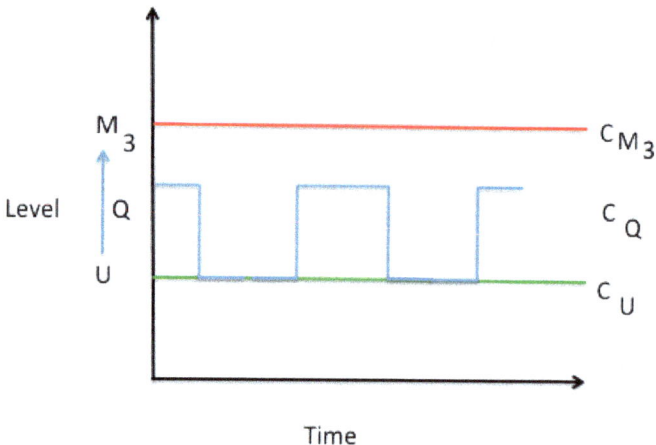

Figure 3.2.1 Speed of Light at Quantum Level

Note that research on the speed of light also indicates that it may go faster than c_U. While the speeds suggested currently through experimental research, and summarized below, may be only incrementally higher than c_U the notion that c_U can be exceeded appears to be put in place:

1. The Heisenberg Uncertainty Principle (to be discussed subsequently in Chapter 3.4) already suggests that photons can travel at any speed, even exceeding c_U, for short periods.

2. Notion of different space-time realities, also known as meta-levels in this treatise, suggests that light can travel differently in a layer different from the four-dimensional space-time that apparently defines our observable world (Hawking, 1988).

3. In his book QED Feynman (Feynman, 1985) says "...there is also an amplitude for light to go faster (or slower) than the conventional speed of light. You found out in the last lecture that light doesn't go only in straight lines; now, you find out that it doesn't go only at the speed of light! It may surprise you that there is an amplitude for a photon to go at speeds faster or slower than the conventional speed, c." In research conducted at Humboldt University (Chown, 1990), Scharnhorst has made calculations using the theory of quantum electrodynamics to reveal the possible existence of "faster-than-light" photons. This is known as the Scharnhorst effect.

4. As reviewed in Chapter 2.1, Perkowitz makes the point that the theory of relativity does not disallow particles already moving at c or greater.

The point is that the reality at Q is going to be different than the reality at U. This should be apparent from considering the relation of c_X to the consequent reality as elaborated in Chapter 2.1. In Q the fundamental lower limit, h, which allows matter to sustain itself, is itself going to fluctuate. Hence, as X tends to M_3, c_X will tend to infinity, and h will become a fraction of itself. As it becomes a fraction of itself the quantization effect will be lowered, and matter will get dispersed more and more easily to in effect take on a wave-like, or field-like, or even block-like appearance as also suggested in Chapter 2.1.

Chapter 3.3: Schrodinger's Equation & Multiple Layers of Light

Schrodinger's equation, which seeks to model how a quantum state of a quantum system changes with time, or in other words seeks to model matter as a wave rather than as a particle (Stewart, 2012), is depicted in Equation 3.3.1:

$$i \frac{h}{2\pi} \frac{\partial}{\partial x} \psi = \hat{H}\psi$$

Eq 3.3.1: Schrodinger's Equation

ψ depicts a wave form and can be thought of as a probable cloud of possible states. $\hat{H}$ is the Hamiltonian operator which is a focusing function, and in its essence what the equation may be suggesting is that the way a wave form changes over time is equivalent to some expressible state of the possibilities inherent in the cloud of possible states.

But the cloud of possible states is another way of saying that behind the layer U, form is represented in another way than at U. If the existence of the meta-levels, and in this case, of Sig_x at M_1, is considered possible, then it is far more reasonable to admit that form is configured by function and the very dynamics of what may appear to have been random, as some interpretations of quantum phenomena suggest (Wimmel, 1992), may now appear to be far more logical. There is now more *context* to interpreting observation at the quantum level.

Interestingly Schrodinger himself had misgivings about the applicability of this equation that seemed to apply at the quantum level, to the macro-world (Stewart, 2012). To bring his misgivings to light he invented a thought experiment concerning a cat. This cat would be in a superposed state in a quantum black-box. A radioactive particle, a decaying-particle detector, and a flask of poison, were the other inhabitants of the black-box. At some point the particle will decay, be detected, and as in the thought-experiment at that point, triggered by the decaying particle, the poison in the flask would be released. The cat would then die. But in the meanwhile the cat would be in a superposed states of being both dead and alive. Only when the box was opened would the wave function collapse and a single definite state emerge. This thought experiment is also considered the origin of the notion of 'superposition'. The mathematical model presented in this book though has already interpreted 'superposition' differently as discussed previously.

Schrodinger was hoping to highlight the absurdity of the application of having a cat in both a dead and alive state at the macro-level. Instead physicists found this thought experiment to be sensible, even at the macro-level, and began to generalize the quantum theory based on this. Hence, for example the idea of superposition at the physical level began to be thought of as real. It is interesting to note that in his lectures on Schrodinger's equation Feynman (Gottlieb, 2013) has stated: "Where did he get that [equation] from? Nowhere. It is not possible to derive it from anything you know. It came out of the mind of Schrödinger".

Considering Schrodinger's equation in the light of the discussion on Q in Chapter 3.2, 'i' is a complex number and suggests the interplay of two dimensions, one being real, and one being 'imaginary'. But the 'imaginary' dimension could be thought of as none other than the meta-levels implicit in the mathematical model in this treatise, and suggested to be real at Q. Further, $\frac{h}{2\pi}$ is in line with the suggestion also made in Chapter 3.2 that h will have to become a fraction of itself as c increases. Hence, the change in the wave function, $\frac{\partial}{\partial x}\psi$, is intimately related to i and $\frac{h}{2\pi}$, and perhaps more fully makes sense when considered in the context of $i \times \frac{h}{2\pi} \times \frac{\partial}{\partial x} \psi$, which has to be the case when dealing with the integration of dynamics of multiple levels of light.

Further, the change in the wave function, $\frac{\partial}{\partial x}\psi$, is related to $\hat{H}\psi$, and suggests that there is some system "energy", represented by the Hamiltonian, $\hat{H}$, that when applied to the existing wave, ψ, will indicate how the wave will be expressed going forward.

But as discussed, at Q the dynamics of Sig_x become real, and in fact is a fundamental organizing principle for all organization at U, and starting at the space-dimension of h.

Chapter 3.4: Heisenberg's Uncertainty Principle & Multiple Layers of Light

In his book, The Little Book of String Theory, Princeton University's Gubser (Gubser, 2010) describes the effect on approaching absolute zero temperature on molecules. He takes the example of water molecules and relates that one cannot make the water molecules colder than absolute zero, -273.15 Celsius, because there is no more thermal energy to suck out at that temperature. However, quantum uncertainty, the phenomenon which relates the momentum and location of electrons in atoms necessitates that the water molecules will still vibrate. Gubser suggests this by considering Heisenberg's uncertainty relation, reproduced in Equation 3.4.1:

$$\Delta p \; X \; \Delta x \; \geq \; \frac{h}{4\pi}$$

Eq 3.4.1: Heisenberg's Uncertainty Relation

In Equation 3.4.1, Δp is the uncertainty in a particle's momentum, Δx is the uncertainty in the particle's location, and h is the Planck's constant. In frozen water crystals it is precisely known where the water molecules are, and therefore Δx is fairly small. This means that Δp has to be considerably larger, and therefore that the water molecules are still vibrating even though they are at absolute zero. This innate vibration, known as 'quantum zero-point' energy, expresses the phenomenon of quantum fluctuations.

The Planck's constant order of magnitude (10^{-34}) though, suggests the boundary between U and M_1 and the quantum fluctuations, the uncertainty relation, and the quantum zero-point energy could be an expression of the essential Signature function, Sig_x, that is posited as a key formative force behind organization at U. In this interpretation the thermal energy describes the essential energy at U, while the uncertainty relation may suggest the phenomenon of function-precipitation from other layers of Light, "physically" linking M_1 and U. In this

case it may be suggested that integration of meta-levels with the surface level, I_U^M, is indicated by the uncertainty relation, as in Equation 3.4.2:

$$I_U^M \;\rightarrow\; \Delta p \; X \; \Delta x \;\geq\; \frac{h}{4\pi}$$

Eq 3.4.2: Integration of Levels (Leveraging Heisenberg's Uncertainty Relation)

But further, it may also be suggested that the uncertainty

principle itself is only valid at U, and that too, because of the finiteness of c. This finiteness as already suggested implies h, which implies that if the position of a particle is going to be observed by shining light on it, the light has to have at least a quantum of energy. But to determine the position of a particle accurately, light of a shorter wavelength would have to be used (Hawking, 1988) which would have to have a minimum amount of energy, which in turn would interfere with the velocity and hence momentum of the particle. The uncertainty in measuring the momentum could therefore be thought of as a consequence of the finiteness of the speed of light, c.

If c_U were to approach c_Q though, which as suggested in Chapter 3.2 could be anywhere between c and ∞ miles per second, the quantum would be smaller and the uncertainty in measuring position or momentum would be reduced. At c_{M_3} there would be no uncertainty since light would accurately tell both position and momentum definitively.

Hence, the uncertainty principle may be further qualified, as in Equations 3.4.3, 3.4.4, and 3.4.5:

$$@C_U: \Delta p \; X \; \Delta x \; \geq \; \frac{h}{4\pi}$$

Eq 3.4.3: Uncertainty Principle at U

$$@C_Q: \Delta p \; X \; \Delta x \; \rightarrow \; 0$$

Eq 3.4.4: Uncertainty Principle at Q

$$@C_{M_3}: \Delta p \; X \; \Delta x \; = 0$$

109

Eq 3.4.5: Uncertainty Principle at M3

The notion of position and momentum becoming finite at U also may imply that space, time, and quanta are emergent rather than absolute properties, as also suggested in Section 1. This is also the conclusion of Arkani-Hamed of the Institute of Advanced Studies in the following thought experiment (Wolchover, 2013):

'Locality says that particles interact at points in space-time. But suppose you want to inspect space-time very closely. Probing smaller and smaller distance scales requires ever higher energies, but at a certain scale, called the Planck length, the picture gets blurry: So much energy must be concentrated into such a small region that the energy collapses the region into a black hole, making it impossible to inspect. "There's no way of measuring space and time separations once they are smaller than the Planck length," said Arkani-Hamed. "So we imagine space-time is a continuous thing, but because it's impossible to talk sharply about that thing, then that suggests it must not be fundamental — it must be emergent."

Unitarity says the quantum mechanical probabilities of all possible outcomes of a particle interaction must sum to one. To prove it, one would have to observe the same interaction over and over and count the frequencies of the different outcomes. Doing this to perfect accuracy would require an infinite number of observations using an infinitely large measuring apparatus, but the latter would again cause gravitational collapse into a black hole. In finite regions of the universe unitarity can therefore only be approximately known.'

Chapter 3.5: A Deeper Look at Quantization of Space, Time, Matter, and Gravity

The Light-Space-Time Emergence equation (3.1.3) derived in Chapter 3.1, and reproduced here for convenience, offers some insight into the process of quantization. Reproducing the equation:

$$Emergence_{light-space-time} =$$

$$\begin{bmatrix}\begin{bmatrix} c_\infty: [Pr, Po, K, H] \\ (\downarrow R_{C_K} = f(R_{C_\infty})) \\ c_K: [S_{Pr}, S_{Po}, S_K, S_H] \\ (\downarrow R_{C_N} = f(R_{C_K})) \\ c_N: f(S_{Pr} \times S_{Po} \times S_K \times S_H) \\ (\downarrow R_{C_U} = f(R_{C_N})) \\ c_U: [P, V, M, C] \\ \Uparrow \\ c_{0:[D,W,I,C]} \end{bmatrix}_{Light} \quad \begin{bmatrix} M_3 \rightarrow System_X \\ (\uparrow F \rightarrow I) \\ M_2 \rightarrow S_{System_X} \\ (\uparrow Sig \rightarrow F) \\ M_1 \rightarrow Sig_X \\ (\uparrow > P_x) \\ U \rightarrow x_U \end{bmatrix}_{Space} \\ \begin{bmatrix} M_3 : -\infty \le t \le \infty \\ \downarrow \\ M_2 : 0 \ge t > \infty \\ \downarrow \\ M_1 : 0 > t > \infty \\ \downarrow \\ U \rightarrow \begin{matrix} t \le E_{Cell}; TC: M_3 \rightarrow U \\ t \sim E_{Human}; TC: U \rightarrow M_3 \end{matrix} \end{bmatrix}_{Time} \quad TC \rightarrow x_T \quad \langle x_U | x_T \rangle \end{bmatrix}$$

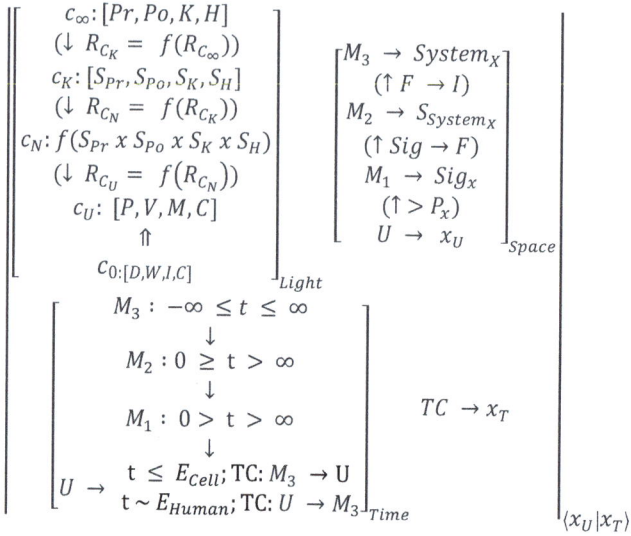

The Light-Matrix, the top left-hand matrix, suggests that there are particular kinds of quantization that could occur. Along the vertical realm these can be thought of as inter-relating one layer of the matrix with the previous layers, or c_0 with c_U. Hence the kind of quantization that is relevant to the physical layer, U, would relate c_U with

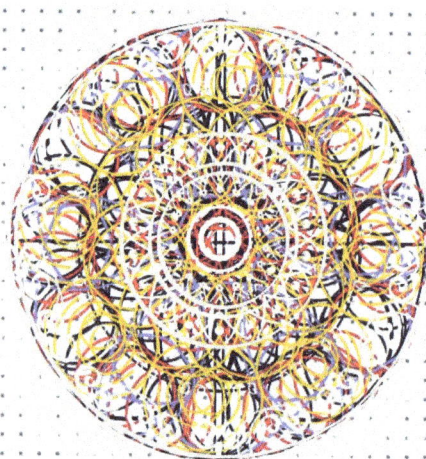

c_x where x $\in$ (N, K, ∞, 0), one of the possible speeds of light as per this mathematical treatise. This quantization can be represented by h_U, where h stands for Planck's constant. Note that in this model there are several fundamental quantization possible that may

inter-relate a specific light-layer with other layers of light. For the sake of simplicity it will be assumed that h_U is the quanta experienced in the inter-relation between U and the collective-set of light-layers behind or above U.

Hence space, time, matter, and gravity, that arise when Light slows down, can be thought of as emergent phenomena. The emergence itself can be thought of as a function of the Light-Matrix. At layer U, and as discussed in Chapter 1.2, Space is related to Light's property

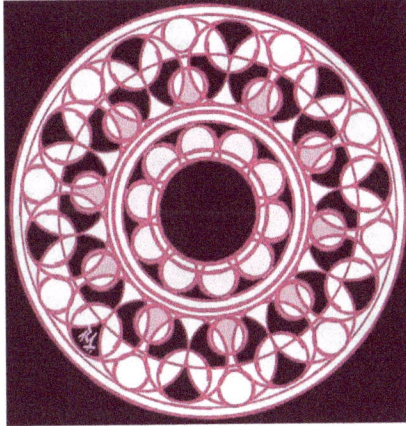

of Knowledge, Time is related to Light's property of Power, Matter or Energy is related to Light's property of Presence, and Gravity is related to Light's Property of Harmony or Nurturing. Each of these will emerge in a particular way and it is possible that there are multiple h_U's.

Hence, the Planck's constant for matter or energy, which we are already familiar with from past discoveries in science, can be depicted as h_{UPr}, since it is related to Presence (Pr). But similarly, quantization for space, time, and gravity could potentially be governed by other similar constants, referred to as h_{UK}, h_{UP}, and h_{UH}, and related to Knowledge (K), Power (P), and Harmony (H), respectively. Since the relationship between these constants, in absolute terms, is uncertain, this can be represented by the equality-inequality as depicted by Equation 3.5.1:

$$h_{UPr} \lesseqgtr h_{UK} \lesseqgtr h_{UP} \lesseqgtr h_{UH}$$

Eq 3.5.1: Equality-Inequality Relationship Between Different 'Planck' Constants

Further, the "quantization-window", that is, the window that quanta provides to layers of light behind U, as it were, potentially allows the precipitation of, or inter-relation with, or creation of a cohesive and compelling meta-function or signature as modeled by the c_N/M_1 layer. This quantization-window is positioned as being key in allowing a phenomenon of "quantum-certainty" to occur. Quantum certainty can be thought of as allowing space-time-energy-gravity quantization to occur and is described in detail in the book Quantum Certainty (Malik, 2017e), part of the 6-book Cosmology of Light series (Malik, 2017-2018).

Subsequent Sections in this book will describe several examples of the activation of the multi-dimensional quantization that ensures change at the material-level initiated at the four-base logic-encoding ecosystems at the quantum-levels. Such change at the four-base logic-encoding ecosystems and material levels is also the basis of genetic mutation. If the mutation is influenced by the layers set up by c_K, c_N, or c_∞, then the mutation is constructive because the material outcomes of Light are being more integrated with Light itself. If the mutation is influenced by c_0, then the mutation is destructive because the material outcomes of Light are being further separated or fragmented from Light itself.

Such quantization requires patterns to be overcome, and so long as these are, this will result in the phenomenon of quantum-certainty. The 'Time' matrix, in (3.1.3) reproduced above, indicates the general default direction under which such quantum-certainty can occur at Layer U. Generally at

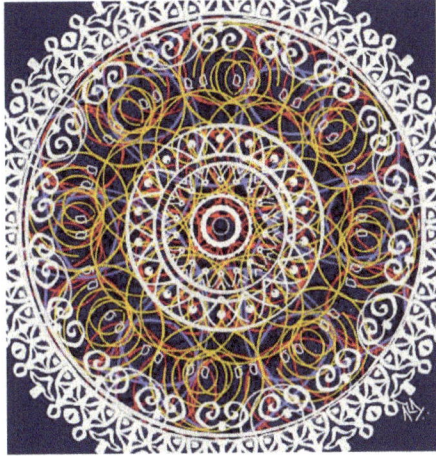

the pre-human level it may proceed more automatically by the dynamics of the system itself, as represented by the segment: $t \sim E_{Cell}$; TC: $M_3 \rightarrow U$. Beyond this level it will generally occur through the action of a cohesive will, as represented by the segment: E_{Human}; TC: $U \rightarrow M_3$. The opening of this quantization-window is modeled by the 'Space' matrix and will happen when habitual patterns are overcome so that 'will' or 'want' becomes cohesive.

Such inter-dependence between human want or will and quantum dynamics is a possible basis of Human-Quantum computational models, as discussed in greater detail in 'The Emperor's Quantum Computer' (Malik, 2018c) and is envisioned to play a critical role in genetic mutation and evolution.

The specific quantization that are occurring along the space, time, energy/matter, and gravity dimensions are modeled by the following equations, that are all based on the Sig_x equations derived in Chapter 2.4. As a reminder (2.4.5) is reproduced here for convenience:

$$Sig = Xa + \overline{Yb_{0-n}}$$

$$where: \begin{bmatrix} X \in [S_{System_{Pr}}, S_{System_P}, S_{System_K}, S_{System_N}] \\ Y \in [S_{System_{Pr}}, S_{System_P}, S_{System_K}, S_{System_N}] \\ a, b \text{ are integers; } a > b \end{bmatrix}$$

The Quantization and Structure of Space

The quantization of Space, which holds the seeds of knowledge of all that will emerge, hence, is modeled in the following way as specified by Equation 3.5.2:

$$Space_{quantization} = h_{UK}(Xa + \overline{Yb_{0-n}})$$

$$where: \begin{bmatrix} X \in [S_{System_K}] \\ Y \in [S_{System_{Pr}}, S_{System_P}, S_{System_K}, S_{System_N}] \\ a, b \text{ are integers; } a > b \end{bmatrix}$$

Eq 3.5.2: Space Quantization

In this model 'space' as an emergent property of Light, is structured by infinite seeds of knowledge. But because the emergence is taking place in a layer of reality generally itself structured by a finite speed of light, c, it has to be quantized. The quantization assures that the knowledge is not dissipated, but can accumulate to create seeds, and therefore, the structure of space itself. Such a structure of space appears consistent with recent models on the structure of space as explored in Rovelli's 'Reality Is Not What It Seems' (Rovelli, 2017). It is assumed that there is some kind of 'Planck's constant' in effect, that is modeled as being specific to the way knowledge or space may be quantized – hence, h_{UK}.

The structure of Space itself would be the summation of infinite seeds, as in Equation 3.5.3, Structure of Space:

$$Space_{Structure} = \sum_{i,j=1}^{\to \infty} h_{UK}\left(X_i a + \overline{Y_j b_{0-n}}\right)$$

$$where: \begin{bmatrix} X_i \in [S_{System_K}] \\ Y_j \in [S_{System_{Pr}}, S_{System_P}, S_{System_K}, S_{System_N}] \\ a, b, i, j \ are \ integers; a > b \end{bmatrix}$$

Eq 3.5.3: Structure of Space

The Structure of Time

The quantization of Time, which holds the inevitability of the seeds or knowledge emerging in a phased maturity, hence, is modeled in the following way as specified by Equation:

$$Time_{quantization} = h_{UP}\left(Xa + \overline{Yb_{0-n}}\right)$$

$$where: \begin{bmatrix} X \in [S_{System_P}] \\ Y \in [S_{System_{Pr}}, S_{System_P}, S_{System_K}, S_{System_N}] \\ a, b \ are \ integers; a > b \end{bmatrix}$$

Eq 3.5.4: Time Quantization

In this model 'time' as an emergent property of Light, is structured by an inevitable process of maturity, which due to its inevitability, is related to power. But because the emergence is taking place in a layer of reality generally itself structured by a finite speed of light, c, it has to be quantized. The quantization assures that the power is not dissipated, but can accumulate to express phased maturity, and therefore, the structure of time itself. It is assumed that there is some kind of 'Planck's constant' in effect, that is modeled as being specific to the way power or time may be quantized – hence, h_{UP}.

The infinity of Time itself would be the summation of the maturation of infinite seeds, as in Equation 3.5.5, Infinity of Time:

$$Time_{Infinity} = \sum_{i,j=1}^{\to\infty} h_{UP}\left(X_i a + \overline{Y_j b_{0-n}}\right)$$

$$where: \begin{bmatrix} X_i \in [S_{System_P}] \\ Y_j \in [S_{System_{Pr}}, S_{System_P}, S_{System_K}, S_{System_N}] \\ a, b, i, j \ are \ integers; a > b \end{bmatrix}$$

Eq 3.5.5: Infinity of Time

Energy & Matter

Energy, which through a process of containment can result in Matter, hence, is modeled in the following way as specified by Equation 3.5.6:

$$Energy_{quantization} = h_{UPr}\left(Xa + \overline{Yb_{0-n}}\right)$$

$$where: \begin{bmatrix} X \in [S_{System_{Pr}}] \\ Y \in [S_{System_{Pr}}, S_{System_P}, S_{System_K}, S_{System_N}] \\ a, b \text{ are integers; } a > b \end{bmatrix}$$

Eq 3.5.6: Energy Quantization

In this model 'energy' as an emergent property of Light, results in the reality of matter. But because the emergence is taking place in a layer of reality generally itself structured by a finite speed of light, c, it has to be quantized. The quantization assures that the energy is not dissipated, but can accumulate to create matter. Planck's constant is referred to as - h_{UPr}.

Gravity

Gravity, which holds seemingly distinct objects in the layer of reality created by c together in a harmony, hence, is modeled in the following way as specified by Equation 3.5.7:

$$Gravity_{quantization} = h_{UH}\left(Xa + \overline{Yb_{0-n}}\right)$$

$$where: \begin{bmatrix} X \in [S_{System_N}] \\ Y \in [S_{System_{Pr}}, S_{System_P}, S_{System_K}, S_{System_N}] \\ a, b \text{ are integers; } a > b \end{bmatrix}$$

Eq 3.5.7: Gravity Quantization

In this model 'gravity' as an emergent property of Light, results in a harmonious collectivity of seemingly independent objects. But because the emergence is taking place in a layer of reality generally itself structured by a finite speed of light, c, it has to be quantized. The quantization assures that the harmony is not dissipated, but can accumulate to express more and more complex collectivities on large-scale. It is assumed that there is some kind of 'Planck's constant' in effect, that is modeled as being specific to the way harmony or gravity may be quantized – hence, h_{UH}.

The preceding analysis also suggests that Space is a repository of pre-genetic information or of pre-genetic libraries. Time, Gravity, and Energy work with Space to bring about material possibilities. Change to any material possibility requires therefore the action of space, time, energy, and gravity. At the margin there is a fourfold quantization that brings about such change. Further, it may very well be the case that the antecedent quantum layer that has been proposed to house the four-base logic-encoding ecosystems is none other than an aspect of, or Space itself.

The Light-Space-Time Emergence equation (3.1.3) being iterative can be used to model emergence from simpler to more complex four-fold manifestations. In other words (3.1.3) suggests a computational approach to the development of the universe. Pre-genetic and genetic information may be thought of as the output of such computation. This computational approach involves multiple layers of reality as suggested by (3.1.3) and is driven more by a process of qualified determinism, to be explored in more detail in the next chapter, than probability and statistics. This notion of qualified determinism is in contrast to the prevalent probabilistic view that has been erected as the cornerstone of quantum theory, and to MIT's Seth Lloyd's Copenhagen-like quantum superposition, probability-based

121

computational approach to the development of the universe as described in his book 'Programming the Universe' (Lloyd, 2007).

An example of the Light-Space-Time based computation will be elaborated in Section 4, to provide an overview of the process by which pre-genetic and genetic information is created. Sections 5, 6, 7, and 8 will further leverage the Light-Space-Time based computation to discuss in more detail how such information changes, or how mutation occurs. However, in keeping with the current quantum-analysis paradigm of using wavefunction to explore outcomes at a particular space and time, this chapter will develop two general approaches summarized by equations, to understanding the effect that different layers or levels of light may have on genetic mutation. While the natural line of development given the mathematical treatise explored in this book is to proceed with such an understanding using (3.1.3), the wavefunction depiction will be elaborated to suggest how (3.1.3) can be leveraged to provide further insight into a probability-based quantum-analytical approach.

(3.1.3) is reproduced below for convenience and a more concise form of it, the Simplified Version of Light-Space-Time Emergence, Equation 3.6.1, shall be utilized through the rest of this chapter.

$$Emergence_{light-space-time} =$$

$$
\left|\left|\begin{array}{l}
\begin{bmatrix}
c_\infty : [Pr, Po, K, H] \\
(\downarrow\ R_{C_K} = f(R_{C_\infty})) \\
c_K : [S_{Pr}, S_{Po}, S_K, S_H] \\
(\downarrow\ R_{C_N} = f(R_{C_K})) \\
c_N : f(S_{Pr} \times S_{Po} \times S_K \times S_H) \\
(\downarrow\ R_{C_U} = f(R_{C_N})) \\
c_U : [P, V, M, C] \\
\Uparrow \\
\quad c_{0:[D,W,I,C]}
\end{bmatrix}_{Light} \\[2em]
\begin{bmatrix}
\qquad M_3 : -\infty \le t \le \infty \\
\qquad\qquad \downarrow \\
\qquad M_2 : 0 \ge t > \infty \\
\qquad\qquad \downarrow \\
\qquad M_1 : 0 > t > \infty \\
\qquad\qquad \downarrow \\
U \to \begin{array}{l} t \le E_{Cell}; TC: M_3 \to U \\ t \sim E_{Human}; TC: U \to M_3 \end{array}
\end{bmatrix}_{Time}
\end{array}\right.
\left.\begin{array}{l}
\begin{bmatrix}
M_3 \to System_X \\
(\uparrow F \to I) \\
M_2 \to S_{System_X} \\
(\uparrow Sig \to F) \\
M_1 \to Sig_x \\
(\uparrow > P_x) \\
U \to x_U
\end{bmatrix}_{Space} \\[4em]
\qquad\qquad TC \to x_T
\end{array}\right|\right|_{\langle x_U | x_T \rangle}
$$

Simplifying each of the main matrices in (3.1.3), hence, yields Equation 3.6.1, the Simplified Version of the Light-Space-Time Emergence equation:

$$Emergence_{light-space-time} = |[L][S][T]TC \to x_T|_{\langle x_U | x_T \rangle}$$

Eq. 3.6.1: Simplified Version of Light-Space-Time Emergence

Effect of Levels of Light on Mutation Using Simplified Version of Light-Space-Time Emergence

123

As detailed in Chapter 2.6 on the systematization of the mutational-sequence, while relatively transformed organizations have opened to the active influence of meta-levels, relatively untransformed organizations remain under the influence of untransformed physical, vital, mental, and integral dynamics at U. This modeling can also be applied to mutation. Hence, as discussed in Chapter 2.6, the notion of relatively transformed organizations will translate to that of constructive-mutation. The notion of relatively untransformed organizations will translate to that of destructive-mutation.

Equations 3.6.2 through 3.6.4 summarize three types of mutation using (3.6.1). These are the destructive-mutation with only untransformed bases active, random-mutation with mixed bases, and constructive-mutation with bases relatively transformed due to the active influence of meta-levels, respectively. Hence:

$$Destructive - Mutation = \left| |[L][S][T]TC \rightarrow x_T|_{\langle x_U | x_T \rangle} \right|_U$$

Eq. 3.6.2: Destructive-Mutation

$$Random - Mutation = \left| |[L][S][T]TC \rightarrow x_T|_{\langle x_U | x_T \rangle} \right|_{U \& M_x}$$

Eq. 3.6.3: Random-Mutation

$$Consgtructive - Mutation = \left| |[L][S][T]TC \rightarrow x_T|_{\langle x_U|x_T \rangle} \right|_{M_x}$$

Eq. 3.6.4: Constructive-Mutation

Now the question is to what extent will the antecedent, and therefore the more permanent four-base logic-encoding ecosystem be altered? For in the model based on Light it is this category of library that is important in heredity or the longer-term scheme of things. Destructive or random mutation may alter local copies of libraries, such as exist in DNA at the cellular-level, but it is only constructive mutation that can alter the deeper level subtle libraries that are preserved in a different kind of way in longevity.

This has to be the case since there is an in-built buffer due to the quantization that occurs as light slows down. The conditions for entering into that region where light is traveling faster must require a special, non-trivial set of conditions to be active. The conditions for traversing this in-built buffer will be explored in greater detail in subsequent chapters.

Traversing the buffer implies that meta-levels are active, and that they have to become active in a special way for four-base logic-encoding ecosystems to be affected.

Equation 3.6.5, Potential Effect of Levels of Light on Pre-Genetic or Genetic Information, summarizes how this may happen:

$Potential\ Effect\ of\ Levels\ of\ Light\ on\ Pre\ Genetic\ or\ Genetic$
$Information =$

$$\left[\begin{array}{c} STATIC\ \langle |[L][S][T]TC \rightarrow x_T|_{\langle x_U | x_T \rangle} \rangle \\ \times \\ \left((Y > U: Z_Q) \vee (Y \le U: Z_F) \vee (Y = U: Z_R) \right) \\ \ni \\ \left(\begin{array}{c} Z \in \mathbb{U}\ (\ Space, Time, Energy, Gravity) \\ Q:\ Quantization;\ F:\ Fragmentation;\ R:\ Random \end{array} \right) \end{array} \right] \rightarrow h \ni$$

$$h \in \left(\begin{array}{c} [Q]:\ Constructive\ zone, \\ [Q]:\ Constructive\ zone \wedge\ Constructive\ mutation, \\ [F]:\ Destructive\ mutation, \\ [R]:\ Random\ mutation \end{array} \right)$$

Eq. 3.6.5: Potential Effect of Levels of Light on Pre-Genetic or Genetic Information

Line 1 from the top in the matrix is simply a static form of (3.6.1) the Simplified Light-Space-Time Emergence equation. The static form is designated by 'STATIC', and implies that fundamental operations true of (3.6.1) are being highlighted in (3.6.5). In other word (3.6.1) already has all the operations highlighted in (3.6.5) in it, but by 'freezing' it by making it static, the essential dynamics leading to possible mutations at the genetic level can more clearly be highlighted.

Line 1 is then subjected (×) to a determination of the dominant levels of light that may be active, designated by Line 2,

'$\left((Y > U: Z_Q) \vee (Y \le U: Z_F) \vee (Y = U: Z_R) \right)$'.   Unpacking this, 'Y > U' implies meta-levels are active and as a result it is possible that Z_Q is going to take place. This also implies activation and potential change of FBLEE.

The call from below, as it were, may invoke some function that already exists in the subtle-libraries 'above', so that some already existing function may influence FBLEE through ∞ - entanglement, K-entanglement, or N-entanglement. This may be thought of as a key-and-lock mechanism, where a deep enough visceral urge from below acting as the key, opens an entangled lock to alter FBLEE as per the visceral urge.

'Y ≤ U' implies that only the untransformed levels are active, and therefore also the sub-level where the speed of light is 0 is active and as a result Z_F is going to take place. 'Y = U' implies that all levels are active and as a result Z_R is going to take place.

Line 3 elaborates the significance of Z_Q, Z_F, and Z_R. Hence Z is the union of potential quantum-operations of space, time, energy, and gravity, designated by '$Z \in \mathbb{U}\,(Space, Time, Energy, Gravity)$'. But the nature of the operations is designated by 'Q: $Quantization$; F: $Fragmentation$; R: $Random$'. Z_Q, then, implies that the full quantization originating from updated four-base logic-encoding ecosystems (FBLEE) can take place, and will result in lasting material change at the pre-genetic or genetic level. Z_F implies that the essential set will be fragmented and that only libraries a the level of local cellular-level DNA or precipitated material-fabric logic can potentially be altered. Z_R also precludes full quantization, and that some partial local-library constructive or destructive mutation may take place.

The '$\rightarrow h \ni h \in$' segment resolves the outcome of the operations implied by Lines 1 – 3, suggesting that the outcome will be 'h' such that ($\ni$) 'h' is an element ($\in$) of the set specified by the members '[Q]: *Constructive zone*', '[Q}: *Constructive zone AND Constructive mutation*', '[F}: *Destructive mutation*', and '{R}: *Random mutation*'. '[Q]: *Constructive zone*' implies that the in-built buffer has been crossed and that access to the deeper four-base logic-encoding ecosystem (FBLEE) has been granted. '[Q}' in this segment implies that there is the possibility that full-quantization as specified by Line 3 of the previous matrix can take place. Access to this zone is a prerequisite for constructive mutation to occur, as designated by the element '$[Q\}$: *Constructive zone* $\cap$ *Constructive mutation* ', which implies that full-quantization is going to take place and will result in material change. The '[F]' specifies the relationship between 'Fragmentation' in Line 3 of the previous matrix and destructive mutation. The '[R]' specifies the relationship between 'Random' in Line 3 of the previous matrix and random mutation.

Effect of Levels of Light on Mutation Using Wavefunction Form of Light-Space-Time Emergence

In Chapter 3.3 we were introduced to Schrodinger's wavefunction equation (3.3.1), reproduced below for convenience, which seeks to model matter as a wave rather than as a particle:

$$i\frac{h}{2\pi}\frac{\partial}{\partial x}\psi = \hbar\psi$$

This wavefunction equation deals with dynamics behind

the surface, and particularly of as many as infinite potential superposed states that then collapse into a material possibility in time and space at U. Such superposition is another way of saying that there is wholeness behind the surface. In (3.6.1) such wholeness is represented by L, S, T,

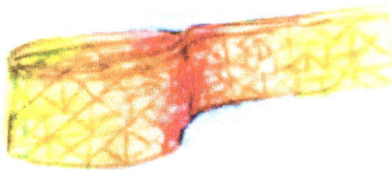

the Light, Space, and Time matrices respectively.

As explained by Neil Turok in his book The Universe Within (Turok, 2012), Euler's formula reproduced below, can be used to model many naturally occurring phenomena because of its sinusoidal oscillation between narrow bounds as x increases. Reproducing Euler's formula:

$$e^{ix} = \cos x + i \sin x$$

A key characteristic of this equation is that the sum of the squares of the ordinary and complex parts, on the right side of the equation, is one. In many contemporary interpretations of quantum theory this ensures that the probabilities for all possible outcomes add up to one. Dealing in probabilities become important when considering as many as infinite superposed states.

Further, in modeling material reality a modified notation of the Schrodinger wavefunction as interpreted by Feynman is leveraged. This version features the integral sign, $\int$, meaning that all terms to the right of it have to be summed up for all space and time till the moment when the wavefunction is required to be known.

Hence, combining (3.6.1) with Feynman's interpretation of the Schrodinger wavefunction, with the Euler formula yields the following equations, 3.6.6-11 for genetic mutation:

$$\psi_{destructive-mutation} = \left| \int e^{i \int |[L][S][T]|} \right|_U$$

Eq. 3.6.6: Wavefunction for Destructive-Mutation

Note that in (3.6.6) the formulation specifying iteration, $\langle x_U | x_T \rangle$, becomes redundant since iteration is implied by the integral across space and time, and hence is removed. Further, (3.6.6) suggests that so long as the basis of an organization is untransformed, specified by the U following the vertical-brackets, the outcome is going to be one of destructive mutation, specified by $\psi_{destructive-mutation}$.

Further, as specified by Equation 3.6.7, the Probability-View of Destructive-Mutation, the driver of such destructive-mutation is going to be either the untransformed physical (P_U), the untransformed vital (V_U), the untransformed mental (M_U), or the untransformed integral (I_U):

$$|\psi_{destructive-mutation}|^2 = P_U^2 + V_U^2 + M_U^2 + I_U^2 = 1$$

Eq. 3.6.7: Probability-View for Destructive-Mutation

As specified by (3.6.7) the probability that any of these bases will be leveraged in destructive mutation adds up to one. Note that the physical, the vital, the mental, and integral were discussed in Chapter 2.6 focusing on the systematization of the mutational-sequence.

Equation 3.6.8, Wavefunction for Random-Mutation, suggests the mixed bases for organizations, as specified by dynamics of both the untransformed (U) and the meta-levels (M_x), which therefore results in random mutation.

$$\psi_{ransom-mutation} = \left| \int e^{i \int |[L][S][T]|} \right|_{U\&M_x}$$

Eq. 3.6.8: Wavefunction for Random-Mutation

As specified by Equation 3.6.9, Probability-View for Random-Mutation, the probability that any of the untransformed (U) and transformed (T) bases will be leveraged in random mutation adds up to one:

$$|\psi_{random-mutation}|^2 = P_U^2 + P_T^2 + V_U^2 + V_T^2 + M_U^2 + M_T^2 + I_U^2 + I_T^2$$
$$= 1$$

Eq. 3.6.9: Probability View for Random-Mutation

Equation 3.6.10, Wavefunction for Constructive-Mutation, suggests some transformed bases for an organization, as specified by dynamics of the meta-levels (M_x), which therefore can result in a constructive mutation.

$$\psi_{constructive-mutation} = \left| \iint e^{i \int |[L][S][T]|} \right|_{M_x}$$

Eq. 3.6.10: Wavefunction for Constructive-Mutation

As specified by Equation 3.6.11, Probability-View for Constructive-Mutation, the probability that the transformed (T) bases will be leveraged in constructive mutation adds up to one:

$$|\psi_{constructive-mutation}|^2 = P_T^2 + V_T^2 + M_T^2 + I_T^2 = 1$$

Eq. 3.6.11: Probability View for Constructive-Mutation

Now the question is how would genetic information be impacted based on the levels of light active, and what might an equation that captured that look like? Equation 3.6.12, Potential Effect of Levels of Light on Pre-Genetic of Genetic Information (Wavefunction Form), is such an equation:

$$\psi_{Potential\ Effect\ of\ Levels\ of\ Light\ on\ Pre\ Genetic\ or\ Genetic\ Information}$$
$$=$$
$$\begin{bmatrix} \left(\left| \iint e^{i \int |[L][S][T]|} \right|_{Y=(U,U \& M_x, M_x)} \right) \\ \times \\ \left((Y > U: Z_Q) \vee (Y \leq U: Z_F) \vee (Y = U: Z_R) \right) \\ \ni \\ \left(\begin{array}{c} Z \in \mathbb{U}\ (Space, Time, Energy, Gravity) \\ Q: Quantization; F: Fragmentation; R: Random \end{array} \right) \end{bmatrix} \rightarrow$$
$$h \ni h \in \left(\begin{array}{c} Constructive\ zone, \\ Constructive\ zone \wedge Constructive\ mutation, \\ Destructive\ mutation, \\ Random\ mutation \end{array} \right)$$

Eq. 3.6.12: Potential Effect of Levels of Light on Pre-Genetic of Genetic Information (Wavefunction Form)

Taking each line in the equation separately: Line 1 from the top combines (3.6.6), (3.6.8), and (3.6.10) to essentially state that the wave-function for effect on genetic information while governed by the Light-Space-Time matrices in (3.1.3) will vary based on the bases that are active, hence $Y = (U, U \& M_x, M_x)$, where Y equates to the family of bases that may be active: untransformed (U), a mix (U & M_x), or M_x. Since the primary part of Line 1 falls under the $\int$ it means that the operation is iterative since dynamics are summed up for space and time to that point where the wavefunction is sought.

Line 1 is then subjected ($\times$) to a determination of the dominant levels of light that may be active, designated by Line 2, $' \left((Y > U: Z_Q) \vee (Y \leq U: Z_F) \vee (Y = U: Z_R) \right) '$. Unpacking this, '$Y > U$' implies meta-levels are active and as a result it is possible that Z_Q is going to take place. This also implies activation and potential change of FBLEE. The call from below, as it were, may invoke some function that already exists in the subtle-libraries 'above', so that some already existing function may influence FBLEE through ∞-entanglement, K-entanglement, or N-entanglement. This may be thought of as a key-and-lock mechanism, where a deep enough visceral urge from below acting as the key, opens an entangled lock to alter FBLEE as per the visceral urge. '$Y \leq U$' implies that only the untransformed levels are active, and therefore also the sub-level where the speed of light is 0 is active and as a result Z_F is going to take place. '$Y = U$' implies that all levels are active and as a result Z_R is going to take place.

Line 3 elaborates the significance of Z_Q, Z_F, and Z_R. Hence Z is the union of potential quantum-operations of space, time, energy, and gravity, designated by '$Z \in$

$\mathbb{U}$ ($Space, Time, Energy, Gravity$)'. But the nature of the operations is designated by ' Q: $Quantization$; F: $Fragmentation$; R: $Random$ '. Z_Q , then, implies that the full quantization originating from updated four-base logic-encoding ecosystems (FBLEE) can take place, and will result in lasting material change at the pre-genetic or genetic level. Z_F implies that the essential set will be fragmented and that only libraries a the level of local cellular-level DNA or precipitated material-fabric logic can potentially be altered. Z_R also precludes full quantization, and that some partial local-library constructive or destructive mutation may take place.

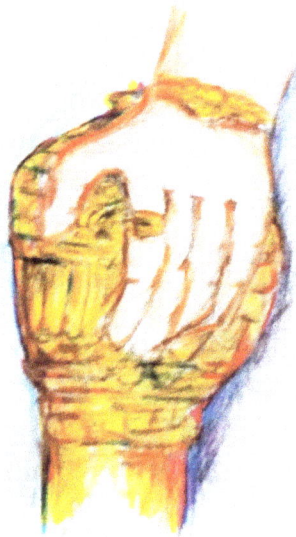

The ' $\rightarrow h \ni h \in$' segment resolves the outcome of the operations implied by Lines 1 – 3, suggesting that the outcome will be 'h' such that ($\ni$) 'h' is an element ($\in$) of the set specified by the members '[Q]: *Constructive zone*', '[Q}: *Constructive zone AND Constructive mutation*', '[F}; *Destructive mutation*', and '{R}: *Random mutation*'. '[Q]: *Constructive zone*' implies that the in-built buffer has been crossed and that access to the deeper four-base logic-encoding ecosystem (FBLEE) has been granted. '[Q}' in this segment implies that there is the possibility that full-quantization as specified by Line 3 of the previous matrix can take place. Access to this zone is a prerequisite for constructive mutation to occur, as designated by the element '   [*Q*}: *Constructive zone* ∩ *Constructive mutation* ', which implies that full-quantization is going to take place and will result in material change. The '[F]' specifies the relationship between 'Fragmentation' in Line 3 of the previous matrix and destructive mutation. The '[R]' specifies the relationship between 'Random' in Line 3 of the previous matrix and random mutation.

Chapter 3.7: Application of Qualified Determinism to Genetic Mutation

The previous chapter explored the possible impact of different layers or levels of light on genetic mutation. The exploration was at a high level and broadly considered a mapping between active layers of light and the type of mutation that may result. Further, the exploration was framed in terms of the light-space-time matrices, central to the mathematics in this treatise, and additionally in terms of a wavefunction form, central to the probabilistic foundation in quantum theory. Consistent with the mathematics in this treatise though, this chapter further elaborates a non-probabilistic approach by application of a process of Qualified Determinism to genetic mutation. In this process structural forces influencing mutation are disaggregated into vertical and horizontal components. The vertical component is resolved by determination of the most active layer of light. The horizontal component is resolved by determination of the most influential base of change amongst the physical, vital, mental, and integral possibilities.

This chapter, hence, explores a mathematical notion of qualified determinism - Dynamic Interaction (DI) that has a 'vertical' and a 'horizontal' component. The vertical component is designated as DI_V and the horizontal component as DI_H. Several equations to capture the inherent dynamism manifest as the likely mutational-sequence along the upward-strand have already been derived in Chapter 2.6. These included equations for the mutational-sequence propelled by the physical, the vital, the mental, and the integral bases. The derived equations propose a model to give insight into how on-going mutation occurs by changing the fundamental states that a microorganism, as an instance of organization in general, is subject to. Several scientists, such as (Prigogine, 1977) and others, are proposing that a system can bifurcate in unpredictable ways to create an emergent property that cannot be predicted. DI is going to propose that in fact there is a 'qualified determinism' as opposed to randomness that occurs (Malik et al, 2017).

DI is going to propose an alternative to the paradigm of statistics and probability thought to govern quantum and other nano- and micro-level objects.

This qualified determinism is the result of the relative strengths of the levels within the core-matrix as summarized by Equation 2.6.7 and reproduced here for convenience:

$$\begin{bmatrix} M_3 & \to & System_x \\ & (\uparrow F & \to I) \\ M_2 & \to & S_{System_x} \\ & (\uparrow Sig \to F) \\ & M_1 & \to Sig_x \\ & (\uparrow > P_x) \\ & U & \to x_U \end{bmatrix}$$

The application of the vertical component of the new function being proposed, DI_V, to this core matrix will yield the nature or 'strength' of the state (x) or orientation under consideration. If the untransformed or U layer is strongest, implying that the habitual patterns that keep an organization locked into its untransformed way of operation are still very active, then the nature of the output of DI_V, notated by x-state, will be x_U. If the habitual patterns have been overcome then the strength of the x-state increases since it is the dynamics of M_1 or Sig_x that are now active. In this case the x-state will be Sig_x. If the unique 'signature' has become a 'force', then the conditions for activation of M_2 have been put in place and the x-state will be even higher, S_{System_x}. The architectural forces active in M_2 are by definition more powerful than Sig_x that is a derivation of a set of such architectural forces. If the 'force' so acting becomes impersonal so that an organizational ego-state is overcome, then the x-state will have the most strength and is characterized by $System_X$ active at M_3. Hence, DI_V applied to a core-matrix will yield the 'strength' in terms of the x-dynamic that is active. Recall also, that each of these upward-strand levels is in fact determined by the parallel light-level that exists in the downward-strand. This is illustrated by the following equation, Equation 3.7.1, which can be considered to be a deductive proof in the context of this model:

$$DI_V \begin{bmatrix} M_3 \rightarrow System_X \\ (\uparrow F \rightarrow I) \\ M_2 \rightarrow S_{System_X} \\ (\uparrow Sig \rightarrow F) \\ M_1 \rightarrow Sig_x \\ (\uparrow > P_x) \\ U \rightarrow x_U \end{bmatrix} =>$$

x-state $\in \left(x_U, Sig_x, S_{System_X}, System_X\right)$

$Where: Strength(System_X) > Strength(S_{System_X})$
$> Strength(Sig_x) > Strength(x_U)$

Eq 3.7.1: Illustrating Action of Dynamic Interaction – vertical component

What is to be noted here is that while the action of DI_V yields a relative strength and therefore a 'single' value for the core- or x-matrix under consideration yet each x-matrix in itself could have an infinite number of possibilities. This should be clear in looking at how x_U, Sig_x, S_{System_X}, and $System_X$, were initially defined.

Hence, taking the example where x = physical:

$Physical_U \ni [inertia, lethargy, status\ quo, ...]$

As can be seen $Physical_U$, defined in Chapter 2.6, is practically speaking already an infinite set with qualities similar to the ones specified.

Similarly, Sig_{Pr}, defined in Chapter 2.4, also has an infinite variation:

$$Sig_{Pr}$$
$$= Xa$$
$$+ \overline{Yb_{0-n}} \quad where \quad \begin{bmatrix} X \in [S_{System_{Pr}}] \\ Y \in [S_{System_{Pr}}, S_{System_P}, S_{System_K}, S_{System_N}] \\ a, b \; are \; integers; a > b \end{bmatrix}$$

$S_{System_{Pr}}$, defined in Chapter 2.3, is also an infinite set with forces of the nature specified in the following equation:

$$S_{System_{Pr}} \ni [Service, Perfection, Diligence, Perseverance, \dots]$$

And recall that in Chapter 2.2, $System_{Pr}$ has been defined as:

$$System_{Pr}$$
$$\equiv TS_{0 \rightarrow N} \begin{bmatrix} Presence \\ \downarrow \\ Opportunity \end{bmatrix} \begin{bmatrix} P_L & \rightarrow & V_L \\ V_L & \rightarrow & M_L \end{bmatrix} \& \; Container \begin{bmatrix} System_P \\ System_K \\ System_N \end{bmatrix}$$

So in essence DI_V is really giving us a summary assessment of the 'level' of the x-matrix under consideration with all its infinite potentiality. An example will follow shortly.

The other component of DI, as suggested earlier in this chapter, is the horizontal component, DI_H. Just as DI_V yields a summary assessment of the level that an x-matrix is operating at, similarly DI_H yields a summary assessment of the direction that a system or organization under consideration is going to continue its development in, considering the physical, the vital, the mental, and the integral bases to be the choices.

Assuming that any organization or system is inherently unique, as this mathematical model proposes, and as the phenomena of entanglement proves, and assuming that the infinite sets of x_U and S_{System_X} applied across the physical, vital, mental, and integral bases or orientations respectively will account for any state that a nano- or micro-organization can experience, then at a certain point in time any such organization under consideration is going to have a direction-bias in one of the possible physical, vital, mental, or integral directions. Hence, DI_H will yield the summary direction that is going to lead an organization into its future given the current states active in it.

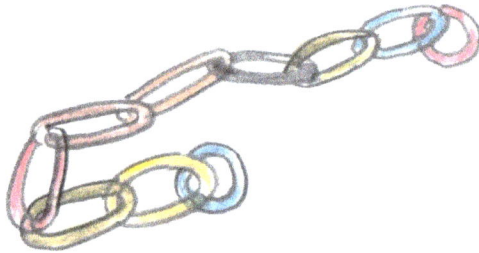

This summary direction is going to be yielded by considering the relative strengths of the separate core x-matrices – the physical, the vital, the mental, and the integral. The assumption is that there will be one core-matrix that will be stronger than the others.

Hence, as an example, first applying DI_V across all four x-matrices may, for example, yield the following results, with the strongest level within each x-matrix highlighted and bolded:

$$\begin{bmatrix} System_{Pr} \\ \mathbf{S_{System_{Pr}}} \\ Sig_P \\ Physical_U \end{bmatrix} \begin{bmatrix} System_P \\ S_{System_P} \\ Sig_V \\ \mathbf{Vital_U} \end{bmatrix} \begin{bmatrix} System_K \\ S_{System_K} \\ Sig_M \\ \mathbf{Mental_U} \end{bmatrix} \begin{bmatrix} System_N \\ S_{System_N} \\ \mathbf{Sig_I} \\ Integral_U \end{bmatrix}$$

Since by definition the strength of $System_x$ is greater than S_{System_x}, which is greater than Sig_x, which is greater than

x_U, applying DI_H, as in Equation 3.7.2, across these x-matrices, as in the example following it will then yield the strongest direction, which in this example is the Physical:

$$DI_H \left(\begin{bmatrix} System_{Pr} \\ S_{System_{Pr}} \\ Sig_P \\ Physical_U \end{bmatrix} \begin{bmatrix} System_P \\ S_{System_P} \\ Sig_V \\ Vital_U \end{bmatrix} \begin{bmatrix} System_K \\ S_{System_K} \\ Sig_M \\ Mental_U \end{bmatrix} \begin{bmatrix} System_N \\ S_{System_N} \\ Sig_I \\ Integral_U \end{bmatrix} \right)$$
$$= Orientation_{Strongest}$$

Eq 3.7.2: Illustrating Action of Dynamic Interaction - horizontal component

Example:

$$DI_H \left(\begin{bmatrix} System_{Pr} \\ \mathbf{S_{System_{Pr}}} \\ Sig_P \\ Physical_U \end{bmatrix} \begin{bmatrix} System_P \\ S_{System_P} \\ Sig_V \\ \mathbf{Vital_U} \end{bmatrix} \begin{bmatrix} System_K \\ S_{System_K} \\ Sig_M \\ \mathbf{Mental_U} \end{bmatrix} \begin{bmatrix} System_N \\ S_{System_N} \\ \mathbf{Sig_I} \\ Integral_U \end{bmatrix} \right)$$
$$= Physical$$

Hence, DI function will yield the following direction of mutation, as in Equation 3.7.3, where 'x_matrix' is used interchangeably with 'orientation' or 'bases':

$$Mutation\_Dir = DI \left(\begin{bmatrix} M_3 \to System_{Pr} \\ (\uparrow F \to I) \\ M_2 \to S_{System_{Pr}} \\ (\uparrow Sig \to F) \\ M_1 \to Sig_P \\ (\uparrow > P_P) \\ U \to Physical_U \end{bmatrix} \begin{bmatrix} M_3 \to System_P \\ (\uparrow F \to I) \\ M_2 \to S_{System_P} \\ (\uparrow Sig \to F) \\ M_1 \to Sig_V \\ (\uparrow > P_V) \\ U \to Vital_U \end{bmatrix} \atop \begin{bmatrix} M_3 \to System_S \\ (\uparrow F \to I) \\ M_2 \to S_{System_S} \\ (\uparrow Sig \to F) \\ M_1 \to Sig_M \\ (\uparrow > P_M) \\ U \to Mental_U \end{bmatrix} \begin{bmatrix} M_3 \to System_N \\ (\uparrow F \to I) \\ M_2 \to S_{System_N} \\ (\uparrow Sig \to F) \\ M_1 \to Sig_I \\ (\uparrow > P_I) \\ U \to Integral_U \end{bmatrix} \right) \to$$

$x\_matrix_{strongest} @ level_{strongest}$

Eq 3.7.3: Direction of Mutation

Generalizing, as in Equation 3.7.4, where Mutation_Dir is direction of mutation:

$$Mutation\_Dir = DI \left(\begin{bmatrix} M_3 \to System_X \\ (\uparrow F \to I) \\ M_2 \to S_{System_X} \\ (\uparrow Sig \to F) \\ M_1 \to Sig_x \\ (\uparrow > P_P) \\ U \to x_U \end{bmatrix}^{x=p,v,m,i} \right) \to$$

$x\_matrix_{strongest} @ level_{strongest}$

Eq 3.7.4: Generalized Equation for Direction of Mutation

Hence, this mathematical model is suggesting that any situation, including micro and nano-level situations, rather than having a random outcome, has a 'qualified deterministic' outcome. In the introduction to his book "Where is Science Going?" (Planck, 1933), James Murphy points out that the reason Planck spent so much of his time giving lectures on causation was because of the trend of physicists at the time, which has continued to the modern day, to overthrowing the principle of causation following the development of quantum theory, which he felt was misplaced. "Planck would claim", he wrote, "and so would Einstein, that it is not the principle of

causation itself which has broken down in modern physics, but rather the traditional formulation of it." Murphy also quotes James Jeans (Jeans, 1932) to suggest the issue associated with causation and determinism: "Einstein showed in 1917 that the theory founded by Planck appeared, at first sight at least, to entail consequences far more revolutionary than mere discontinuity", and here he is referring to the finding that radiant energy is not emitted in a continuous flow, but in integral quantities, or quanta, which can be expressed in integral numbers. Continuing: "It appeared to dethrone the law of causation from the position it had therefore held as guiding the course of the natural world. The old science had confidently proclaimed that nature could follow only one road, the road which was mapped out from the beginning of time to its end by the continuous chain of cause and effect; state A was inevitably succeeded by state B. So far the new science has only been able to say that state A may be followed by state B or C or D or by innumerable other states. It can, it is true, say that B is more likely than C, C than D, and so on; it can even specify the relative probabilities of B, C, and D. But, just because it has to speak in terms of probabilities, it cannot speak with certainty which state will follow which; this is

143

a matter which lies on the knees of the gods – whatever gods there may be."

While under the apparent dynamics at the quantum, micro, or nano level there may appear to be randomness and a dethroning of the principle of causation, the notion of a multiplicity of levels, each having its impact on the strength of an orientation or base and further on the consequent direction from a multiplicity of possible orientations, is being suggested

here as determining the direction of any system, while still allowing infinite variation in the details that may define its. Hence, the positions of Planck and Einstein are vindicated when considering Equation 3.7.4.

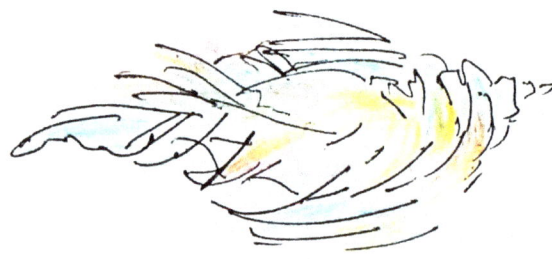

Further, assuming any system where multiple elements are active, connected, interdependent, and emergent, it

may be possible to understand, through application of calculus, as to which level is the source for change.

Hence, where N may be source of change, the rate of change of N will resolve into one of P_U, V_U, M_U, I_U, P_T, V_T, M_T, or I_T. This may be summarized by Equation 3.7.5, where y is either U or T:

$$\frac{dN}{dt} \rightarrow \begin{bmatrix} P_U & P_T \\ V_U & V_T \\ M_U & M_T \\ I_U & I_T \end{bmatrix} \rightarrow x_y, where\ y \in (U,T)$$

Eq 3.7.5: Establishing the Nature of the Change

If T, implying that the action of one of the meta-levels has caused the transformation or mutation, then application of one of the following integrals will determine which level is the likely source for change.

Hence, for M_1, if the integral of $\frac{\partial(x_U \rightarrow x_T)}{\partial t}$ across a limited area 'a' in the vicinity of the change, is greater than some threshold value $Threshold_{Signature}$, then the signature dynamics are likely the source of change, where 'a' can be thought of as some kind of gene-footprint. This is summarized by Equation 3.7.6:

$$\int_0^a \frac{\partial(x_U \rightarrow x_T)}{\partial t} > Threshold_{Signature}$$

Eq 3.7.6: Signature Dynamics as the Source of Change

For M_2, if the integral of $\frac{\partial(x_U \rightarrow x_T)}{\partial t}$ across a larger area 'b' extending beyond the vicinity of the change, is greater than some threshold value $Threshold_{Architectural\ Forces}$, then the architectural forces are likely the source of change, , where 'b' can be thought of as some kind of gene-footprint larger than 'a'. This is summarized by Equation 3.7.7:

$$\int_0^b \frac{\partial(x_U \to x_T)}{\partial t} > Threshold_{ArchitecturalForces}$$

Eq 3.7.7: Architectural Forces as the Source of Change

For M_3, if the double integral of $\frac{\partial(x_U \to x_T)}{\partial t}$ across the system specified by 'A', and across some time 't', is greater than some threshold value $Threshold_{System\ Property}$, then the system properties are likely the source of change, , where 'A' can be thought of as some kind of gene-footprint larger than both 'a' and 'b'. This is summarized by Equation 3.7.8:

$$\int_0^t \int_0^A \frac{\partial(x_U \to x_T)}{\partial t} > Threshold_{SystemProperty}$$

Eq 3.7.8: System Properties as the Source of Change

SECTION 4: OVERVIEW OF REAL-TIME LIGHT-SPACE-TIME MATRIX COMPUTATION IN DEVELOPMENT OF PRE-GENETIC CODE, MATERIAL-FABRIC, GENETIC CODE, AND POST-GENETIC CODE

If the mathematics presented in the previous sections is true, then the pre-genetic and genetic code that imbues everything becomes the result of continuous, real-time Light-Space-Time Matrix based computations implicitly involving quantum-levels.

This section therefore briefly outlines how it may be that Light-Space-Time Matrix based computations generates a range of code from an era pre-dating the Big Bang to modern day Global Civilization, to whatever else may emerge.

The implications of this are that everything that exists is imbued with code. It is just the way the code is housed that changes in forms recognized as living. This also implies that there is likely code within all living cells that has to do with space, time, energy, gravity, the functioning of the electromagnetic spectrum, and all the stages of material elaboration that precede the emergence of a living cell.

Note that the sections following this will go into more detail to elaborate the process by which quantization incrementally changes material reality through alteration of pre-material-fabric, four-base logic encoding ecosystem, pre-genetic material-fabric, genetic, and post-genetic code.

Chapter 4.1: Generation of Pre-Big Bang to Big Bang Pre-Genetic Code and Activation of Material-Fabric

The Light-Matrix component of the Light-Space-Time Emergence equation (3.1.3), reproduced below for convenience, essentially provides an algorithm by which pre-Big Bang existence proceeds to a reality of the Big Bang due to precipitating light. The process generates pre-Big Bang pre-genetic code, and with the Big Bang culminates in the activation of the material-fabric through the initialization of space-time-energy-gravity pre-genetic code.

$Light - Space - Time\ Emergence_T =$

$$\left| \begin{bmatrix} \begin{matrix} c_\infty : [Pr, Po, K, H] \\ (\downarrow R_{C_K} = f(R_{C_\infty})) \\ c_K : [S_{Pr}, S_{Po}, S_K, S_H] \\ (\downarrow R_{C_N} = f(R_{C_K})) \\ c_N : f(S_{Pr} \times S_{Po} \times S_K \times S_H) \\ (\downarrow R_{C_U} = f(R_{C_N})) \\ c_U : [P, V, M, C] \\ \Uparrow \\ c_{0:[D,W,I,C]} \end{matrix} \end{bmatrix}_{Light} \begin{bmatrix} M_3 \rightarrow System_x \\ (\uparrow F \rightarrow I) \\ M_2 \rightarrow S_{System_x} \\ (\uparrow Sig \rightarrow F) \\ M_1 \rightarrow Sig_x \\ (\uparrow > P_x) \\ U \rightarrow x_U \end{bmatrix}_{Space} \\ \begin{bmatrix} U \rightarrow \begin{matrix} M_3 : -\infty \le t \le \infty \\ \downarrow \\ M_2 : 0 \ge t > \infty \\ \downarrow \\ M_1 : 0 > t > \infty \\ \downarrow \\ t \le E_{Cell}; TC: M_3 \rightarrow U \\ t \sim E_{Human}; TC: U \rightarrow M_3 \end{matrix} \end{bmatrix}_{Time} \begin{matrix} \\ TC \rightarrow x_T \\ \end{matrix} \right| \langle x_U | x_T \rangle$$

Pre-Big Bang Pre-Precipitation ∞-Entanglement Pre-Material-Fabric Pre-Genetic Code

Before the Big Bang, as per discussions in Section 1 and 2, it can be assumed that Light precipitated in stages to the point where it slowed down to c. Equation 4.1.1 highlights the pre-precipitation pre-material-fabric pre-genetic code. Hence (3.1.3) collapses to the active highlighted portion only.

$$Light - Space - Time\ Emergence_T =$$

$$\begin{bmatrix} \begin{bmatrix} \begin{array}{c} \boldsymbol{c_\infty}:[\boldsymbol{Pr,Po,K,H}] \\ \left(\downarrow R_{C_K} = f(R_{C_\infty})\right) \\ c_K:[S_{Pr},S_{Po},S_K,S_H] \\ \left(\downarrow R_{C_N} = f(R_{C_K})\right) \\ c_N: f(S_{Pr} \times S_{Po} \times S_K \times S_H) \\ \left(\downarrow R_{C_U} = f(R_{C_N})\right) \\ c_U:[P,V,M,C] \\ \Uparrow \\ C_0:[D,W,I,C] \end{array} \end{bmatrix}_{Light} \begin{bmatrix} M_3 \rightarrow System_X \\ (\uparrow F \rightarrow I) \\ M_2 \rightarrow S_{System_X} \\ (\uparrow Sig \rightarrow F) \\ M_1 \rightarrow Sig_X \\ (\uparrow > P_x) \\ U \rightarrow x_U \end{bmatrix}_{Space} \\ \begin{bmatrix} M_3: -\infty \leq t \leq \infty \\ \downarrow \\ M_2: 0 \geq t > \infty \\ \downarrow \\ M_1: 0 > t > \infty \\ \downarrow \\ U \rightarrow \begin{array}{c} t \leq E_{Cell}; TC: M_3 \rightarrow U \\ t \sim E_{Human}; TC: U \rightarrow M_3 \end{array} \end{bmatrix}_{Time} \quad TC \rightarrow x_T \end{bmatrix} \langle x_U | x_T \rangle$$

$$\Rightarrow$$

$$\left(\begin{array}{l} System_{Pr} \equiv TS_{0 \rightarrow N} \begin{bmatrix} Presence \\ \downarrow \\ Opportunity \end{bmatrix} \begin{bmatrix} P_L & \rightarrow & V_L \\ V_L & \rightarrow & M_L \end{bmatrix} \& Container_{\begin{bmatrix} System_P \\ System_K \\ System_N \end{bmatrix}} \\ \\ System_P \equiv power > \sum_{TS=0}^{N} P_R * V_R * M_R \\ \\ System_K \equiv TS_{0 \rightarrow N} \begin{bmatrix} Knowledge \\ \downarrow \\ Leverage \end{bmatrix} \begin{bmatrix} P_{I,C} \\ V_{I,C} \\ M_{I,C} \end{bmatrix} \rightarrow \begin{bmatrix} P_L & \rightarrow & V_L \\ V_L & \rightarrow & M_L \end{bmatrix} \\ \\ System_N \equiv TS_{0 \rightarrow N} \left(\coprod_{Nurturing} \begin{pmatrix} P_- & M_+ \\ V_- & V_+ \\ M_- & P_+ \end{pmatrix} mod\ (r,\theta) \right) \end{array} \right)$$

Eq 4.1.1: Pre-Big Bang Pre-Precipitation ∞-Entanglement Pre-Material-Fabric Pre-Genetic Code

Note the equation-segments following the '⇒' were derived in Chapter 2.2, Ubiquitous-Point-Instant Pre-

Genetic ∞-Entanglement Library. The generated pre-genetic code belongs to the pre-material-fabric genre since it will remain in a subtle domain exercising its influence through ∞-entanglement where possible.

Pre-Big Bang Pre-Precipitation Zero-Speed Pre-Material-Fabric Pre-Genetic Code

As discussed in the Chapter 2.1, Light-Matrix & The Ubiquitous Downward-Strand, Light could project itself at speed zero. This would create a reality opposite to that in its native state. This reality could have been projected prior to the gradual precipitation that is the corner stone of this treatise on Light. The reality created when light moves at zero speed, as already discussed in Sections 2 and 3, can be seen to play a significant part in destructive mutation, and is seen to influence the reality modeled by the layer U, where light moves at c. Because obstinate patterns can be better explained with a 'speed zero' reality in the background, it will be assumed that this pre-exists the precipitation of Light down to c.

Equation 4.1.2 highlights the pre-precipitation zero-speed pre-material-fabric pre-genetic code. Hence (3.1.3) collapses to the active highlighted portion only.

$Light - Space - Time\ Emergence_T =$

$$\left\|\begin{bmatrix} \begin{array}{c} c_\infty : [Pr, Po, K, H] \\ \left(\downarrow R_{C_K} = f(R_{C_\infty}) \right) \\ c_K : [S_{Pr}, S_{Po}, S_K, S_H] \\ \left(\downarrow R_{C_N} = f(R_{C_K}) \right) \\ c_N : f(S_{Pr} \times S_{Po} \times S_K \times S_H) \\ \left(\downarrow R_{C_U} = f(R_{C_N}) \right) \\ c_U : [P, V, M, C] \\ \Uparrow \\ \boldsymbol{c_{0:[D,W,I,C]}} \end{array} \end{bmatrix}_{Light} \begin{bmatrix} M_3 \to \text{System}_X \\ (\uparrow F \to I) \\ M_2 \to S_{\text{System}_X} \\ (\uparrow Sig \to F) \\ M_1 \to Sig_x \\ (\uparrow > P_x) \\ U \to x_U \end{bmatrix}_{Space} \right.$$

$$\left. \begin{bmatrix} U \to \begin{array}{c} M_3 : -\infty \le t \le \infty \\ \downarrow \\ M_2 : 0 \ge t > \infty \\ \downarrow \\ M_1 : 0 > t > \infty \\ \downarrow \\ t \le E_{Cell}; TC: M_3 \to U \\ t \sim E_{Human}; TC: U \to M_3 \end{array} \end{bmatrix}_{Time} \quad TC \to x_T \right\| \langle x_U | x_T \rangle$$

$$\Rightarrow$$

$$\left\langle \begin{array}{c} Pre - Big - Bang\ Pre - Precipitation\ \infty - Entanglement \\ Pre - Material - Fabric\ Pre - Genetic\ Code \end{array} \right\rangle +$$

$$(c_0 : [Darkness, Weakness, Ignorance, Chaos])$$

Eq 4.1.2:   Pre-Big Bang Pre-Precipitation Zero-Speed Pre-Material-Fabric Pre-Genetic Code

Note that the last equation-segment following the '⇒' was derived in Chapter 2.1, Light-Matrix & The Ubiquitous Downward-Strand. This equation is stating that when light is projected at zero-speed the only active code-segments are due to ∞-entanglement as discussed in the previous sub-section and zero-speed of light.

Pre-Big Bang K-Entanglement Pre-Material-Fabric Pre-Genetic Code

The first stage of precipitation prior to the Big Bang is modeled by K-entanglement as discussed in Chapter 2.3, The Architectural Forces K-Entanglement Pre-Genetic

Library. Equation 4.1.3 highlights the pre-material-fabric pre-genetic code active at this point. Hence (3.1.3) collapses to the active highlighted portion only.

$$Light - Space - Time\ Emergence_T =$$

$$
\left[
\begin{array}{l}
\left[
\begin{array}{c}
c_\infty: [Pr, Po, K, H] \\
\left(\downarrow R_{C_K} = f\left(R_{C_\infty}\right)\right) \\
c_K: [S_{Pr}, S_{Po}, S_K, S_H] \\
\left(\downarrow R_{C_N} = f\left(R_{C_K}\right)\right) \\
c_N: f(S_{Pr} \times S_{Po} \times S_K \times S_H) \\
\left(\downarrow R_{C_U} = f\left(R_{C_N}\right)\right) \\
c_U: [P, V, M, C] \\
\Uparrow \\
c_{0: [D,W,I,C]}
\end{array}
\right]_{Light}
\begin{array}{c}
\left[
\begin{array}{c}
M_3 \rightarrow System_X \\
(\uparrow F \rightarrow I) \\
M_2 \rightarrow S_{System_X} \\
(\uparrow Sig \rightarrow F) \\
M_1 \rightarrow Sig_x \\
(\uparrow > P_x) \\
U \rightarrow x_U
\end{array}
\right]_{Space}
\end{array} \\[1cm]
\left[U \rightarrow
\begin{array}{c}
M_3 : -\infty \le t \le \infty \\
\downarrow \\
M_2 : 0 \ge t > \infty \\
\downarrow \\
M_1 : 0 > t > \infty \\
\downarrow \\
t \le E_{Cell}; TC: M_3 \rightarrow U \\
t \sim E_{Human}; TC: U \rightarrow M_3
\end{array}
\right]_{Time}
\begin{array}{c}
TC \rightarrow x_T
\end{array}
\end{array}
\right]
\langle x_U | x_T \rangle
$$

$$\Rightarrow$$

$$\left\langle
\begin{array}{c}
Pre - Big - Bang\ Pre - Precipitation\ \infty - Entanglement \\
Pre - Material - Fabric\ Pre - Genetic\ Code
\end{array}
\right\rangle +$$

$$\left\langle
\begin{array}{c}
Pre - Big - Bang\ Pre - Precipitation\ Zero - Speed \\
Pre - Material - Fabric\ Pre - Genetic\ Code
\end{array}
\right\rangle +$$

$$\left(
\begin{array}{l}
\sum S_{System_{Pr}} \ni [Service, Perfection, Diligence, Perseverance, \dots] \\
\sum S_{System_P} \ni [Power, Courage, Adventure, Justice, \dots] \\
\sum S_{System_K} \ni [Wisdom, Law\ Making, Spread\ of\ Knowledge \dots] \\
\sum S_{System_N} \ni [Love, Compassion, Harmony, Relationship \dots]
\end{array}
\right)$$

Eq 4.1.3: Pre-Big Bang K-Entanglement Pre-Material-Fabric Pre-Genetic Code

Note that the last equation-segment following the '⇒' was derived in Chapter 2.3, The Architectural Forces K-Entanglement Pre-Genetic Library. The '$\sum x$' signifies all the possibilities for the 'x' equation-segment. All the segments following the '⇒' are active as pre-material-fabric pre-genetic code at this point.

Pre-Big Bang N-Entanglement Pre-Material-Fabric Pre-Genetic Code

The second stage of precipitation prior to the Big Bang is modeled by N-entanglement as discussed in Chapter 2.4, The Organizational Uniqueness N-Entanglement Pre-Genetic Library. Equation 4.1.4 highlights the pre-material-fabric pre-genetic code active at this point. Hence (3.1.3) collapses to the active highlighted portion only.

$$Light - Space - Time\ Emergence_T =$$

$$\left[\left[\begin{array}{c} c_\infty: [Pr, Po, K, H] \\ \left(\downarrow R_{C_K} = f(R_{C_\infty})\right) \\ c_K: [S_{Pr}, S_{Po}, S_K, S_H] \\ \left(\downarrow R_{C_N} = f(R_{C_K})\right) \\ c_N: f(S_{Pr} \times S_{Po} \times S_K \times S_H) \\ \left(\downarrow R_{C_U} = f(R_{C_N})\right) \\ c_U: [P, V, M, C] \\ \Uparrow \\ c_{0:[D,W,I,C]} \end{array}\right]_{Light} \quad \begin{array}{c} M_3 \rightarrow \text{System}_X \\ (\uparrow F \rightarrow I) \\ M_2 \rightarrow S_{System_X} \\ (\uparrow Sig \rightarrow F) \\ M_1 \rightarrow Sig_x \\ (\uparrow > P_x) \\ U \rightarrow x_U \end{array}\right]_{Space}$$

$$\left[U \rightarrow \begin{array}{c} M_3 : -\infty \le t \le \infty \\ \downarrow \\ M_2 : 0 \ge t > \infty \\ \downarrow \\ M_1 : 0 > t > \infty \\ \downarrow \\ t \le E_{Cell}; TC: M_3 \rightarrow U \\ t \sim E_{Human}; TC: U \rightarrow M_3 \end{array}\right]_{Time} \quad TC \rightarrow x_T \qquad \langle x_U | x_T \rangle$$

$$\Rightarrow$$

$$\left\langle \begin{array}{c} Pre - Big - Bang\ Pre - Precipitation\ \infty - Entanglement \\ Pre - Material - Fabric\ Pre - Genetic\ Code \end{array} \right\rangle +$$

$$\left\langle \begin{array}{c} Pre - Big - Bang\ Pre - Precipitation\ Zero - Speed \\ Pre - Material - Fabric\ Pre - Genetic\ Code \end{array} \right\rangle +$$

$$\left\langle \begin{array}{c} Pre - Big - Bang\ K - Entanglement\ Pre - Material - Fabric \\ Pre - Genetic\ Code \end{array} \right\rangle$$
$$+$$

$$\left(where \begin{array}{c} \sum Sig = Xa + \overline{Yb_{0-n}} \\ \left[\begin{array}{c} X \in [S_{System_{Pr}}, S_{System_P}, S_{System_K}, S_{System_N}] \\ Y \in [S_{System_{Pr}}, S_{System_P}, S_{System_K}, S_{System_N}] \\ a, b\ are\ integers; a > b \end{array}\right] \end{array} \right)$$

Eq 4.1.4: Pre-Big Bang N-Entanglement Pre-Material-Fabric Pre-Genetic Code

Note that the last equation-segment following the '⇒' was derived in Chapter 2.4, The Organizational Uniqueness

N-Entanglement Pre-Genetic Library. The '$\sum x$' signifies all the possibilities for the 'x' equation-segment.

Big Bang Pre-Material-Fabric and Material-Fabric Pre-Genetic Code

Assuming Light then projects itself from its state of ∞ to that of c, this creates the Big Bang. The process is modeled as yielding two equations.

Equation 4.1.5, Big Bang Pre-Material-Fabric Pre-Genetic Code, highlights the operative portions. The important point is that this equation generates the additional space-time-energy-gravity code-segment as a four-base logic-encoding ecosystem depicted by the sub-script 'FBLEE' at the right-base of the newly generated code-segment, which then immediately activates the material-fabric in Equation 4.1.6.

Hence (4.1.5):

$$Light - Space - Time\ Emergence_T =$$

$$\left[\begin{array}{l}\left[\begin{array}{c}c_\infty:[Pr,Po,K,H] \\ \left(\downarrow R_{C_K} = f(R_{C_\infty})\right) \\ c_K:[S_{Pr},S_{Po},S_K,S_H] \\ \left(\downarrow R_{C_N} = f(R_{C_K})\right) \\ c_N:f(S_{Pr}\,x\,S_{Po}\,x\,S_K\,x\,S_H) \\ \left(\downarrow R_{C_U} = f(R_{C_N})\right) \\ c_U:[P,V,M,C] \\ \Uparrow \\ c_{0:[D,W,I,C]}\end{array}\right]_{Light} \begin{bmatrix}M_3 \to System_X \\ (\uparrow F \to I) \\ M_2 \to S_{System_X} \\ (\uparrow Sig \to F) \\ M_1 \to Sig_x \\ (\uparrow > P_x) \\ U \to x_U\end{bmatrix}_{Space} \\ \left[\begin{array}{c}M_3 : -\infty \le t \le \infty \\ \downarrow \\ M_2 : 0 \ge t > \infty \\ \downarrow \\ M_1 : 0 > t > \infty \\ \downarrow \\ U \to \begin{array}{l}t \le E_{Cell};\text{TC}:M_3 \to U \\ t \sim E_{Human};\text{TC}:U \to M_3\end{array}\end{array}\right]_{Time} \quad TC \to x_T\end{array}\right] \langle x_U|x_T\rangle$$

$\Rightarrow$

$\left\langle\begin{array}{c}Pre - Big - Bang\ Pre - Precipitation\ \infty - Entanglement \\ Pre - Material - Fabric\ Pre - Genetic\ Code\end{array}\right\rangle +$

$\left\langle\begin{array}{c}Pre - Big - Bang\ Pre - Precipitation\ Zero - Speed \\ Pre - Material - Fabric\ Pre - Genetic\ Code\end{array}\right\rangle +$

$\left\langle\begin{array}{c}Pre - Big - Bang\ K - Entanglement\ Pre - Material - Fabric \\ Pre - Genetic\ Code\end{array}\right\rangle +$

$\left\langle\begin{array}{c}Pre - Big - Bang\ N - Entanglement \\ Pre - Material - Fabric\ Pre - Genetic\ Code\end{array}\right\rangle +$

$$\left(\begin{array}{c} Space_{quantization} \ = h_{UK}\left(Xa + \overline{Yb_{0-n}}\ \right) \\ \left[\begin{array}{c} X \in [S_{System_K}] \\ where: \quad \left[\begin{array}{c} Y \in [S_{System_{Pr}}, S_{System_P}, S_{System_K}, S_{System_N}] \\ a, b\ are\ integers; a > b \end{array}\right] \end{array}\right] \\ Time_{quantization} \ = h_{UP}\left(Xa + \overline{Yb_{0-n}}\ \right) \\ \left[\begin{array}{c} X \in [S_{System_P}] \\ where: \quad \left[\begin{array}{c} Y \in [S_{System_{Pr}}, S_{System_P}, S_{System_K}, S_{System_N}] \\ a, b\ are\ integers; a > b \end{array}\right] \end{array}\right] \\ Energy_{quantization} \ = h_{UPr}\left(Xa + \overline{Yb_{0-n}}\ \right) \\ \left[\begin{array}{c} X \in [S_{System_{Pr}}] \\ where: \quad \left[\begin{array}{c} Y \in [S_{System_{Pr}}, S_{System_P}, S_{System_K}, S_{System_N}] \\ a, b\ are\ integers; a > b \end{array}\right] \end{array}\right] \\ Gravity_{quantization} \ = h_{UH}\left(Xa + \overline{Yb_{0-n}}\ \right) \\ \left[\begin{array}{c} X \in [S_{System_N}] \\ where: \quad \left[\begin{array}{c} Y \in [S_{System_{Pr}}, S_{System_P}, S_{System_K}, S_{System_N}] \\ a, b\ are\ integers; a > b \end{array}\right] \end{array}\right] \end{array}\right)_{FBLEE}$$

Eq 4.1.5: Big Bang Pre-Material-Fabric Pre-Genetic Code

Equation 4.1.5 may be thought of as being active as light approaches c. Hence the generated FBLEE-segment exists at the quantum-levels and can be thought of as exhibiting entanglement, referred to as FBLEE-entanglement (FBLEEE).

As light slows down to c the multiple quantization discussed in Section 3 becomes active and spawns the material-fabric, which may be thought of as the gestalt or the emergent property due to the combined space-time-energy-gravity precipitation.

Space, Time Energy, and Gravity have therefore a specific meaning and significance in the context of the material universe. At the macro-level as discussed in Chapter 3.5 through acting as the mechanisms by which potentially infinite number of seeds are interrelated and mature, they define essential cosmic parameters by which the universe operates.

At the micro-level their action is essential in materializing pre-genetic code, genetic code, and other codes that may also cause different kinds of materialization in the future. In living cells such code is known to exist in DNA. In forms prior to the living cell, while Science may not yet have acknowledged that such code exists and similarly must be housed in a material structure, yet this must be the case.

In this treatise such a material structure housing pre-genetic code is referred to as the 'material-fabric'. Material-fabric is envisioned as existing as or at the interface of the antecedent quantum-layer and matter. Note though that the Big Bang space-time-energy-gravity pre-genetic code is still created first as a four-base logic encoding ecosystem assumed to exist in the antecedent quantum-layer and depicted by the FBLEE subscript in (4.1.5) as just discussed.

Through further application of (3.1.3) this FBLEE pre-genetic code will be housed in the material-fabric as illustrated by Equation 4.1.6:

$$Light - Space - Time\ Emergence_{Big\ Bang} =$$

$$\left\| \begin{bmatrix} c_\infty : [Pr, Po, K, H] \\ \left(\downarrow R_{C_K} = f(R_{C_\infty}) \right) \\ c_K : [S_{Pr}, S_{Po}, S_K, S_H] \\ \left(\downarrow R_{C_N} = f(R_{C_K}) \right) \\ c_N : f(S_{Pr} \times S_{Po} \times S_K \times S_H) \\ \left(\downarrow R_{C_U} = f(R_{C_N}) \right) \\ c_U : [P, V, M, C] \\ \Uparrow \\ C_{0:[D,W,I,C]} \end{bmatrix}_{Light} \begin{bmatrix} M_3 \rightarrow System_x \\ (\uparrow F \rightarrow I) \\ M_2 \rightarrow S_{System_x} \\ (\uparrow Sig \rightarrow F) \\ M_1 \rightarrow Sig_x \\ (\uparrow > P_x) \\ U \rightarrow x_{Big\ Bang[FBLEE]} \end{bmatrix}_{Space} \right.$$

$$U \rightarrow \begin{bmatrix} M_3 : -\infty \leq t \leq \infty \\ \downarrow \\ M_2 : 0 \geq t > \infty \\ \downarrow \\ M_1 : 0 > t > \infty \\ \downarrow \\ t \leq E_{Cell}; TC: M_3 \rightarrow U \\ t \sim E_{Human}; TC: U \rightarrow M_3 \end{bmatrix}_{Time} \qquad TC \rightarrow x_{Big\ Bang} \qquad \left. \vphantom{X} \right\| \langle x_U | x_T \rangle$$

$$\Rightarrow$$

$$\left(\begin{array}{l} Space_{quantization} = h_{UK}\left(Xa + \overline{Yb_{0-n}} \right) \\ where: \begin{bmatrix} X \in [S_{System_K}] \\ Y \in [S_{System_{Pr}}, S_{System_P}, S_{System_K}, S_{System_N}] \\ a, b\ are\ integers; a > b \end{bmatrix} \\ Time_{quantization} = h_{UP}\left(Xa + \overline{Yb_{0-n}} \right) \\ where: \begin{bmatrix} X \in [S_{System_P}] \\ Y \in [S_{System_{Pr}}, S_{System_P}, S_{System_K}, S_{System_N}] \\ a, b\ are\ integers; a > b \end{bmatrix} \\ Energy_{quantization} = h_{UPr}\left(Xa + \overline{Yb_{0-n}} \right) \\ where: \begin{bmatrix} X \in [S_{System_{Pr}}] \\ Y \in [S_{System_{Pr}}, S_{System_P}, S_{System_K}, S_{System_N}] \\ a, b\ are\ integers; a > b \end{bmatrix} \\ Gravity_{quantization} = h_{UH}\left(Xa + \overline{Yb_{0-n}} \right) \\ where: \begin{bmatrix} X \in [S_{System_N}] \\ Y \in [S_{System_{Pr}}, S_{System_P}, S_{System_K}, S_{System_N}] \\ a, b\ are\ integers; a > b \end{bmatrix} \end{array} \right)_{MF}$$

Eq 4.1.6: Big Bang Material-Fabric Pre-Genetic Code

Not the change in several notations in (4.1.6) as compared with previous iterations of (3.1.3). These include the highlighted portions in 4.1.6, the change in subscripts T

160

and U, and the additional subscript 'MF' after the generated code-segment.

Hence, all the portions of the Light-Matrix are highlighted as in (4.1.5). But in addition in the Time-Matrix, the highlighted 't $\leq E_{Cell}$; TC: $M_3 \rightarrow$ U' indicates that since the space-time-energy-gravity construct precedes the emergence of the cell, all meta-layers will be active in the Space-Matrix. Hence, the entire space-matrix is also highlighted indicating that quantization is active and therefore precipitation into a material-fabric is inevitable. This materialization into the material-fabric is depicted by the subscript 'MF' following the generated code-segment. Specifically also, the U has been changed to 'Big Bang[FBLEE]' and the T to 'Big Bang'.

161

Note that in future sections the actions specific to (4.1.5) and (4.1.6) will generally be combined into one equation only. This will signify that FBLEE code generation is implicit and will always precede MF code generation. The combined equation will always have the U and T subscripts named.

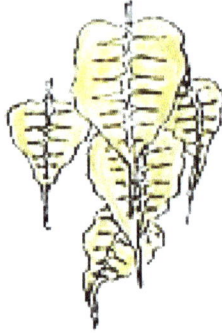

An observation of (4.1.5) and (4.1.6) suggest the vast amount of pre-genetic information that has already been generated, and also suggests the means by which this information can be materialized. The Big Bang is therefore a watershed event that puts into play a potentially endless evolution materializing more and more possibility that exists in Light.

While (4.1.6) does not explicitly show all the pre-material-fabric code-segments, these exist in antecedent layers of Light and will continue to exercise influence through both entanglement and through being part of the source material that generates FBLEE code-segments. The means for materialization of any generated FBLEE code-segment is the fourfold space-time-gravity-energy quantization that is itself generated as pre-genetic code that must exist in every iota of the universe as part of its material-fabric.

Note also that the Big-Bang is a significant event that in the model of emergence depicted in this book alters the possibilities of evolution. If the only operative layer besides c_∞ was c_0 then the nature of emergence and evolution would be radically different than with all the layers as depicted by the Light-Matrix in (3.1.3). As

162

discussed each of these light layers implies the presence of libraries of pre-genetic pre-material-fabric information, and a much quicker and potentially richer evolution than would be possible if only c_0 existed. Ironically, though, it appears that in the contemporary rendering of physics and the notion of emergence as described in mainstream complex adaptive systems literature evolution is viewed as originating from c_0 only.

Chapter 4.2: Post-Big Bang Generation of Material-Fabric, Genetic Code, and Post-Genetic Code

As implied in Chapter 4.1 it is only with (4.1.6) that the emergence through matter, life, and mega-organization can proceed. The occurrence of space-time-energy-gravity quantization means that for pre-cellular structures the mechanism for affecting the fabric of material existence, or material-fabric, and for cellular structures and beyond, the means for affecting genetic code and any possible post-genetic code is now in place. Such quantization puts into place the mechanism for the evolution of the universe as will be elaborated in subsequent sections.

This chapter will offer a high-level overview of various fourfold code projections that accelerate material evolution. Hence, if we trace the potential in Light working from the basis of an initial electromagnetic spectrum to the current basis of a possible sustainable global civilization, and beyond, some of the iterations the equation may have gone through can be envisioned as follows in Equation 4.2.1 – 8. Each iteration generates a new code segment and in totality these are added to existing code segments in the material-fabric in the case of pre-genetic information, to DNA in the case of genetic information, and to some hybrid or unknown structure in the case of post-genetic information.

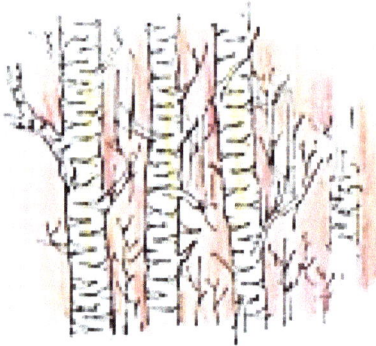

Note that the following equations and their outputs are only snapshots and always there will be a vaster number of iterations that will be required to capture the enormity of any kind of genetic-type information that exists in reality.

Further, these iterations do not necessarily imply that one code-segment output is dependent on a previous one. Space-Time-Energy-Gravity logic embedded in the material-fabric allows quantization to take place. Similarly the emergence of other code-segments allows other necessary operations to take place and accelerates what is possible in the material universe. But emergence itself may equally be dictated by the different kinds of entanglement and the vast variety of information in pre-material-fabric subtle-libraries.

Quantum Particle Material-Fabric Pre-Genetic Code Iteration

In Equation 4.2.1, the starting point, x_U, is 'EM Spectrum', and the ending point, x_T, is 'Quantum Particle'. x_U is highlighted, and x_T is bolded. As suggested by (4.1.6) the naming of subscripts implies that the code generated is actually housed in the material-fabric through a similar 2-step process as captured by (4.1.5) and (4.1.6). Hence the FBLEE code-segments for quantum particles will already have been generated at the quantum-levels. The output is the pre-genetic code for quantum particles that will be added to the existing space-time-energy-gravity and electromagnetic spectrum

165

pre-genetic code in the material-fabric. In this model such code embedded into the material-fabric is assumed to be operative and binding on constructs that arise within the universe. Such code also changes the possibilities that can arise within the material universe.

Hence:

$$Light - Space - Time\ Emergence_{Quantum\ Particles} =$$

$$\left\Vert \begin{bmatrix} \begin{bmatrix} c_{\infty} : [Pr, Po, K, H] \\ \left(\downarrow R_{C_K} = f(R_{C_\infty}) \right) \\ c_K : [S_{Pr}, S_{Po}, S_K, S_H] \\ \left(\downarrow R_{C_N} = f(R_{C_K}) \right) \\ c_N : f(S_{Pr} \times S_{Po} \times S_K \times S_H) \\ \left(\downarrow R_{C_U} = f(R_{C_N}) \right) \\ c_U : [P, V, M, C] \\ \Uparrow \\ c_{0:[D,W,I,C]} \end{bmatrix}_{Light} \begin{bmatrix} M_3 \rightarrow System_X \\ (\uparrow F \rightarrow I) \\ M_2 \rightarrow S_{System_X} \\ (\uparrow Sig \rightarrow F) \\ M_1 \rightarrow Sig_X \\ (\uparrow > P_x) \\ U \rightarrow x_{EM\ Spectrum} \end{bmatrix}_{Space} \\ \begin{bmatrix} U \rightarrow \begin{matrix} M_3 : -\infty \le t \le \infty \\ \downarrow \\ M_2 : 0 \ge t > \infty \\ \downarrow \\ M_1 : 0 > t > \infty \\ \downarrow \\ t \le E_{Cell}; TC: M_3 \rightarrow U \\ t \sim E_{Human}; TC: U \rightarrow M_3 \end{matrix} \end{bmatrix}_{Time} \quad TC \rightarrow x_{Quantum\ Particles} \end{bmatrix} \right\Vert_{\langle x_U | x_T \rangle}$$

$$\Rightarrow$$

$$\langle Space - Time - Energy - Gravity\ Material - Fabric\ Pre - Genetic\ Code \rangle +$$

$$\langle Electromagnetic\ Spectrum\ Material - Fabric\ Pre - Genetic\ Code \rangle +$$

$$Quantum - Particle\ Material - Fabric\ Pre - Genetic\ Code$$

Eq 4.2.1: Quantum Particle Material-Fabric Pre-Genetic Code

Note that Section 5 will discuss the computation leading to the emergence of the electromagnetic spectrum and its associated pre-genetic code segments, while Section 6 will discuss the computations leading to the emergence of matter, including quantum particles, and the associated pre-genetic code segments.

Atoms Material-Fabric Pre-Genetic Code Iteration

In Equation 4.2.2, Atoms Material-Fabric Pre-Genetic Code, the starting point, x_U, is 'Quantum Particles', and the ending point, x_T, is 'Atoms'. This iteration of the Light-Space Time Emergence equation suggests therefore that the generated code-segments for atoms will be added to the pre-existing material-fabric code-segments or "laws" for space-time-energy-gravity, the electromagnetic spectrum, and quantum particles.

Hence:

$$Light - Space - Time\ Emergence_{Atoms} =$$

$$\left| \begin{bmatrix} \begin{bmatrix} c_\infty : [Pr, Po, K, H] \\ \left(\downarrow R_{C_K} = f\!\left(R_{C_\infty}\right) \right) \\ c_K : [S_{Pr}, S_{Po}, S_K, S_H] \\ \left(\downarrow R_{C_N} = f\!\left(R_{C_K}\right) \right) \\ c_N : f(S_{Pr} \times S_{Po} \times S_K \times S_H) \\ \left(\downarrow R_{C_U} = f\!\left(R_{C_N}\right) \right) \\ c_U : [P, V, M, C] \\ \Uparrow \\ C_{0:[D,W,I,C]} \end{bmatrix}_{Light} \\ \begin{bmatrix} U \to \begin{array}{c} M_3 : -\infty \le t \le \infty \\ \downarrow \\ M_2 : 0 \ge t > \infty \\ \downarrow \\ M_1 : 0 > t > \infty \\ \downarrow \\ t \le E_{Cell}; TC: M_3 \to U \\ t \sim E_{Human}; TC: U \to M_3 \end{array} \end{bmatrix}_{Time} \\ \Rightarrow \end{bmatrix} \quad \begin{bmatrix} M_3 \to System_X \\ (\uparrow F \to I) \\ M_2 \to S_{System_X} \\ (\uparrow Sig \to F) \\ M_1 \to Sig_x \\ (\uparrow > P_x) \\ U \to x_{Quantum\ Particles} \end{bmatrix}_{Space} \\ TC \to x_{Atoms} \end{matrix} \right| \langle x_U | x_T \rangle$$

$\langle Space - Time - Energy - Gravity\ Material - Fabric\ Pre - Genetic\ Code \rangle +$

$\langle Electromagnetic\ Spectrum\ Material - Fabric\ Pre - Genetic\ Code \rangle +$

$\langle Quantum - Particle\ Material - Fabric\ Pre - Genetic\ Code \rangle +$

$Atoms\ Material - Fabric\ Pre - Genetic\ Code$

Eq 4.2.2: Atoms Material-Fabric Pre-Genetic Code

The computation leading to the emergence of atoms and its associated pre-genetic code segments is discussed in greater detail in Chapter 6.3.

168

In Equation 4.2.3, Molecules Material-Fabric Pre-Genetic Code, the starting point, x_U, is 'Atoms', and the ending point, x_T, is 'Molecules':

$$Light - Space - Time\ Emergence_{Molecules} =$$

$$\left\|\begin{bmatrix} \begin{bmatrix} c_\infty\colon [Pr, Po, K, H] \\ \left(\downarrow R_{C_K} = f(R_{C_\infty})\right) \\ c_K\colon [S_{Pr}, S_{Po}, S_K, S_H] \\ \left(\downarrow R_{C_N} = f(R_{C_K})\right) \\ c_N\colon f(S_{Pr} \times S_{Po} \times S_K \times S_H) \\ \left(\downarrow R_{C_U} = f(R_{C_N})\right) \\ c_U\colon [P, V, M, C] \\ \Uparrow \\ c_{0\colon[D,W,I,C]} \end{bmatrix}_{Light} \begin{bmatrix} M_3 \to System_x \\ (\uparrow F \to I) \\ M_2 \to S_{System_x} \\ (\uparrow Sig \to F) \\ M_1 \to Sig_x \\ (\uparrow > P_x) \\ U \to x_{Atoms} \end{bmatrix}_{Space} \\ \begin{bmatrix} M_3 \colon -\infty \le t \le \infty \\ \downarrow \\ M_2 \colon 0 \ge t > \infty \\ \downarrow \\ M_1 \colon 0 > t > \infty \\ \downarrow \\ U \to \begin{array}{l} t \le E_{Cell}; TC\colon M_3 \to U \\ t \sim E_{Human}; TC\colon U \to M_3 \end{array} \end{bmatrix}_{Time} \quad TC \to x_{Molecules} \end{bmatrix}\right\|_{\langle x_U | x_T \rangle}$$

$$\Rightarrow$$

$\langle Space - Time - Energy - Gravity\ Material - Fabric\ Pre$
$\qquad - Genetic\ Code \rangle +$

$\langle Electromagnetic\ Spectrum\ Material - Fabric\ Pre - Genetic\ Code \rangle +$

$\langle Quantum - Particle\ Material - Fabric\ Pre - Genetic\ Code \rangle +$

$\langle Atoms\ Material - Fabric\ Pre - Genetic\ Code \rangle +$

$Molecules\ Material - Fabric\ Pre - Genetic\ Code$

Eq 4.2.3: Molecules Material-Fabric Pre-Genetic Code

Cells Genetic Code Iteration

In Equation 4.2.4, Cells Genetic Code, the starting point, x_U, is 'Molecules', and the ending point, x_T, is 'Cells':

$Light - Space - Time\ Emergence_{Cells} =$

170

$$\begin{bmatrix} \begin{bmatrix} c_\infty : [Pr, Po, K, H] \\ \left(\downarrow R_{C_K} = f(R_{C_\infty}) \right) \\ c_K : [S_{Pr}, S_{Po}, S_K, S_H] \\ \left(\downarrow R_{C_N} = f(R_{C_K}) \right) \\ c_N : f(S_{Pr} \times S_{Po} \times S_K \times S_H) \\ \left(\downarrow R_{C_U} = f(R_{C_N}) \right) \\ c_U : [P, V, M, C] \\ \Uparrow \\ c_{0:[D,W,I,C]} \end{bmatrix}_{Light} \begin{bmatrix} M_3 \rightarrow System_X \\ (\uparrow F \rightarrow I) \\ M_2 \rightarrow S_{System_X} \\ (\uparrow Sig \rightarrow F) \\ M_1 \rightarrow Sig_x \\ (\uparrow > P_x) \\ U \rightarrow x_{Molecules} \end{bmatrix}_{Space} \\ \\ U \rightarrow \begin{bmatrix} M_3 : -\infty \le t \le \infty \\ \downarrow \\ M_2 : 0 \ge t > \infty \\ \downarrow \\ M_1 : 0 > t > \infty \\ \downarrow \\ t \le E_{Cell}; TC: M_3 \rightarrow U \\ t \sim E_{Human}; TC: U \rightarrow M_3 \end{bmatrix}_{Time} \quad TC \rightarrow x_{Cells} \quad \langle x_U | x_T \rangle \\ \Rightarrow \end{bmatrix}$$

$\langle Space - Time - Energy - Gravity \; Material - Fabric \; Pre$
$\qquad - Genetic \; Code \rangle +$

$\langle Electromagnetic \; Spectrum \;\; Material - Fabric \; Pre - Genetic \; Code \rangle$
$\qquad\qquad +$

$\langle Quantum - Particle \; Material - Fabric \; Pre - Genetic \; Code \rangle +$

$\langle Atoms \; Material - Fabric \; Pre - Genetic \; Code \rangle +$

$\langle Molecules \; Material - Fabric \; Pre - Genetic \; Code \rangle + LSTE \; \langle ... \rangle$
$\qquad\qquad + Cells \; Genetic \; Code$

Eq 4.2.4: Cells Genetic Code

The computation leading to the emergence of cells and associated code segments is discussed in greater detail in Chapter 7.2. Note that there are likely many iterations of this equation between molecules and cells and this in general is depicted by 'LSTE ⟨...⟩' in (4.2.4), where LSTE signifies iterations(s) of the Light-Space-Time Emergence equation. Also note that an assumption is made that all the previous material-fabric code-segments are intimately active at the level of cellular genetic code. This may be through the device of entanglement, or perhaps even by the material-fabric code segments being embedded in cellular genetic code.

Humans Genetic Code Iteration

In Equation 4.2.5, Humans Genetic Code, the starting point, x_U, is 'Cells', and the ending point, x_T, is 'Humans'. Clearly there will be a vast number of iterations before humans emerge which also covers vast tracts of the plant and animal kingdoms that will cause an appropriate divergence in genetic code. Further the development of microbiomes – that contain a multitude of bacteria and

other microbes – will have been developed and shared between a vast number of species (Young, 2016).

Further, somewhere between the cell and the human, the emergence of vision is said to have sparked an explosion of evolution (Parker, 2003). This is not surprising if in fact we exist due to a Cosmology of Light as is being suggested in this book. It should be natural that instrumentation to perceive light, such as vision, would then generate an explosion in evolution.

Since any genetic code will be the materialization of pre-material-fabric or more accurately four-base logic-encoding ecosystems (FBLEE) as a result of the Light-Matrix computation, there will also be a degree of entanglement to such ecosystems. Such entanglement will be different from ∞-entanglement, K-entanglement, and N-entanglement, and as suggested in the last chapter, will be referred to as FBLEE-entanglement. Such entanglement is due to four-base logic-encoding ecosystems existing at the quantum level. FBLEE-entanglement will allow code-segment developments to be shared or accessed on an inter-species level, and are depicted by 'FBLEEE $\langle ... \rangle$', as in Equation 4.2.5.

Hence:

$$Light - Space - Time\ Emergence_{Humans} =$$

$$\left[\left[\begin{array}{c} c_\infty: [Pr, Po, K, H] \\ \left(\downarrow R_{C_K} = f(R_{C_\infty})\right) \\ c_K: [S_{Pr}, S_{Po}, S_K, S_H] \\ \left(\downarrow R_{C_N} = f(R_{C_K})\right) \\ c_N: f(S_{Pr} \times S_{Po} \times S_K \times S_H) \\ \left(\downarrow R_{C_U} = f(R_{C_N})\right) \\ c_U: [P, V, M, C] \\ \Uparrow \\ c_{0:[D,W,I,C]} \end{array}\right]_{Light} \quad \left[\begin{array}{c} M_3 \to System_X \\ (\uparrow F \to I) \\ M_2 \to S_{System_X} \\ (\uparrow Sig \to F) \\ M_1 \to Sig_X \\ (\uparrow > P_X) \\ U \to x_{Cells} \end{array}\right]_{Space}\right.$$

$$\left.\left[U \to \begin{array}{c} M_3: -\infty \le t \le \infty \\ \downarrow \\ M_2: 0 \ge t > \infty \\ \downarrow \\ M_1: 0 > t > \infty \\ \downarrow \\ t \le E_{Cell}; TC: M_3 \to U \\ t \sim E_{Human}; TC: U \to M_3 \end{array}\right]_{Time} \quad TC \to x_{Humans} \right]_{\langle x_U | x_T \rangle}$$

$$\Rightarrow$$

$\langle Space - Time - Energy - Gravity\ Material - Fabric\ Pre$
$- Genetic\ Code \rangle +$

$\langle Electromagnetic\ Spectrum\ Material - Fabric\ Pre - Genetic\ Code \rangle +$

$\langle Quantum - Particle\ Material - Fabric\ Pre - Genetic\ Code \rangle +$

$\langle Atoms\ Pre - Genetic\ Code \rangle +$

$\langle Molecules\ Material - Fabric\ Pre - Genetic\ Code \rangle + LSTE\ \langle ... \rangle +$

$\langle Cells\ Genetic\ Code \rangle + LSTE\ \langle ... \rangle + FBLEEE\ \langle ... \rangle$
$+ Humans\ Genetic\ Code$

Eq 4.2.5: Humans Genetic Code

If we follow the lines of emergence as will be described in the following subsection then (4.2.5) will be the starting point for potentially different trajectories of development. In one trajectory 'Humans' will evolve 'Basic Capacities of Self' and then 'Truer Individuality'. In another line of development 'Humans' will be the starting point for 'Mega-Organization' culminating in 'Sustainable Global Civilization'. The gist of these can be represented by Equations 4.2.6 and 4.2.7 respectively. As in (4.2.5) FBLEEE-segments depict the sharing of code segments between these different lines of development through the process of four-base logic-encoding ecosystem entanglement at the quantum level.

Truer Individuality Genetic Code Iteration

Hence in Equation 4.2.6, Truer Individuality Genetic Code Emergence, the starting point, x_U, is 'Humans', and the ending point, x_T, is 'Truer Individuality'. Clearly there will be a vast number of iterations before 'Truer Individuality' emerges, that will contain 'Basic Capacities of Self' as a milestone along the way:

$$Light - Space - Time\ Emergence_{Truer\ Individuality} =$$

$$
\left| \left[\begin{array}{c}
\begin{array}{c}
c_\infty : [Pr, Po, K, H] \\
\left(\downarrow R_{C_K} = f(R_{C_\infty}) \right) \\
c_K : [S_{Pr}, S_{Po}, S_K, S_H] \\
\left(\downarrow R_{C_N} = f(R_{C_K}) \right) \\
c_N : f(S_{Pr} \times S_{Po} \times S_K \times S_H) \\
\left(\downarrow R_{C_U} = f(R_{C_N}) \right) \\
c_U : [P, V, M, C] \\
\Uparrow \\
c_{0:[D,W,I,C]}
\end{array}
\end{array} \right]_{Light}
\begin{bmatrix}
M_3 \rightarrow System_X \\
(\uparrow F \rightarrow I) \\
M_2 \rightarrow S_{System_X} \\
(\uparrow Sig \rightarrow F) \\
M_1 \rightarrow Sig_x \\
(\uparrow > P_x) \\
U \rightarrow x_{Humans}
\end{bmatrix}_{Space}
\right.
$$

$$
\left. \begin{array}{c}
\left[U \rightarrow \begin{array}{c}
M_3 : -\infty \le t \le \infty \\
\downarrow \\
M_2 : 0 \ge t > \infty \\
\downarrow \\
M_1 : 0 > t > \infty \\
\downarrow \\
t \le E_{Cell}; TC: M_3 \rightarrow U \\
t \sim E_{Human}; TC: U \rightarrow M_3
\end{array} \right]_{Time}
\end{array} \quad TC \rightarrow x_{Truer\ Individuality} \right| \langle x_U | x_T \rangle
$$

$\Rightarrow$

$\langle Space - Time - Energy - Gravity\ Material - Fabric\ Pre$
$\qquad - Genetic\ Code \rangle +$

$\langle Electromagnetic\ Spectrum\ Material - Fabric\ Pre - Genetic\ Code \rangle +$

$\langle Quantum - Particle\ Material - Fabric\ Pre - Genetic\ Code \rangle +$

$\langle Atoms\ Pre - Genetic\ Code \rangle +$

$\langle Molecules\ Material - Fabric\ Pre - Genetic\ Code \rangle + LSTE\ \langle ... \rangle +$

$\langle Cells\ Genetic\ Code \rangle + LSTE\ \langle ... \rangle + FBLEEE\ \langle ... \rangle$
$\qquad + Humans\ Genetic\ Code +$

$LSTE\ \langle ... \rangle + FBLEEE\ \langle ... \rangle + Truer\ Individuality\ Genetic\ Code$

Eq 4.2.6: Truer Individuality Genetic Code Iteration

Sustainable Global Civilization Genetic Code Iteration

In equation 4.2.7, Sustainable Global Civilization Genetic Code Iteration, the starting point, x_U, is 'Humans', and the ending point, x_T, is 'Sustainable Global Civilization'. Clearly there will be a vast number of iterations before 'Sustainable Global Civilization' emerges, that will contain 'Mega-Organization' as a milestone along the way:

176

*Light − Space − Time Emergence*_{Sustainable Global Civilization} =

$$\left| \begin{array}{l} \begin{bmatrix} \begin{array}{l} c_{\infty}: [Pr, Po, K, H] \\ \left(\downarrow R_{C_K} = f(R_{C_\infty}) \right) \\ c_K: [S_{Pr}, S_{Po}, S_K, S_H] \\ \left(\downarrow R_{C_N} = f(R_{C_K}) \right) \\ c_N: f(S_{Pr} \; x \; S_{Po} \; x \; S_K \; x \; S_H) \\ \left(\downarrow R_{C_U} = f(R_{C_N}) \right) \\ c_U: [P, V, M, C] \\ \Uparrow \\ c_{0:[D,W,I,C]} \end{array} \end{bmatrix}_{Light} \quad \begin{bmatrix} M_3 \; \to \; System_X \\ (\uparrow F \to I) \\ M_2 \; \to \; S_{System_X} \\ (\uparrow Sig \to F) \\ M_1 \; \to \; Sig_x \\ (\uparrow > P_x) \\ U \; \to \; x_{Humans} \end{bmatrix}_{Space} \\[2em] \begin{bmatrix} M_3 : -\infty \leq t \leq \infty \\ \downarrow \\ M_2 : 0 \geq t > \infty \\ \downarrow \\ M_1 : 0 > t > \infty \\ \downarrow \\ U \; \to \; \begin{array}{l} t \leq E_{Cell}; TC: M_3 \to U \\ t \sim E_{Human}; TC: U \to M_3 \end{array} \end{bmatrix}_{Time} \quad TC \to x_{Sust.Global\;Civilization} \end{array} \right| \langle x_U | x_T \rangle$$

$\Rightarrow$

$\langle Space - Time - Energy - Gravity \; Material - Fabric \; Pre - Genetic \; Code \rangle +$

$\langle Electromagnetic \; Spectrum \; Material - Fabric \; Pre - Genetic \; Code \rangle +$

$\langle Quantum - Particle \; Material - Fabric \; Pre - Genetic \; Code \rangle +$

$\langle Atoms \; Pre - Genetic \; Code \rangle + \langle Molecules \; Material - Fabric \; Pre - Genetic \; Code \rangle +$

$LSTE \langle \dots \rangle + \langle Cells \; Genetic \; Code \rangle + LSTE \langle \dots \rangle + FBLEEE \langle \dots \rangle +$

$Humans \; Genetic \; Code \; + LSTE \langle \dots \rangle + FBLEEE \langle \dots \rangle +$

$Sustainable \; Global \; Civilization \; Code$

Eq 4.2.7: *Sustainable Global Civilization Genetic Code*

178

The computation leading to the emergence of a sustainable global civilization is discussed in greater detail in Chapter 8.3.

Super-Matter Post-Genetic Code Iteration

In Equation 4.2.8, Super-Matter Post-Genetic Code, the starting point, x_U, is 'Molecules, and the ending point, x_T, is 'Super-Matter'. As per the mathematical model presented here all the preceding Light-Space-Time Emergence equation iterations change the reality of matter through subtle quantization that allows space, time, energy, and gravity to operate differently as will be further explored in subsequent section. The following equation is therefore one approximation in stating how super-matter may emerge:

$$Light - Space - Time\ Emergence_{Suoer-Matter} =$$

$$\left| \left| \left[\begin{array}{c} c_\infty : [Pr, Po, K, H] \\ \left(\downarrow R_{C_K} = f(R_{C_\infty}) \right) \\ c_K : [S_{Pr}, S_{Po}, S_K, S_H] \\ \left(\downarrow R_{C_N} = f(R_{C_K}) \right) \\ c_N : f(S_{Pr} \times S_{Po} \times S_K \times S_H) \\ \left(\downarrow R_{C_U} = f(R_{C_N}) \right) \\ c_U : [P, V, M, C] \\ \Uparrow \\ c_{0:[D,W,I,C]} \end{array} \right]_{Light} \left[\begin{array}{c} M_3 \to System_X \\ (\uparrow F \to I) \\ M_2 \to S_{System_X} \\ (\uparrow Sig \to F) \\ M_1 \to Sig_x \\ (\uparrow > P_x) \\ U \to x_{Molecules} \end{array} \right]_{Space} \right. \right.$$

$$\left. \left. \left[\begin{array}{c} M_3 : -\infty \le t \le \infty \\ \downarrow \\ M_2 : 0 \ge t > \infty \\ \downarrow \\ M_1 : 0 > t > \infty \\ \downarrow \\ U \to \begin{array}{l} t \le E_{Cell}; TC: M_3 \to U \\ t \sim E_{Human}; TC: U \to M_3 \end{array} \end{array} \right]_{Time} \quad TC \to x_{Super-Matter} \right| \langle x_U | x_T \rangle \right|$$

$\Rightarrow$

$$\langle Space - Time - Energy - Gravity \; Material - Fabric \; Pre \\ - Genetic \; Code \rangle \; +$$

$$\langle Electromagnetic \; Spectrum \;\; Material - Fabric \; Pre - Genetic \; Code \rangle \\ +$$

$$\langle Quantum - Particle \; Material - Fabric \; Pre - Genetic \; Code \rangle \; +$$

$$\langle Atoms \;\; Material - Fabric \; Pre - Genetic \;\; Code \rangle \; +$$

$$\langle Molecules \;\; Material - Fabric \; Pre - Genetic \; Code \rangle + LSTE \; \langle \dots \rangle \\ + \; FBLEEE \; \langle \dots \rangle \; +$$

$$LSTE \; \langle \dots \rangle \; + \; Super - Matter \; Post - Genetic \; Code$$

Eq 4.2.8: Super-Matter Post-Genetic Code

The emergence of super matter is also discussed in the book Super-Matter (Malik, 2018a), with (4.2.8) included here to suggest computational realities yet to emerge. The essential action of space-time-energy-gravity quantization can cause the materialization of potentially infinite four-base logic-encoding ecosystems. It is conceivable that such a variation of space-time-energy-gravity quantization coupled with the ability to easily move back and forth between the material and antecedent realms makes the need to house genetic information in a form such as DNA incomplete. It is conceivable that four-base logic-encoding ecosystems, or even the ∞ - entanglement, K-entanglement, and N-entanglement libraries may be able to more directly act at the material level in some composite subtle-material post-genetic form. This possibility will be referred to as Post-Genetic Code.

SECTION 5: GENERATING THE MATERIAL-FABRIC, PRE-GENETIC CODE FOR THE ELECTROMAGNETIC SPECTRUM

Section 5 will explore the computation involved in the creation of the pre-genetic, material-fabric electromagnetic spectrum code.

The electromagnetic spectrum is a technical way to refer to light, essentially because amongst its range of properties it displays a tangible and simultaneous electric and magnetic or "electromagnetic" reality as well. So as light becomes more concrete to us or as the possibilities within it begin to emerge, one of the first forms it takes is as the electromagnetic spectrum. It must be the case that the previously surfaced properties of light – Presence, Power, Knowledge, and Harmony – emerge so as to define the very architecture of the electromagnetic spectrum.

Chapter 5.1, Generation of Pre-Genetic, Material-Fabric Electromagnetic Spectrum Code, summarizes the emergence of the electromagnetic spectrum in terms of the underlying Light-Space-Time Emergence equation and the process of quantization that must occur to create the logic of the electromagnetic spectrum ecosystem that precipitates into the material-fabric. So we find that the four underlying properties of Light that we call Harmony, Knowledge, Power, and Presence are of the essence of the speed with which the electromagnetic spectrum moves, the wave-range within the electromagnetic spectrum, the energy-gradient within the electromagnetic spectrum, and the mass-possibilities due to the electromagnetic spectrum, respectively. Further the description – electromagnetic – seems to have captured the Power-Harmony aspects implicit in light. In reality the electro-magnetic spectrum can likely be more completely described as electro-magnetic-wavearchetype-masspotential spectrum.

Chapter 5.1: Generation of Material-Fabric, Pre-Genetic Electromagnetic Spectrum Code

As elaborated in Section 4, the Light-Space-Time Emergence equation (3.1.3) can be used to model emergence as it proceeds from simpler four-fold to more complex four-fold manifestation. Further, being iterative, it can be used to understand the generation of code-segments that will continue to add to pre-genetic or genetic code. (3.6.5), the Potential Effect of Levels of Light on Pre-Genetic of Genetic Information equation, basically sheds light on specific types of mutation that can occur and therefore on the nature of the code-segment that is being generated by (3.1.3). (3.6.5) shows a particular working of (3.1.3) but is wholly contained within it.

This chapter elaborates the action of (3.1.3) and (3.6.5) in the creation of the code-segments that define the electromagnetic spectrum ecosystem logic.

Application of Light-Space-Time Emergence Equation

As suggested by (3.1.3) the architecture and details of the electromagnetic spectrum, and the resulting material-fabric, pre-genetic code can be seen to be the result of the application of the Light, Space, and Time matrices. This is illustrated in Equation 5.1.1, Electromagnetic Spectrum Material-Fabric Pre-Genetic Code:

$$Light - Space - Time\ Emergence_{EM\ Spectrum} =$$

183

$$\Rightarrow \begin{vmatrix} \begin{bmatrix} c_\infty : [Pr, Po, K, H] \\ \left(\downarrow R_{C_K} = f(R_{C_\infty}) \right) \\ c_K : [S_{Pr}, S_{Po}, S_K, S_H] \\ \left(\downarrow R_{C_N} = f(R_{C_K}) \right) \\ c_N : f(S_{Pr} \times S_{Po} \times S_K \times S_H) \\ \left(\downarrow R_{C_U} = f(R_{C_N}) \right) \\ c_U : [P, V, M, C] \\ \Uparrow \\ c_{0:[D,W,I,C]} \end{bmatrix}_{Light} \begin{bmatrix} M_3 \to System_X \\ (\uparrow F \to I) \\ M_2 \to S_{System_X} \\ (\uparrow Sig \to F) \\ M_1 \to Sig_X \\ (\uparrow > P_x) \\ U \to x_{Space-Time-Energy-Gravity} \end{bmatrix}_{Space} \\ \begin{bmatrix} M_3 : -\infty \le t \le \infty \\ \downarrow \\ M_2 : 0 \ge t > \infty \\ \downarrow \\ M_1 : 0 > t > \infty \\ \downarrow \\ U \to \begin{array}{l} t \le E_{Cell}; TC: M_3 \to U \\ t \sim E_{Human}; TC: U \to M_3 \end{array} \end{bmatrix}_{Time} \qquad TC \to x_{EM\ Spectrum} \end{vmatrix} \langle x_U | x_T \rangle$$

$$\langle Space - Time - Energy - Gravity\ Material - Fabric\ Pre - Genetic\ Code \rangle +$$

$$Electromagnetic\ Spectrum\ Material - Fabric\ Pre - Genetic\ Code$$

Eq 5.1.1: *Electromagnetic Spectrum Material-Fabric Pre-Genetic Code*

Starting with the Light-Matrix, the top left-hand matrix in (5.1.5), the first line from the top, $C_\infty : [Pr, Po, K, H]$, specifies the fundamental architecture of the electromagnetic spectrum. As will be elaborated in the subsections on Harmony, Knowledge, Power, and Presence, each of these aspects are an emergence of the fundamental properties of Light at ∞.

Line 3 in the Light-Matrix, $C_K : [S_{Pr}, S_{Po}, S_K, S_H]$, elaborates the sets for Presence, Power, Knowledge, and Harmony, each containing multiple elements. For example, as will be explored in the subsection on Harmony, the following elements: 'connection', 'growing into one's own', 'form bonds', are suggested to exist in the Set of Harmony or Nurturing, and are accessed to contribute to the operative reality so set up by virtue of light traveling at c.

Specifically, Line 5, $C_{N:} f(S_{Pr} \times S_{Po} \times S_K \times S_H)$, suggests that unique seeds are created from a combination of the elements from all four sets, with a particular element leading, that in effect creates the distinctness of the vast variety of waves possible in the electromagnetic spectrum.

Line 6, $(\downarrow R_{C_U} = f(R_{C_N}))$, specifies quantization between the layer where the seeds are formed, and the physical layer we are familiar with, and as explored in Chapter 3.5 and 3.6, will result in Line 7, C_U: $[P, V, M, C]$, hence changing the material-fabric of existence. Note that as in the process describing the generation of the code-segments for space-time-energy-gravity quantization at the time of the Big Bang, the generation of the FBLEE (four-base logic-encoding ecosystem) code-segment as specified by (4.1.5) and the MF (material-fabric) code segment as specified by (4.1.6) are combined together into one equation so that the FBLEE process is apparently transparent.

The possibilities represented by Lines 1 through 5 hence concretize through the quantization represented by Line 6 to become the electromagnetic spectrum with its physical (related to Presence), vital (related to Power), mental (related to Knowledge), and connection (related to Harmony) aspects now existing in material reality typified by Light moving at c. Note that just as Line 6 represents a process of quantization relating the layer of reality created by Light traveling at c with the antecedent layers, so too Lines 2 and 4 as previously discussed, also represent quantization of a more subtle kind that ultimately plays a critical part in allowing the material-fabric to express infinite diversity.

Typically it is the process as captured by the Space-Matrix that will determine if Line 6 is activated. Specifically patterns at the untransformed layer, U, will need to be overcome, as specified by the second-line from the bottom of the Space-Matrix: $(\uparrow > P_x)$. But as specified by the bottom-line of the Time-Matrix, reproduced below, it is assumed that only with the advent of the human-system that the automaticity of the action of meta-levels is reversed:

$$U \to \begin{array}{l} \text{t} \le E_{Cell}; \text{TC: } M_3 \to \text{U} \\ \text{t} \sim E_{Human}; \text{TC: } U \to M_3 \end{array}$$

Hence in the case of the electromagnetic spectrum system, which in this emergence is a pre-human system, the fact that patterns do not need to be overcome means that quantization happens automatically.

From the point of view of pre-genetic information this means that constructive mutation has occurred and that the available code-segments that impact the material operation of the universe has been changed. But 'constructive' implies that the operation, materially, allows the manifested universe to 'materially' come closer to the integrated nature of Light. Of course the emergence of the electromagnetic spectrum in this rendering occurs after the emergence of space-time-energy-gravity, and marks the beginning stages, relatively, of the long curve that evolution will go through as more and more possibility emerges from Light.

Application of Potential Effect of Levels of Light on Pre-Genetic or Genetic Information Equation

Even though the electromagnetic spectrum is a pre-human system and therefore the action of meta-levels are modeled as being automatic, it is nonetheless useful to review (3.6.5), Potential Effect of Levels of Light on Pre-Genetic or Genetic Information, to understand how the relationship with different forms of mutation has been specified:

Potential Effect of Levels of Light on Pre Genetic or Genetic Information =

$$
\begin{bmatrix}
STATIC \ \langle |[L][S][T]TC \ \to \ x_T|_{\langle x_U|x_T\rangle}\rangle \\
\times \\
\left((Y > U: Z_Q) \lor (Y \le U: Z_F) \lor (Y = U: Z_R) \right) \\
\ni \\
\left(\begin{array}{c} Z \in \mathbb{U} \ (Space, Time, Energy, Gravity) \\ Q: Quantization; F: Fragmentation; R: Random \end{array} \right)
\end{bmatrix} \to h \ni
$$

$$
h \in \left(\begin{array}{c} [Q]: Constructive\ zone, \\ [Q]: Constructive\ zone \land Constructive\ mutation, \\ [F]: Destructive\ mutation, \\ [R]: Random\ mutation \end{array} \right)
$$

Line 1 from the top in the matrix is simply a static form of (3.6.1) the Simplified Light-Space-Time Emergence equation. The static form is designated by 'STATIC', and implies that fundamental operations true of (3.6.1) are being highlighted in (3.6.5). In other word (3.6.1) already has all the operations highlighted in (3.6.5) in it, but by 'freezing' it by making it static, the essential dynamics leading to possible mutations at the genetic level can more clearly be highlighted.

Line 1 is then subjected ($\times$) to a determination of the dominant levels of light that may be active, designated by Line 2, ' $\left((Y > U: Z_Q) \lor (Y \le U: Z_F) \lor (Y = U: Z_R) \right)$ '. Unpacking this, 'Y > U' implies meta-levels are active and as a result it is possible that Z_Q is going to take place. This also implies activation and potential change of FBLEE. The call from below, as it were, in this case the rendering of the space-time-energy-gravity quadrumvirate that gives rise to the material-fabric and the potential seed-activity that determines the cosmic quadrumvirate, may invoke some function that already exists in the subtle-libraries 'above', so that some already existing function may influence FBLEE through ∞ -entanglement, K-entanglement, or N-entanglement. This may be thought of as a key-and-lock mechanism, where a fundamental enough need from below acting as the key, opens an entangled lock to alter FBLEE as per the need. In this case the need may be for something like the electromagnetic spectrum to stimulate activity of the seeds embedded in space.

'$Y \leq U$' implies that only the untransformed levels are active, and therefore also the sub-level where the speed of light is 0 is active and as a result Z_F is going to take place. '$Y = U$' implies that all levels are active and as a result Z_R is going to take place.

Line 3 elaborates the significance of Z_Q, Z_F, and Z_R. Hence Z is the union of potential quantum-operations of space, time, energy, and gravity, designated by '$Z \in \mathbb{U}\,(Space, Time, Energy, Gravity)$'. But the nature of the operations is designated by '$Q:\ Quantization;\ F:\ Fragmentation;\ R:\ Random$'.

Z_Q, then, implies that the full quantization originating from updated four-base logic-encoding ecosystems (FBLEE) can take place, and will result in lasting material change at the pre-genetic. Z_F implies that the essential set will be fragmented and that only local libraries a the level of local material-fabric can potentially be altered. Z_R also precludes full quantization, and that some partial local-library constructive or destructive mutation may take place.

The '$\rightarrow h \ni h \in$' segment resolves the outcome of the operations implied by Lines 1 – 3, suggesting that the outcome will be 'h' such that ($\ni$) 'h' is an element ($\in$) of the set specified by the members '[Q]: *Constructive zone*', '[Q]: *Constructive zone AND Constructive mutation*', '[F]; *Destructive mutation*', and '{R}: *Random mutation*'. '[Q]: *Constructive zone*' implies that the in-built buffer has been crossed and that access to the deeper four-base logic-encoding ecosystem (FBLEE) has been granted. '[Q}' in this segment implies that there is the possibility that full-quantization as specified by Line 3 of the previous matrix can take place. Access to this zone is a prerequisite for constructive mutation to occur, as designated by the element '   [Q}: *Constructive zone* ∩ *Constructive mutation* ', which implies that full-quantization is going to take place and will result in material change. The '[F]' specifies the relationship between 'Fragmentation' in Line 3 of the previous matrix and destructive mutation. The '[R]' specifies the relationship between 'Random' in Line 3 of the previous matrix and random mutation.

But as just summarized in the Time-Matrix in (5.1.1) Y is by definition greater than U and hence quantization is automatic. In terms of the electromagnetic spectrum such quantization implies that wholeness becomes fully active through specific space, time, energy, and gravity quantization to create an holistic "ecosystem" with its own "electromagnetic spectrum logic" as it were. The wholeness has now precipitated into the material-fabric and is available to be consciously and unconsciously tapped into. This "logic" or more specifically pre-genetic code is elaborated in the following sub-sections.

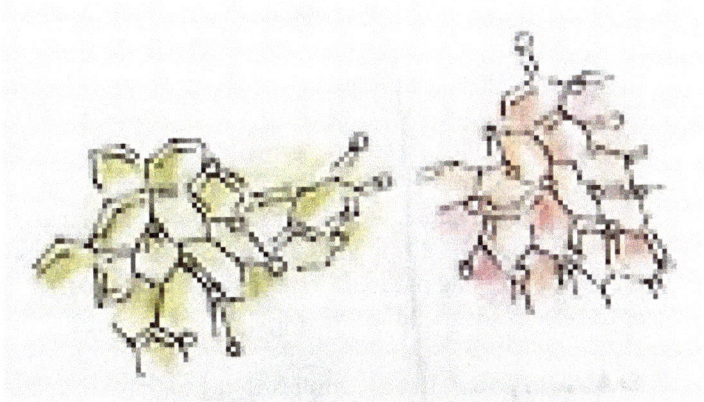

Generation of 'Magnetic' Material-Fabric Pre-Genetic Code of Electromagnetic Spectrum

To begin with, we had already looked at how the speed of light sets up the nature of reality because of its speed. So for light traveling at c, past, present, future, and the notion of separation due to creation of islands of matter is the reality. It seems then that c architects the possibility of interaction in our system and can therefore be thought of as a projection from the property of Harmony to create the basis for a matter-based harmony. So being, it can be suggested that the nature of the resultant interactions allows matter-based organizations, regardless of scale, to come into their own, to grow into their boundaries, and to form bonds based on the sense of being separated from other perceived organizations. This notion of forming bonds seems also to be related to the "magnetic" in electromagnetic.

In equation form, as in Equation 5.1.2 it would therefore be possible to specify the nature of reality so set up by the electromagnetic spectrum (EM_Spectrum) moving at c:

$$EM\_Spectrum_{Speed} = Xa +$$
$$\overline{Yb_{0-n}} \quad where \left[\begin{array}{c} X \in [S_{System_N}] \\ Y \in [S_{System_{Pr}}, S_{System_P}, S_{System_K}, S_{System_N}] \\ a, b \ are \ integers; a > b \end{array} \right]$$

Eq 5.1.2: Speed of EM Spectrum

The notion of 'connection', 'growing into one's own', 'form bonds', amongst other attributes of such a reality can be seen as elements of the four sets of architectural forces. Hence (5.1.2) suggests the mathematical equation that so defines the nature of reality due to the speed of c of the electromagnetic spectrum.

Note that (5.1.2) already implies that Lines 1 − 5 in the Light Matrix of (5.1.1) have been activated, and that the

190

logic of the "magnetic" in the electro-magnetic-ecosystem will automatically precipitate into the material-fabric through the action of Line 6-7 of (5.1.1). This logic is none other than the pre-genetic code as specified by (5.1.2).

This also implies that as the speed of light changes, and as discussed in Chapter 2.1, the nature of relationship in such a reality will also change.

Generation of 'Wave Archetype' Material-Fabric Pre-Genetic Code of Electromagnetic Spectrum

The electromagnetic spectrum contains a range of waves embedded in it. These waves enable many different applications with practical utility apparent in every day life. So for example there is a region in the electromagnetic spectrum that we call radio waves, and others that we know as microwave, infrared visible light, ultraviolet, x-rays and gamma rays that each make possible many technologies that we use every single day. These ranges essentially code a range of technological-possibility, and therefore the wave-range within the electromagnetic spectrum can be thought of as encoding or of expressing some kind of precipitation or projection or emergence of the property of Knowledge. Or put another way, the property of Knowledge that is found in light, emerges as the wave-range implicit in the electromagnetic spectrum.

The EM spectrum itself, with its vast range of natures from gamma rays through visible light through radio waves, with its implicitness of time-space possibility as suggested by frequency (v) and wavelength (λ), may be thought of as an arrangement of archetypes of what is possible in systems, and therefore is perhaps a precipitation of system-knowledge, as in Equation 5.1.3:

$$EM_{Spectrum_{Structure}} = Xa + \overline{Yb_{0-n}}$$

$$where \begin{bmatrix} X \in [S_{System_K}] \\ Y \in [S_{System_{Pr}}, S_{System_P}, S_{System_K}, S_{System_N}] \\ a, b \ are \ integers; a > b \end{bmatrix}$$

Eq 5.1.3: Structure of EM Spectrum

Note that (5.1.3) already implies that Lines 1 – 5 in the Light Matrix of (5.1.1) have been activated, and that the logic of the "wave-archetype" in the electro-magnetic-wavearchetype-masspotential ecosystem will automatically precipitate into the material-fabric through the action of Line 6-7 of (5.1.1). This logic is none other than the pre-genetic code as specified by (5.1.3).

What this also implies is that the significance or intent of the different types of waves that exist can also be expressed by this general equation where the X and Y elements will vary. What precisely these elements are will need to be worked out. A few representative equations, Equations 5.1.4 through 5.1.7 follow:

$$Gamma \ Rays \ _{Intent} = Xa +$$
$$\overline{Yb_{0-n}} \ \ where \begin{bmatrix} X \in [S_{System_K}] \\ Y \in [S_{System_{Pr}}, S_{System_P}, S_{System_K}, S_{System_N}] \\ a, b \ are \ integers; a > b \end{bmatrix}$$

Eq 5.1.4: Gamma Rays Intent

Knowing that some of the applications of Gamma Rays are in sterilizing and radiotherapy, these can be attributed as elements to the architectural sets. The totality of the use of can be thought of as the 'intent' of Gamma Rays. An understanding of the uses will allow a full set of the secondary elements to be mapped thus allowing (5.1.4) to be reverse-engineered.

Similarly each of the archetypes present in the electromagnetic spectrum can be mapped out. Sample mapping of intent include x-rays, infrared, and microwaves:

$$X - Rays_{Intent} = Xa +$$

$$\overline{Yb_{0-n}} \quad where \quad \left[\begin{array}{c} X \in [S_{System_K}] \\ Y \in [S_{System_{Pr}}, S_{System_P}, S_{System_K}, S_{System_N}] \\ a, b \text{ are integers}; a > b \end{array} \right]$$

Eq 5.1.5: X-Rays Intent

$$Infrared_{Intent} = Xa +$$

$$\overline{Yb_{0-n}} \quad where \quad \left[\begin{array}{c} X \in [S_{System_K}] \\ Y \in [S_{System_{Pr}}, S_{System_P}, S_{System_K}, S_{System_N}] \\ a, b \text{ are integers}; a > b \end{array} \right]$$

Eq 5.1.6: Infrared Intent

$$Microwaves_{Intent} = Xa +$$

$$\overline{Yb_{0-n}} \quad where \quad \left[\begin{array}{c} X \in [S_{System_K}] \\ Y \in [S_{System_{Pr}}, S_{System_P}, S_{System_K}, S_{System_N}] \\ a, b \text{ are integers}; a > b \end{array} \right]$$

Eq 5.1.7: Microwaves Intent

Generation of 'Electro' Material-Fabric Pre-Genetic Code of Electromagnetic Spectrum

The range of different wave-types or wavelengths implicit in the electromagnetic spectrum also moves with different frequencies. Since the speed of light is a constant, and it is known that speed is the product of frequency and wavelength, the greater the wavelength the lower will be the frequency of the wave-type. Conversely the less the wavelength the higher will be the frequency of the wave-type. And energy or power of a wave-type will depend on its frequency.

Hence, gamma rays that have a lower wavelength will have a higher frequency, and higher energy associated with it. Radio waves on the other hand that have a higher wavelength will have a lower frequency and therefore lower energy associated with it. So implicit in the electromagnetic spectrum is a gradient of energy. But it

193

is also known that the penetration power is dependent on frequency. Therefore, the higher the frequency or energy, the higher will be the power of the wave-type. Another way to say this is that the property of Power in Light emerges as the energy-gradient in the electromagnetic spectrum. This Power aspect seems to have been captured by the "electro" in electromagnetic.

The energy-gradient implicit in the EM spectrum suggests the power and energy with which knowledge moves and is perhaps a precipitation of system-power, as in Equation 5.1.8:

$$EM\_Spectrum_{Energy} = Xa +$$

$$\overline{Yb_{0-n}} \quad where \begin{bmatrix} X \in [S_{System_P}] \\ Y \in [S_{System_{Pr}}, S_{System_P}, S_{System_K}, S_{System_N}] \\ a, b \ are \ integers; a > b \end{bmatrix}$$

Eq 5.1.8: EM Spectrum Energy

Note that (5.1.8) already implies that Lines 1 – 5 in the Light Matrix of (5.1.1) have been activated, and that the logic of the "electro" in the electro-magnetic-wavearchetype-masspotential ecosystem will automatically precipitate into the material-fabric through the action of Line 6-7 of (5.1.1). This logic is none other than the pre-genetic code as specified by (5.1.8).

This electromagnetic energy, as suggested, is directly proportional to the frequency, of which there is an infinite range as predicted by Maxwell's equations. Different frequencies have different penetration profiles (HyperPhysics, 2016) and it may be suggested that the nature of the energy is also a precipitation of a range of to be determined meta-functions as suggested in some sample Equations 5.1.9 through 5.1.13. Hence:

Gamma Rays $_{Nature\ of\ Energy}$ =

$$Xa +$$

$$\overline{Yb_{0-n}} \quad where \quad \left[\begin{array}{c} X \in [S_{System_P}] \\ Y \in [S_{System_{Pr}}, S_{System_P}, S_{System_K}, S_{System_N}] \\ a, b \ are\ integers; a > b \end{array} \right]$$

Eq 5.1.9: Gamma Rays Nature of Energy

$X - Rays_{Nature\ of\ Energy} \ =$

$$Xa +$$

$$\overline{Yb_{0-n}} \quad where \quad \left[\begin{array}{c} X \in [S_{System_P}] \\ Y \in [S_{System_{Pr}}, S_{System_P}, S_{System_K}, S_{System_N}] \\ a, b \ are\ integers; a > b \end{array} \right]$$

Eq 5.1.10: X-Rays Nature of Energy

$Ultraviolet\ _{Nature\ of\ Energy} \ =$

$$Xa +$$

$$\overline{Yb_{0-n}} \quad where \quad \left[\begin{array}{c} X \in [S_{System_P}] \\ Y \in [S_{System_{Pr}}, S_{System_P}, S_{System_K}, S_{System_N}] \\ a, b \ are\ integers; a > b \end{array} \right]$$

Eq 5.1.11: Ultraviolet Nature of Energy

$White\ Light_{Nature\ of\ Energy} \ =$

$$Xa +$$

$$\overline{Yb_{0-n}} \quad where \quad \left[\begin{array}{c} X \in [S_{System_P}] \\ Y \in [S_{System_{Pr}}, S_{System_P}, S_{System_K}, S_{System_N}] \\ a, b \ are\ integers; a > b \end{array} \right]$$

Eq 5.1.12: White Light Nature of Energy

$AM\ Radio\ Waves_{Nature\ of\ Energy} \ =$

$$Xa +$$

$$\overline{Yb_{0-n}} \quad where \quad \left[\begin{array}{c} X \in [S_{System_P}] \\ Y \in [S_{System_{Pr}}, S_{System_P}, S_{System_K}, S_{System_N}] \\ a, b \ are \ integers; a > b \end{array} \right]$$

Eq 5.1.13: AM Radio Waves Nature of Energy

Generation of 'Mass Potential' Material-Fabric Pre-Genetic Code of Electromagnetic Spectrum

If there is a large range of frequencies implicit in the electromagnetic spectrum then there is also the possibility of different types of masses implicit in the electromagnetic spectrum. Frequency determines energy, and mass and energy are related through Einstein's famous MC-squared equation. So pushing a little further it is not just that mass and energy are related, but a different kind of frequency or wave-type potentially allows a different type of mass to emerge. So the possibility of different types of mass seems to be related to the property of Presence in Light. In other words the property of Presence emerges as the possibility of different types of masses as suggested by the range of mass-possibilities that can emerge from the electromagnetic spectrum.

Mass can be thought of as a container at U within which all possibility happens. In other words it can be thought of as a precipitation or emergence of system-presence as depicted in Equation 5.1.14:

$$EM_{Spectrum_{Mass_{Possibility}}} = Xa + \overline{Yb_{0-n}}$$

$$where \quad \left[\begin{array}{c} X \in [S_{System_{Pr}}] \\ Y \in [S_{System_{Pr}}, S_{System_P}, S_{System_K}, S_{System_N}] \\ a, b \ are \ integers; a > b \end{array} \right]$$

Eq 5.1.14: EM Spectrum Mass Possibility

Note that (5.1.14) already implies that Lines 1 – 5 in the Light Matrix of (5.1.1) have been activated, and that the logic of the "masspotential" in the electro-magnetic-wavearchetype-masspotential ecosystem will automatically precipitate into the material-fabric through the action of Line 6-7 of (5.1.1). This logic is none other than the pre-genetic code as specified by (5.1.14).

Further, if the frequencies are infinite, then the possibility of the 'types' of masses or matter is also infinite. The mystery of 'Dark Matter' suggested by scientists to be 27% of our universe, as opposed to 5% of visible matter (NASA-darkmatter, 2016) may have some relation to this. This aspect is also explored in the book Cosmology of Light (Malik, 2018b).

Hypothetically the Equations 5.1.15 through 5.1.19 depict a range of mass possibilities:

$Gamma\ Rays\ _{Mass_{Possibility}} =$

$Xa +$

$\overline{Yb_{0-n}}$ $where$ $\begin{bmatrix} X \in [S_{System_{Pr}}] \\ Y \in [S_{System_{Pr}}, S_{System_P}, S_{System_K}, S_{System_N}] \\ a, b\ are\ integers; a > b \end{bmatrix}$

Eq 5.1.15: Gamma Rays Mass Possibility

$Ultraviolet\ _{Mass_{Possibility}} =$

$Xa +$

$\overline{Yb_{0-n}}$ $where$ $\begin{bmatrix} X \in [S_{System_{Pr}}] \\ Y \in [S_{System_{Pr}}, S_{System_P}, S_{System_K}, S_{System_N}] \\ a, b\ are\ integers; a > b \end{bmatrix}$

Eq 5.1.16: Ultraviolet Mass Possibility

$Blue\ Light\ _{Mass_{Possibility}} =$

$$Xa +$$

$$\overline{Yb_{0-n}} \ \ where \begin{bmatrix} X \in [S_{System_{Pr}}] \\ Y \in [S_{System_{Pr}}, S_{System_P}, S_{System_K}, S_{System_N}] \\ a, b \ are \ integers; a > b \end{bmatrix}$$

Eq 5.1.17: Blue Light Mass Possibility

$$Microwaves_{Mass_{Possibility}} =$$

$$Xa +$$

$$\overline{Yb_{0-n}} \ \ where \begin{bmatrix} X \in [S_{System_{Pr}}] \\ Y \in [S_{System_{Pr}}, S_{System_P}, S_{System_K}, S_{System_N}] \\ a, b \ are \ integers; a > b \end{bmatrix}$$

Eq 5.1.18: Microwaves Mass Possibility

$$FM \ Radio \ Waves_{Mass_{Possibility}} =$$

$$Xa +$$

$$\overline{Yb_{0-n}} \ \ where \begin{bmatrix} X \in [S_{System_{Pr}}] \\ Y \in [S_{System_{Pr}}, S_{System_P}, S_{System_K}, S_{System_N}] \\ a, b \ are \ integers; a > b \end{bmatrix}$$

Eq 5.1.19: FM Radio Waves Mass Possibility

Summary of Material-Fabric Pre-Genetic Code at Electromagnetic Spectrum Stage

Summarizing, after the electromagnetic spectrum stage iterations of the Light-Space-Time Emergence equation are complete, the following code-segments will have been generated as specified by Equation 5.1.20, Active Material-Fabric Pre-Genetic Code Segments at Electromagnetic Spectrum Stage:

$$Light - Space - Time \ Emergence_{EM \ Spectrum} =$$

198

$$\left\lVert \begin{bmatrix} c_\infty: [Pr, Po, K, H] \\ \left(\downarrow R_{C_K} = f\left(R_{C_\infty}\right)\right) \\ c_K: [S_{Pr}, S_{Po}, S_K, S_H] \\ \left(\downarrow R_{C_N} = f\left(R_{C_K}\right)\right) \\ c_N: f(S_{Pr} \times S_{Po} \times S_K \times S_H) \\ \left(\downarrow R_{C_U} = f\left(R_{C_N}\right)\right) \\ c_U: [P, V, M, C] \\ \Uparrow \\ c_{0:[D,W,I,C]} \end{bmatrix}_{Light} \begin{bmatrix} M_3 \;\rightarrow\; System_X \\ (\uparrow F \rightarrow I) \\ M_2 \;\rightarrow\; S_{System_X} \\ (\uparrow Sig \rightarrow F) \\ M_1 \;\rightarrow\; Sig_x \\ (\uparrow > P_x) \\ U \;\rightarrow\; x_{Space-Time-Energy-Gravity} \end{bmatrix}_{Space} \right.$$

$$\left. \begin{bmatrix} M_3 : -\infty \le t \le \infty \\ \downarrow \\ M_2 : 0 \ge t > \infty \\ \downarrow \\ M_1 : 0 > t > \infty \\ \downarrow \\ U \rightarrow \begin{matrix} t \le E_{Cell}; TC: M_3 \rightarrow U \\ t \sim E_{Human}; TC: U \rightarrow M_3 \end{matrix} \end{bmatrix}_{Time} \quad TC \;\rightarrow\; x_{EM\ Spectrum} \right\rVert \langle x_U | x_T \rangle$$

$$\Rightarrow$$

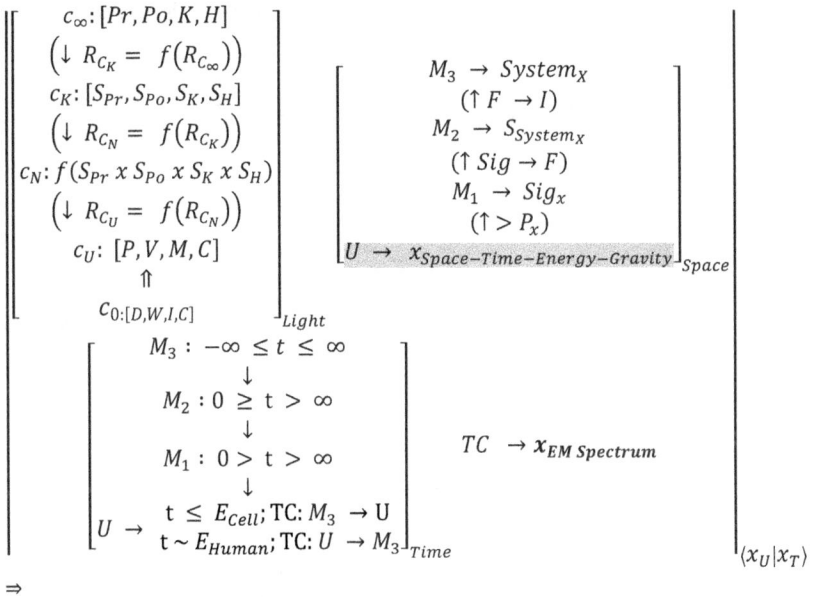

$$\langle Space - Time - Energy - Gravity\ Material - Fabric\ Pre - Genetic\ Code \rangle +$$

$$\left(\begin{matrix} \sum EM\_Spectrum_{Speed} = Xa + \overline{Yb_{0-n}} \\ where \begin{bmatrix} X \in [S_{System_N}] \\ Y \in [S_{System_{Pr}}, S_{System_P}, S_{System_K}, S_{System_N}] \\ a, b\ are\ integers; a > b \end{bmatrix} \\ \sum EM\_Spectrum_{Structure} = Xa + \overline{Yb_{0-n}} \\ where \begin{bmatrix} X \in [S_{System_K}] \\ Y \in [S_{System_{Pr}}, S_{System_P}, S_{System_K}, S_{System_N}] \\ a, b\ are\ integers; a > b \end{bmatrix} \\ \sum EM\_Spectrum_{Energy} = Xa + \overline{Yb_{0-n}} \\ where \begin{bmatrix} X \in [S_{System_P}] \\ Y \in [S_{System_{Pr}}, S_{System_P}, S_{System_K}, S_{System_N}] \\ a, b\ are\ integers; a > b \end{bmatrix} \\ \sum EM_{Spectrum_{Mass_{Possibility}}} = Xa + \overline{Yb_{0-n}} \\ where \begin{bmatrix} X \in [S_{System_{Pr}}] \\ Y \in [S_{System_{Pr}}, S_{System_P}, S_{System_K}, S_{System_N}] \\ a, b\ are\ integers; a > b \end{bmatrix} \end{matrix} \right)$$

Eq. 5.1.20, Active Material-Fabric Pre-Genetic Code Segments at Electromagnetic Spectrum Stage

The '$\sum x$' signifies all the possibilities for the 'x' equation-segment.

SECTION 6:   GENERATING THE PRE-GENETIC CODE FOR MATTER

So we find that the four underlying properties of Light that we call Harmony, Knowledge, Power, and Presence are of the essence of the speed with which the electromagnetic spectrum moves, the wave-range within the electromagnetic spectrum, the energy-gradient within

the electromagnetic spectrum, and the mass-possibilities due to the electromagnetic spectrum, respectively. Further the description – electromagnetic – seems to have captured the Power-Harmony aspects implicit in light. In reality the electro-magnetic spectrum can likely be more completely described as electro-magnetic-wavearchetype-masspotential spectrum.

But further, we will find that layers of matter – quantum

particles, which include bosons, and atoms – are also

201

structured or emerge along the same property-lines or property-families of Light.

There is similarly a continuous process of computation that involves the quantum-realms and quantization to create the realities of quantum particles, including bosons, and atoms, to generate the material-fabric, pre-genetic code for matter.

However, there is an apparent difference in the way matter materializes, in contrast to the electromagnetic spectrum. Chapter 6.1 explores this in greater detail.

Chapters, 6.2, 6.3, and 6.4, then, describe a process of computation by which the material-fabric, pre-genetic code for quantum particles, bosons, and atoms is generated respectively.

Chapter 6.1: Containerization of Matter

So far the action of the Light-Space-Time Emergence has generated pre-genetic code. After the Big Bang this pre-genetic code is housed in the material-fabric, which has a universal action on all constructs arising in the universe. Previous chapters explored such actions at the macro space-time-energy-gravity level, and the electromagnetic spectrum level. The emergence of matter however precipitates a phenomenon of distinct material containerization. It is interesting to speculate as to why this may be, especially in reference to the math already developed previously. What is it about the pre-genetic code that in the case of the electromagnetic spectrum maintains its action more as a wave or a field, and what

is it about matter, to be discussed in more detail in this section, that causes it to containerize? In this chapter it will be proposed that it is an essential action of the space-time-energy-gravity quadrumvirate that curves on itself so as to speak, to thereby allow such individual containers to be generated. Further, this containerizing action can occur at multiple scale as will also be explored in this chapter.

Universality Versus Localization of Space, Time, Energy, and Gravity

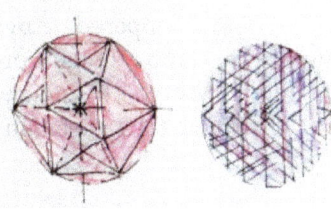

In Chapter 3.5, A Deeper Look at Quantization of Space, Time, Energy, and Gravity, (3.5.3) proposed a model for the structure of space. This is reproduced here for convenience:

$$Space_{Structure} = \sum_{i,j=1}^{\to\infty} h_{UK}\left(X_i a + \overline{Y_j b_{0-n}}\right)$$

$$where: \begin{bmatrix} X_i \in [S_{System_K}] \\ Y_j \in [S_{System_{Pr}}, S_{System_P}, S_{System_K}, S_{System_N}] \\ a, b, i, j \ are \ integers; a > b \end{bmatrix}$$

Summarizing, space, consisting of a vast array of seeds derived from the properties of Light, is itself an expression of Light's property of Knowledge. It is this essential action of a vast number of seeds that gives space its apparent never-ending-ness. In fact as previously discussed, time, energy and gravity can also be interpreted in terms of these 'seeds'. Equations 6.1.1-4

205

restate Space, Time, Energy, and Gravity, in terms of this relationship to seeds. Hence, Equation 6.1.1, Relation of Seeds to Space:

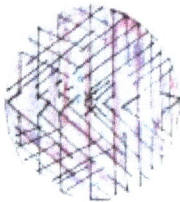

$$Space = f(\#seeds)$$
$$= f(Light's property of Knowledge)$$

Eq 6.1.1: Relation of Seeds to Space

Time, bringing forth the meaning contained in the seeds, regardless of circumstance, and even being opposed by circumstance, can be thought of as Light's property of Power. Hence, Equation 6.1.2, Relation of Seeds to Time:

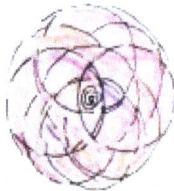

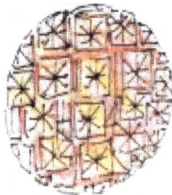

$$Time = f(maturity\ of\ seeds) = f(Light's\ property\ of\ Power)$$

Eq 6.1.2: Relation of Seeds to Time

Matter or Energy itself, being a container in which space and time can allow deeper properties of Light to become materially tangible, must be an expression of Light's property of Presence. Hence, Equation 6.1.3, Relation of Seeds to Energy/Matter:

$$Matter = f(materialization\ of\ seeds)$$
$$= f(Light's\ property\ of\ Presence)$$

Eq 6.1.3: Relation of Seeds to Energy/Matter

But it is also known from Einstein's General Theory of Relativity that gravity is associated with mass and space, in that it is none other than a mass's instruction telling space how to curve, and again is nothing else that space's instruction telling mass how to move through it (Wheeler, 2000). As such, where mass and space exist, there gravity has to exist as well. Hence it may be inferred that gravity is none other than an expression of Light's property of Harmony, which fixes the collective relationship between object and object. Hence, Equation 6.1.4, Relation of Seeds to Gravity:

$$Gravity = f(cohesion\ of\ seeds)$$
$$= f(Light's\ property\ of\ Harmony)$$

Eq 6.1.4: Relation of Seeds to Gravity

If the number of seeds is large, then the collective action of space, time, energy, and gravity is going to create an apparent universality and the reality of 'never-ending-ness'. By contrast if the number of seeds is smaller, then the action of space, time, energy, and gravity will be relatively localized, and create a more apparent reality of containerization.

Such containerization is likely spurred by the action of quantization already captured by (3.5.2), (3.5.4), (3.5.6), and (3.5.7) reproduced here for convenience.

Hence (3.5.2):

$$Space_{quantization} = h_{UK}\left(Xa + \overline{Yb_{0-n}}\right)$$

$$\text{where:} \quad \begin{bmatrix} X \in [S_{System_K}] \\ Y \in [S_{System_{Pr}}, S_{System_P}, S_{System_K}, S_{System_N}] \\ a, b \text{ are integers}; a > b \end{bmatrix}$$

(3.5.4):

$$Time_{quantization} = h_{UP}\left(Xa + \overline{Yb_{0-n}}\,\right)$$

$$\text{where:} \quad \begin{bmatrix} X \in [S_{System_P}] \\ Y \in [S_{System_{Pr}}, S_{System_P}, S_{System_K}, S_{System_N}] \\ a, b \text{ are integers}; a > b \end{bmatrix}$$

(3.5.6):

$$Energy_{quantization} = h_{UPr}\left(Xa + \overline{Yb_{0-n}}\,\right)$$

$$\text{where:} \quad \begin{bmatrix} X \in [S_{System_{Pr}}] \\ Y \in [S_{System_{Pr}}, S_{System_P}, S_{System_K}, S_{System_N}] \\ a, b \text{ are integers}; a > b \end{bmatrix}$$

(3.5.7):

$$Gravity_{quantization} = h_{UH}\left(Xa + \overline{Yb_{0-n}}\,\right)$$

$$\text{where:} \quad \begin{bmatrix} X \in [S_{System_N}] \\ Y \in [S_{System_{Pr}}, S_{System_P}, S_{System_K}, S_{System_N}] \\ a, b \text{ are integers}; a > b \end{bmatrix}$$

Universality of Electromagnetic Spectrum

As discussed in Chapter 5, the range of frequencies in the electromagnetic spectrum is infinite. This has been predicted by Maxwell's equations. (5.1.8), the equation for the infinite energy-gradient, (5.1.13) the equation that relates the infinite range of wavelengths to archetypes, and (5.1.14) that relates the infinite range of mass-potential, each essentially capture the infinite seed-range held by the electromagnetic spectrum. The original forms of the equations follow for convenience.

Hence, (5.1.8):

$$EM_{Spectrum_{Energy}} = Xa + \overline{Yb_{0-n}}$$

$$where \begin{bmatrix} X \in [S_{System_P}] \\ Y \in [S_{System_{Pr}}, S_{System_P}, S_{System_K}, S_{System_N}] \\ a, b \ are \ integers; a > b \end{bmatrix}$$

(5.1.13):

$$EM_{Spectrum_{Structure}} = Xa + \overline{Yb_{0-n}}$$

$$where \begin{bmatrix} X \in [S_{System_K}] \\ Y \in [S_{System_{Pr}}, S_{System_P}, S_{System_K}, S_{System_N}] \\ a, b \ are \ integers; a > b \end{bmatrix}$$

(5.1.14):

$$EM_{Spectrum_{Mass_{Possibility}}} = Xa + \overline{Yb_{0-n}}$$

$$where \begin{bmatrix} X \in [S_{System_{Pr}}] \\ Y \in [S_{System_{Pr}}, S_{System_P}, S_{System_K}, S_{System_N}] \\ a, b \ are \ integers; a > b \end{bmatrix}$$

Recall that the infinite frequency-wavelength range becomes possible because of the constancy of the speed of light at c, which creates time-matter reality that marks this universe.

It is because of the infinite seed-range that the action of the electromagnetic spectrum is apparently universalized. Hence when quantization does take place as per the light-space-time emergence equation, (3.1.3), and the potential effects of levels of light on pre-genetic and genetic information equation, (3.6.5), because there are a larger range of seeds the effect of the material-fabric and subsequently on materialization is distributed and spread-out.

Material Localization at Multiple Scale

Since quanta hold within themselves the essence of what must be projected from properties or function in higher-velocity light to lower-velocity more-material light, and since the four-fold properties are a representation of the inherent oneness of Light, it is reasonable to expect that such four-foldness will continue to have a bias to uphold that oneness in its more material manifestation. In other words, it is reasonable to expect that space-quanta, time-quanta, energy-quanta, and gravity-quanta related to a single particle will operate as a composite fourfold quantum. This means that four quantum fields operate together or as one composite field in the emergence and possibilities represented by a particle and that perhaps is the reason for this apparently intense material localization.

The following subsections explore the notion of intense material localization at different scale – specifically at the atomic-particle, unit-space, big-planet, expanding universe, black hole, and cosmic bounce levels. In the possible iterations of the Light-Space-Time Emergence Equation (3.1.3) the generated code can apply to organizations of increasing size, including those just specified, as discussed in Section 4.

Quanta at Atomic-Particle Level

Figure 6.1.1 below, Composite-Quantum at Atomic-Particle Level, depicts a composite-quantum along the space, time, gravity, and energy dimensions. Essentially the graphic is illustrating that all things being equal there is a balanced accumulation of anterior possibility or

information along the space, time, gravity, and energy dimensions that plays a primary role in orchestrating the possibility contained in the atomic-particle. Recall that the space dimension contains information to do with unique seeds, the time dimension information to do with the maturity of seeds, the gravity dimension information to do with collectivities of seeds, and the energy dimension information to do with the materialization of seeds. Hence:

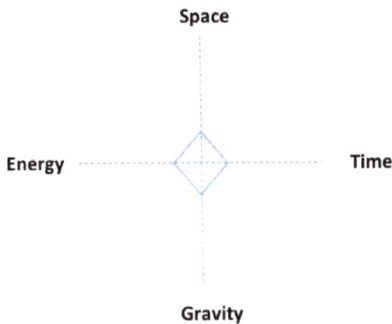

Figure 6.1.1: Composite-Quantum at Atomic-Particle Level

Quanta at Unit-Space Level

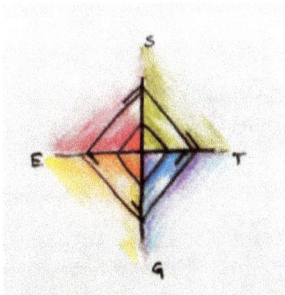

The small rhombus in Figure 6.1.2 represents a quantum at the atomic-particle level, while the larger rhombus is an extrapolation of that in some unit space. All things being equal it is the same fourfold accumulation of anterior possibility or information captured by the quantum that will govern particle behavior at the unit-space level. Hence:

211

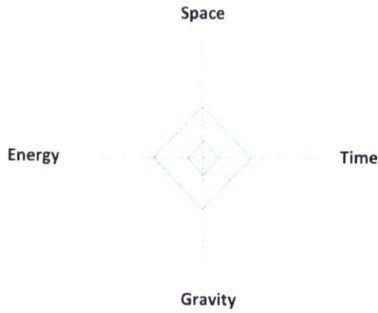

Space

Energy   Time

Gravity

Figure 6.1.2: Composite-Quantum at Unit-Space Level

Further, if Q_A represents a composite-quantum at the atomic-particle level, then Q_{US} will be a summation of such quanta as modeled by Equation 6.1.5:

$$Q_{US} = \Sigma Q_A$$

Eq 6.1.5 Composite-Quantum and Unit-Space Level

Quanta at Level of Big-Planet

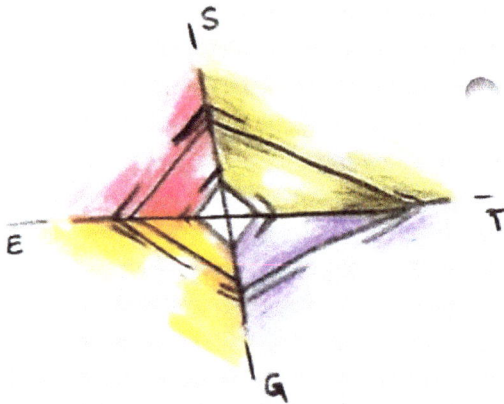

In Figure 6.1.3, Composite-Quantum at Level of Big-Planet, the smaller rhombus represents a quantum at the unit-space level, while the larger kite-quadrilateral represents a quantum at the Big-Planet level:

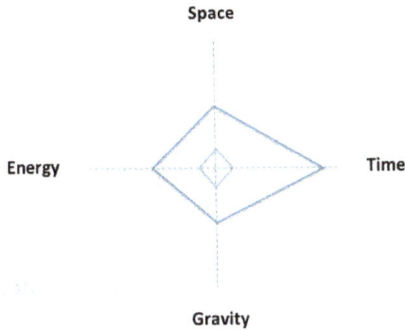

Space

Energy   **Time**

Gravity

Figure 6.1.3: Composite-Quantum at Level of Big-Planet

The kite-quadrilateral is elongated along the time dimension implying that the larger collectivity of seeds represented by a space such as a Big-Planet will take longer to reach maturity, relative to a seed contained in a unit-space. This time-elongation is also modeled by the '$mod(Time_{quantization})$' component of Equation 6.1.6:

$$Q_{BP} = \Sigma\, Q_{US} \times mod(Time_{quantization})$$

Eq 6.1.6 Composite-Quantum at the Level of Big-Planet

Note that in (6.1.6), Q_{BP} represents a composite-quantum at the level of a Big-Planet, while Q_{US} represents a composite-quantum at the unit-space level as modeled by (6.1.5). Equation (6.1.6) illustrates the notion of composite-quantum as creating the field within which a Big-Planet can materialize.

The time-elongation captures the underlying dynamic of many more finite steps to maturity for the vaster collectivity of seeds, that therefore manifests as time being experienced more rapidly: hence time speeds-up.

213

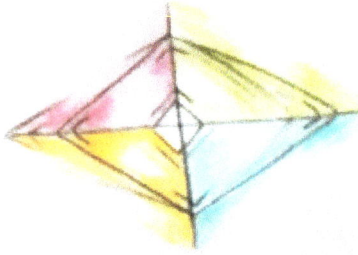

Figure 6.1.4, Composite-Quantum in an Expanding Universe, depicts the relative relationship between the smaller unit-space composite-quantum rhombus and the scaled up expanding-universe composite-quantum rhombus, in which it is estimated that the space, time, gravity, and energy components all go through some related increase:

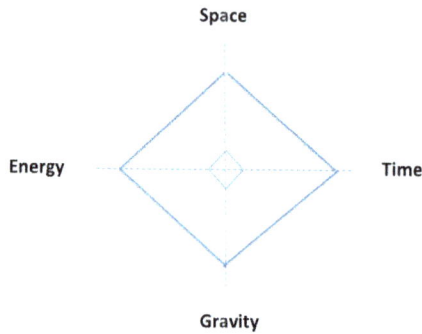

Figure 6.1.4: Composite-Quantum in an Expanding Universe

Equation (26) models such expanding-universe composite-quantum:

$$Q_{EU} = \Sigma Q_{US}$$

Eq 6.1.7 Composite-Quantum in an Expanding Universe

214

Figure 6.1.5, Composite-Quantum at the Black Hole Level, illustrates the relatively different quadrilateral, as compared with previously considered composite-quantum in Figures (6.1.1-4):

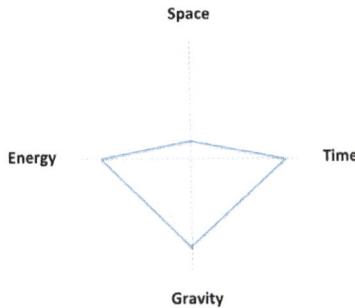

Space

Energy   Time

Gravity

Figure 6.1.5: Composite-Quantum at Black Hole Level

This composite-quantum is hypothesized as being the result of a large number of seeds previously spread in space, coming together to be reformulated as a smaller set

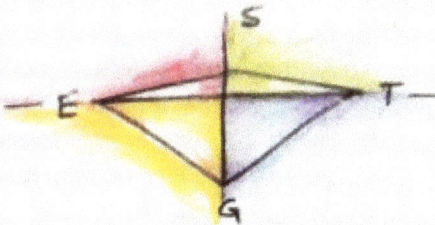

or a new seed. The new seed(s) have a different intent than the spread out un-mashed seeds, will have a very different and relatively elongated maturity-dynamic as depicted along the time-dimension, will have a very different collectivity-relationship between other seeds as represented by the gravity-dimension, and will similarly also have a very different energy-

215

materialization dynamic as represented by the energy-dimension.

Centered on the black hole, space-time-gravity-energy is going to proceed differently than may have existed prior to the formation of the black hole. Such modification is modeled by Equation (6.1.8), where as can be seen there is a summation of previous seeds along each dimension, captured by $\sum Seed_i$, and a holding of the new essence, captured by h_{Ui}, which together create the new black hole (BH) composite-quantum, Q_{BH}:

$$
Q_{BH} =
\begin{bmatrix}
mod(Space_{quantization}) \rightarrow h_{UK} \sum Seed_K \\
mod(Time_{quantization}) \rightarrow h_{UP} \sum Seed_P \\
mod(Gravity_{quantization}) \rightarrow h_{UH} \sum Seed_H \\
mod(Energy_{quantization}) \rightarrow h_{UPr} \sum Seed_{Pr}
\end{bmatrix}
$$

Eq 6.1.8 Composite-Quantum at Black-Hole Level

Note that consistent with General Relativity, the dynamic of gravity intensifies, space is compressed, time slows down, and energy-potential increases.

Quanta of Cosmic Bounce

Figure 6.1.6, Composite-Quantum at Cosmic Bounce, depicts the composite-quantum at a possible Cosmic Bounce event. The smaller quadrilateral represents the black hole composite-quantum, while the larger quadrilateral is a scaled up Cosmic Bounce version of Q_{BH}:

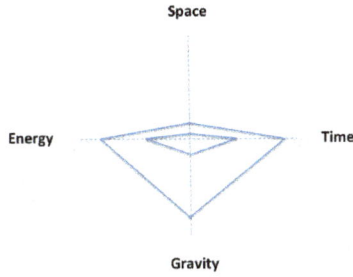

Figure 6.1.6: Composite-Quantum at Cosmic Bounce

The notion of scaling up, is captured by e^M where M is some very large number, in the Cosmic Bounce composite-quantum as modeled by Equation (6.1.9):

$$Q_{CB} = Q_{BH}e^M$$

Eq 6.1.9 Composite-Quantum at Cosmic Bounce

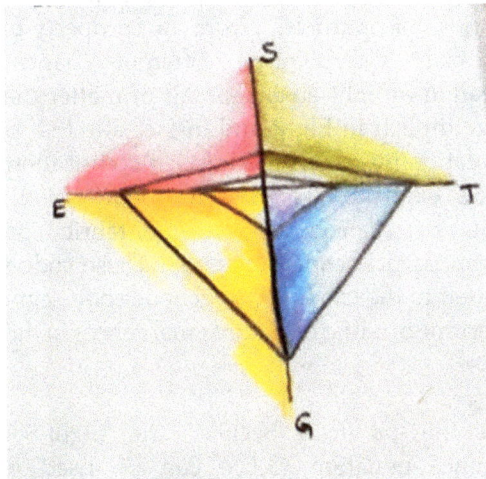

Hence, this brief seed-based analyses suggests that even as organizations scale, there will still be a containerization of matter albeit, possibly, with different quadrumvirate balance.

217

Chapter 6.2: Generation of Material-Fabric, Pre-Genetic Quantum Particle Code

The electromagnetic spectrum or perhaps more accurately, the electro-magnetic-wavearchetype-masspotential spectrum, describes one of the first layers or translation of light into pre-genetic material manifestation.

As already explored the finite speed of such spectrums allows the build-up of energies or quanta, and this in turn may express itself as a series or as an array of quantum particles. The vast number of quantum particles that have so far been discovered has in turn yielded what is known as the Standard Model. This model is made up of what is called Quarks, Leptons, and Bosons, and a Higgs-Boson (Cottingham, 2007).

But if we look at these four fundamental quantum categories of particles from a property-based, or "functional" viewpoint a different chapter in the exploration of light emerges. All of matter then can be seen as implicit in Light and in fact due to a process of computation by which the implicit codification in Light becomes explicit. The implicit codification becomes explicit by generating material-fabric pre-genetic quantum particle code-segments. These code-segments are added to the existing pre-genetic code-segments and also become binding on future emergences in the material universe.

As in the previous Section, the Light-Space-Time Emergence equation (3.1.3) can be used to model emergence as it proceeds to this relatively more complex form of fourfold manifestation. (3.6.5), the Potential Effect of Levels of Light on Pre-Genetic of Genetic Information equation, basically sheds light on specific types of mutation that can occur and therefore on the nature of the

218

code-segment that is being generated by (3.1.3). As discussed before, (3.6.5) shows a particular working of (3.1.3) but is wholly contained within it.

This chapter elaborates the action of (3.1.3) and (3.6.5) in the creation of the code-segments that define the quantum particle ecosystem logic.

Light's Emergence as Quantum Particles

As discussed previously the Light-Space-Time Emergence equation (3.1.3) being iterative, can be used to model emergence as it proceeds from simpler four-fold to more complex four-fold manifestations. But further, as implied by (3.6.5), the Potential Effect of Levels of Light on Pre-Genetic of Genetic Information equation, any process of organization has the possibility of altering the material-fabric or fabric of existence so long as the bases involved are driven primarily by a meta-level.

As suggested by Equation 6.2.1, Quantum Particles Material-Fabric Pre-Genetic Code, the architecture and details of quantum particles can be seen to be the result of the application of the Light, Space, and Time matrices as will be elaborated:

$$Light - Space - Time\ Emergence_{Quantum\ Particles} =$$

$$\left| \begin{array}{l} \left| \begin{bmatrix} c_\infty : [Pr, Po, K, H] \\ \left(\downarrow R_{C_K} = f\left(R_{C_\infty}\right) \right) \\ c_K : [S_{Pr}, S_{Po}, S_K, S_H] \\ \left(\downarrow R_{C_N} = f\left(R_{C_K}\right) \right) \\ c_N : f(S_{Pr} \times S_{Po} \times S_K \times S_H) \\ \left(\downarrow R_{C_U} = f\left(R_{C_N}\right) \right) \\ c_U : [P, V, M, C] \\ \Uparrow \\ c_{0 : [D, W, I, C]} \end{bmatrix}_{Light} \\ \begin{bmatrix} M_3 : -\infty \leq t \leq \infty \\ \downarrow \\ M_2 : 0 \geq t > \infty \\ \downarrow \\ M_1 : 0 > t > \infty \\ \downarrow \\ U \rightarrow \begin{array}{l} t \leq E_{Cell};\ TC : M_3 \rightarrow U \\ t \sim E_{Human};\ TC : U \rightarrow M_3 \end{array} \end{bmatrix}_{Time} \end{array} \right.$$

$$\begin{bmatrix} M_3 \rightarrow System_X \\ (\uparrow F \rightarrow I) \\ M_2 \rightarrow S_{System_X} \\ (\uparrow Sig \rightarrow F) \\ M_1 \rightarrow Sig_X \\ (\uparrow > P_X) \\ U \rightarrow x_{EM\ Spectrum} \end{bmatrix}_{Space}$$

$$TC \rightarrow x_{Quantum\ Particles}$$

$$\langle x_U | x_T \rangle$$

$$\Rightarrow$$

$$\langle Space - Time - Energy - Gravity\ Material - Fabric\ Pre \\ - Genetic\ Code \rangle +$$

$$(Electromagnetic\ Spectrum\ Material - Fabric\ Pre - Genetic\ Code) +$$

$$Quantum\ Particles\ Material - Fabric\ Pre - Genetic\ Code$$

Eq 6.2.1: Quantum Particles Material-Fabric Pre-Genetic Code

Starting with the Light-Matrix, the top left-hand matrix in (6.2.1), the first line from the top, $C_\infty : [Pr, Po, K, H]$, specifies the fundamental architecture of quantum particles. In this rendering, and as elaborated in subsequent subsections, quarks are an emergence of Light's property of Knowledge, leptons are an emergence

220

of Light's property of Power, bosons are an emergence of Light's property of Harmony, and the Higgs-boson is an emergence of Light's property of Presence. The fundamental architecture of these aspects hence, is an

emergence of the properties of Light at ∞.

Line 3 in the Light-Matrix, C_K: $[S_{Pr}, S_{Po}, S_K, S_H]$, elaborates the sets for Presence, Power, Knowledge, and Harmony, each containing multiple elements. For example, as will be explored in greater detail in the section on quarks, various elements derived from the four sets define the behavior of quarks and could be functions such as 'composite-arrangements', 'specifying attributes' amongst others, hence collectively describing quarks' way of being. Specifically, Line 5, $C_{N:} f(S_{Pr} \times S_{Po} \times S_K \times S_H)$, suggests that unique seeds are created from a combination of such elements from all four sets, with a particular element leading, that in effect creates the distinctness possible at the level of quantum particles.

Line 6, $(\downarrow R_{C_U} = f(R_{C_N}))$, specifies quantization between the layer where the seeds are formed, and the physical layer we are familiar with, and as explored in Chapter 3.5 and 3.6, will result in Line 7, C_U: $[P, V, M, C]$, hence changing the material-fabric of existence. Note that as in the process describing the generation of the code-segments for space-time-energy-gravity quantization at the time of the Big Bang, the generation of the FBLEE (four-base logic-encoding ecosystem) code-segment as specified by (4.1.5) and the MF (material-fabric) code segment as specified by (4.1.6) are combined together into one equation so that the FBLEE process is apparently transparent.

The possibilities represented by Lines 1 through 5 hence concretize through the quantization represented by Line 6 to become the quantum particles with its physical (related to Presence), vital (related to Power), mental (related to Knowledge), and connection (related to Harmony) aspects now existing in material reality typified by Light moving at c. Note that just as Line 6 represents a process of quantization relating the layer of reality created by Light traveling at c with the antecedent layers, so too Lines 2 and 4 as previously discussed, also

represent quantization of a more subtle kind that ultimately plays a critical part in allowing the material-fabric to express infinite diversity.

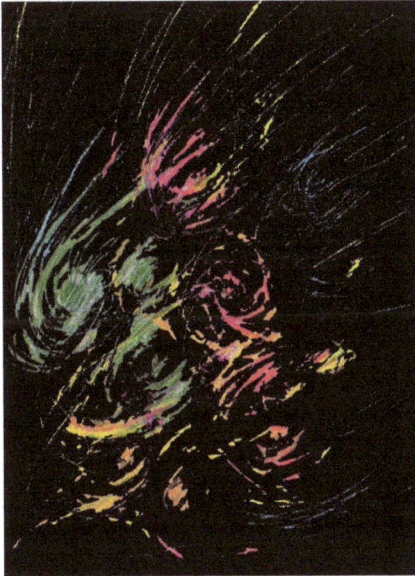

Typically it is the process as captured by the Space-Matrix that will determine if Line 6 is activated. Specifically patterns at the untransformed layer, U, will need to be overcome, as specified by the second-line from the bottom of the Space-Matrix: $(\uparrow > P_x)$. But as specified by the bottom-line of the Time-Matrix, reproduced below, it is only with the advent of the human-system that the automaticity of the action of meta-levels is reversed:

$$U \rightarrow \begin{array}{l} t \leq E_{Cell}; \text{TC}: M_3 \rightarrow U \\ t \sim E_{Human}; \text{TC}: U \rightarrow M_3 \end{array}$$

Hence in the case of the quantum particle system, which in this emergence is a pre-human system, the fact that patterns do not need to be overcome means that quantization happens automatically.

Even though quantum particles are a pre-human system and therefore the action of meta-levels are modeled as being automatic, it is nonetheless useful to review (3.6.5), Potential Effect of Levels of Light on Pre-Genetic or Genetic Information, to understand how the relationship with different forms of mutation has been specified:

Potential Effect of Levels of Light on Pre Genetic or Genetic Information =

$$
\begin{bmatrix}
STATIC \ \langle |[L][S][T]TC \ \rightarrow x_T | \langle x_U | x_T \rangle \rangle \\
\times \\
\left((Y > U\colon Z_Q) \ \vee (Y \leq U\colon Z_F) \ \vee (Y = U\colon Z_R) \right) \\
\ni \\
\begin{pmatrix}
Z \in \mathbb{U} \ (Space, Time, Energy, Gravity) \\
Q\colon Quantization; F\colon Fragmentation; R\colon Random
\end{pmatrix}
\end{bmatrix} \rightarrow h \ni
$$

$$
h \in \begin{pmatrix}
[Q]\colon Constructive \ zone, \\
[Q]\colon Constructive \ zone \ \wedge \ Constructive \ mutation, \\
[F]\colon Destructive \ mutation, \\
[R]\colon Random \ mutation
\end{pmatrix}
$$

Line 1 from the top in the matrix is simply a static form of (3.6.1) the Simplified Light-Space-Time Emergence equation. The static form is designated by 'STATIC', and implies that fundamental operations true of (3.6.1) are being highlighted in (3.6.5). In other word (3.6.1) already has all the operations highlighted in (3.6.5) in it, but by 'freezing' it by making it static, the essential dynamics leading to possible mutations at the genetic level can more clearly be highlighted.

Line 1 is then subjected ($\times$) to a determination of the dominant levels of light that may be active, designated by Line 2, ' $\left((Y > U\colon Z_Q) \ \vee (Y \leq U\colon Z_F) \ \vee (Y = U\colon Z_R) \right)$ '. Unpacking this, 'Y > U' implies meta-levels are active and as a result it is possible that Z_Q is going to take place. This also implies activation and potential change of FBLEE. The call from below, as it were, may invoke some function that already exists in the subtle-libraries 'above', so that some already existing function may influence FBLEE through ∞ -entanglement, K-entanglement, or N-entanglement. Perhaps it is that the existing electro-magnetic-wavearchetype-masspotential spectrum yearns for objects to interact with, which formulates itself as a call from below, as it were. The call-and-response dynamic may be thought of as a key-and-lock mechanism, where a deep enough visceral urge from below acting as the key, opens an entangled lock to alter

FBLEE as per the visceral urge. In the alteration of FBLEE the response from 'above' may precipitate some already existing meta-function from a subtle-library such as results in the functional-elements that comprise a quantum particle ecosystem. Such a call-response system at the pre-human level is considered to be relatively automatic as compared with the emergence of human-type systems because of the complexity of many more sources of action in the case of the latter.

'$Y \leq U$' implies that only the untransformed levels are active, and therefore also the sub-level where the speed of light is 0 is active and as a result Z_F is going to take place. '$Y = U$' implies that all levels are active and as a result Z_R is going to take place.

Line 3 elaborates the significance of Z_Q, Z_F, and Z_R. Hence, as already discussed in the previous chapter, Z is the union of potential quantum-operations of space, time, energy, and gravity, designated by ' $Z \in U\,(Space, Time, Energy, Gravity)$'. But the nature of the operations is designated by ' $Q: Quantization; F: Fragmentation; R: Random$ '. Z_Q, then, implies that the full quantization originating from updated four-base logic-encoding ecosystems (FBLEE) can take place, and will result in lasting material change at the pre-genetic, material-fabric level. Z_F implies that the essential set will be fragmented and that only local libraries a the level of the local material-fabric can potentially be altered. Z_R also precludes full quantization, and that some partial local-library constructive or destructive mutation may take place.

The '$\rightarrow h \ni h \in$' segment resolves the outcome of the operations implied by Lines 1 – 3, suggesting that the outcome will be 'h' such that ($\ni$) 'h' is an element ($\in$) of the set specified by the members '[Q]: *Constructive zone*', '[Q]: *Constructive zone AND Constructive mutation*', '[F]; *Destructive mutation*', and '{R}: *Random mutation*'.

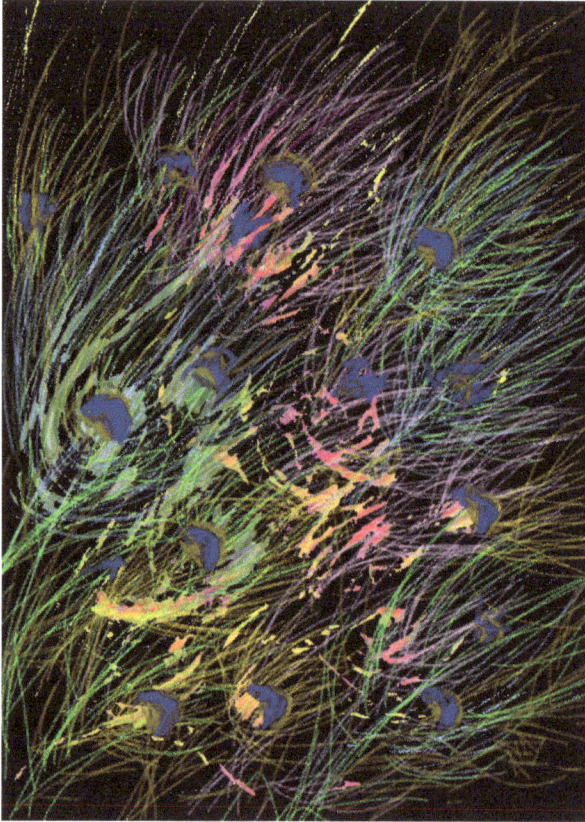

'[Q]: *Constructive zone*' implies that the in-built buffer has been crossed and that access to the deeper four-base logic-encoding ecosystem (FBLEE) has been granted. '[Q]' in this segment implies that there is the possibility that full-quantization as specified by Line 3 of the previous matrix can take place. Access to this zone is a prerequisite for constructive mutation to occur, as designated by the

226

element   '   [Q]: *Constructive zone* ∩
Constructive mutation ', which implies that full-quantization is going to take place and will result in material change. The '[F]' specifies the relationship between 'Fragmentation' in Line 3 of the previous matrix and destructive mutation. The '[R]' specifies the relationship between 'Random' in Line 3 of the previous matrix and random mutation.

But as just summarized in the Time-Matrix in (6.2.1) Y is by definition greater than U and hence quantization is automatic. In terms of quantum particles such quantization implies that wholeness becomes fully active through specific space, time, energy, and gravity quantization to create an holistic "ecosystem" with its own "quantum particles logic" as it were. The wholeness has now precipitated into the material-fabric and is available to be consciously and unconsciously tapped into. This "logic" or more specifically pre-genetic code is elaborated in the following sub-sections.

Generation of Material-Fabric Pre-Genetic Quark Code

It can be seen first that the nucleus of an atom is made up of a combination of quarks. Specifically, a proton is composed of two "up" quarks and one "down" quark. Quarks have unusual names – up, down, charm, strange, top, bottom, with each subsequent pair belonging to a different "generation" of quarks. A neutron is composed of two "down" quarks and one "up" quark. Protons and neutrons together make up the nucleus of an atom. But we also know that the number of protons in the nucleus specifies the Atomic Number of an atom. Atomic number in turn uniquely identifies the element from the periodic table. Hence, an atomic number of 47, for example, specifies that the element is Silver. In other words it can be suggested that the unique properties of an element, the knowledge of what it is and how it will behave in the universe, is related to the quark. It may be suggested that quarks, therefore, are associated with the precipitation or emergence of Light's property of Knowledge in the quantum world.

Hence, it could be that the signature or code-segment for the family of quarks, as in Equation 6.2.2, is:

$$Sig_{quarks} = Xa +$$

$$\overline{Yb_{0-n}} \quad where \begin{bmatrix} X \in [S_{System_K}] \\ Y \in [S_{System_{Pr}}, S_{System_P}, S_{System_K}, S_{System_N}] \\ a, b \ are \ integers; a > b \end{bmatrix}$$

Eq 6.2.2: Generalized Signature or Code-Segment for Quarks

As can be seen the primary element X is derived from the set of knowledge, S_{System_K}. Various elements, derived from the four sets would define the behavior of quarks and could be functions such as 'composite-arrangements', 'specifying attributes' amongst others, hence collectively describing quarks' way of being.

Note that (6.2.2) already implies that Lines 1 – 5 in the Light Matrix (6.2.1) have been activated, and that the logic of the quark-ecosystem will automatically precipitate into the material-fabric through the action of Line 6-7 of (6.2.1). This logic is none other than the pre-genetic code as specified by (6.2.2).

Generation of Material-Fabric Pre-Genetic Lepton Code

When considering leptons it is useful to know that unlike quarks that only exist in composite arrangements with other quarks, leptons are solitary, point-like particles without internal structure (Arabatzis, 2006). The best-known lepton is the electron. So the electron may be considered as a surrogate for the lepton class. The electron appears to be associated with the flow of energy and power. Further they appear to be the adventurers easily leaving the atom they are a part of. They also form locks or bonds with other atoms through the force of attraction and repulsion. In some sense they seem to be a representation or precipitation or emergence of Light's property of Power.

The signature for the family of leptons, as in Equation 6.2.3, is:

$$Sig_{leptons} = Xa + \overline{Yb_{0-n}} \quad where \begin{bmatrix} X \in [S_{System_P}] \\ Y \in [S_{System_{Pr}}, S_{System_P}, S_{System_K}, S_{System_N}] \\ a, b \ are \ integers; a > b \end{bmatrix}$$

Eq 6.2.3: Generalized Signature or Code-Segment for Leptons

As can be seen the primary element X is derived from the set of power, S_{System_P}. Various elements, derived from the four sets would define the behavior of leptons and could be functions such as 'adventurer', 'solitary', 'create attraction', amongst others, hence collectively describing leptons' way of being.

Note that (6.2.3) already implies that Lines 1 – 5 in the Light Matrix (6.2.1) have been activated, and that the logic of the lepton-ecosystem will automatically precipitate into the material-fabric through the action of Line 6-7 of (6.2.1). This logic is none other than the pre-genetic code as specified by (6.2.3).

Generation of Material-Fabric Pre-Genetic Boson Code

The bosons on the other hand are thought of as force-carriers. They are what allow all known matter particles to interact. The three fundamental bosons in this category are the photon, the W and Z bosons, and the gluon. The carrier particle of the electromagnetic force or spectrum is the photon. The carrier particle of the strong nuclear force that holds quarks together is the gluon. The carrier particle for the weak interactions, responsible for the decay of massive quarks and leptons into lighter quarks and leptons, are the W and Z bosons.

Bosons can be thought of as the precipitation of what creates relationship and harmony at the quantum level. Hence they can be thought of as the precipitation or emergence of Light's property of Harmony at the quantum level.

The signature for the family of gauge bosons, as in Equation 6.2.4, is:

$$Sig_{bosons} = Xa +$$

$$\overline{Yb_{0-n}} \quad where \begin{bmatrix} X \in [S_{System_N}] \\ Y \in [S_{System_{Pr}}, S_{System_P}, S_{System_K}, S_{System_N}] \\ a, b \ are \ integers; a > b \end{bmatrix}$$

Eq 6.2.4: Generalized Signature or Code-Segment for Bosons

As can be seen the primary element X is derived from the set of nurturing, S_{System_N}. Various elements, derived from the four sets would define the behavior of bosons and could be functions such as 'creating interaction', 'holding together', amongst others, hence collectively describing bosons' way of being.

Note that (6.2.4) already implies that Lines 1 – 5 in the Light Matrix (6.2.1) have been activated, and that the logic of the boson-ecosystem will automatically precipitate into the material-fabric through the action of Line 6-7 of (6.2.1). This logic is none other than the pre-genetic code as specified by (6.2.4).

Equations in this form give a clue as to how to think about the fundamental particles. While the default approach is always to think about physical qualities, equations in the preceding format allow us to think about detailed functions or qualities associated with particles and give insight into how the process of emergence or complexification of fourfold functionality takes place.

Generation of Material-Fabric Pre-Genetic Higgs-Boson Code

This leaves the other discovered fundamental particle the Higgs-Boson. In ordinary matter, most of the mass is contained in atoms, and the majority of the mass of an atom resides in the nucleus, made of protons and neutrons. Protons and neutrons are each made of three quarks. But it is the quarks that get their mass by interacting with what is known as the Higgs field (Olive, 2014). Hence the Higgs-Boson can be thought of as the mass-giver. In other words it is what gives presence to the quarks and it can be thought of as the precipitation or emergence of Light's property of Presence at the quantum level. Just as there are multiple particles in each of the other 'families' it is likely that there will be multiple particles in the Higgs-Boson family. Recent research at CERN indicates that the Higgs-Boson may have a cousin.

The signature for the Higgs-boson and any other similar particle, as in Equation 6.2.5, is:

$$Sig_{Higgs-boson} = Xa + \overline{Yb_{0-n}}$$

$$where \begin{bmatrix} X \in [S_{System_{Pr}}] \\ Y \in [S_{System_{Pr}}, S_{System_P}, S_{System_K}, S_{System_N}] \\ a, b \ are \ integers; a > b \end{bmatrix}$$

Eq 6.2.5: Generalized Signature or Code-Segment for Higgs-Boson

As can be seen the primary element X is derived from the set of presence, $S_{System_{Pr}}$. Various elements, derived from the four sets would define the behavior of Higgs-bosons and could be a function such as 'creating mass', amongst others, hence collectively describing Higgs-bosons' way of being.

Note that (6.2.5) already implies that Lines 1 – 5 in the Light Matrix (6.2.1) have been activated, and that the logic of the Higgs-boson-ecosystem will automatically precipitate into the material-fabric through the action of Line 6-7 of (6.2.1). This logic is none other than the pre-genetic code as specified by (6.2.5).

Combining all the preceding particle equations it is possible to create a generalized particle equation, as in Equation 6.2.6:

$$Sig_{particle} = Xa + \overline{Yb_{0-n}} \; where \; \begin{bmatrix} X \in [S_{System_{Pr}}, S_{System_P}, S_{System_K}, S_{System_N}] \\ Y \in [S_{System_{Pr}}, S_{System_P}, S_{System_K}, S_{System_N}] \\ a, b \; are \; integers; a > b \end{bmatrix}$$

Eq 6.2.6: Generalized Signature of Particle

So we see that again the underlying properties of light - Knowledge, Power, Harmony, and Presence - emerge as quarks, leptons, bosons, and Higgs-bosons respectively.

Summary of Material-Fabric Pre-Genetic Code at Quantum Particle Stage

Summarizing, after the quantum particle stage iterations of the Light-Space-Time Emergence equation are complete, the following code-segments will have been generated as specified by Equation 6.2.7, Active Material-Fabric Pre-Genetic Code Segments at Quantum Particle Stage:

$$Light - Space - Time\ Emergence_{Quantum\ Particle} =$$

$$\left\lVert \begin{bmatrix} c_\infty : [Pr, Po, K, H] \\ \left(\downarrow R_{C_K} = f\left(R_{C_\infty}\right) \right) \\ c_K : [S_{Pr}, S_{Po}, S_K, S_H] \\ \left(\downarrow R_{C_N} = f\left(R_{C_K}\right) \right) \\ c_N : f(S_{Pr} \times S_{Po} \times S_K \times S_H) \\ \left(\downarrow R_{C_U} = f\left(R_{C_N}\right) \right) \\ c_U : [P, V, M, C] \\ \Uparrow \\ c_{0:[D,W,I,C]} \end{bmatrix}_{Light} \begin{bmatrix} M_3 \rightarrow System_X \\ (\uparrow F \rightarrow I) \\ M_2 \rightarrow S_{System_X} \\ (\uparrow Sig \rightarrow F) \\ M_1 \rightarrow Sig_x \\ (\uparrow > P_x) \\ U \rightarrow x_{EM\ Spectrum} \end{bmatrix}_{Space} \right.$$

$$\left. \begin{bmatrix} M_3 : -\infty \leq t \leq \infty \\ \downarrow \\ M_2 : 0 \geq t > \infty \\ \downarrow \\ M_1 : 0 > t > \infty \\ \downarrow \\ U \rightarrow \begin{array}{l} t \leq E_{Cell}; \text{TC}: M_3 \rightarrow \text{U} \\ t \sim E_{Human}; \text{TC}: U \rightarrow M_3 \end{array} \end{bmatrix}_{Time} \quad TC \rightarrow x_{Quantum\ Particles} \right\rVert \langle x_U | x_T \rangle$$

$\Rightarrow$

$\langle Space - Time - Energy - Gravity\ Material - Fabric\ Pre$
$\qquad - Genetic\ Code \rangle +$

$(Electromagnetic\ Spectrum\ Material - Fabric\ Pre - Genetic\ Code) +$

$$\left(where \begin{bmatrix} \sum Sig_{bosons} = Xa + \overline{Yb_{0-n}} \\ X \in [S_{System_N}] \\ Y \in [S_{System_{Pr}}, S_{System_P}, S_{System_K}, S_{System_N}] \\ a, b\ are\ integers; a > b \end{bmatrix} \atop where \begin{bmatrix} \sum Sig_{quarks} = Xa + \overline{Yb_{0-n}} \\ X \in [S_{System_K}] \\ Y \in [S_{System_{Pr}}, S_{System_P}, S_{System_K}, S_{System_N}] \\ a, b\ are\ integers; a > b \end{bmatrix} \atop where \begin{bmatrix} \sum Sig_{leptons} = Xa + \overline{Yb_{0-n}} \\ X \in [S_{System_P}] \\ Y \in [S_{System_{Pr}}, S_{System_P}, S_{System_K}, S_{System_N}] \\ a, b\ are\ integers; a > b \end{bmatrix} \atop where \begin{bmatrix} \sum Sig_{Higgs-boson} = Xa + \overline{Yb_{0-n}} \\ X \in [S_{System_{Pr}}] \\ Y \in [S_{System_{Pr}}, S_{System_P}, S_{System_K}, S_{System_N}] \\ a, b\ are\ integers; a > b \end{bmatrix}} \right)$$

Eq. 6.2.7, Active Material-Fabric Pre-Genetic Code Segments at Quantum Particle Stage

The '$\sum x$' signifies all the possibilities for the 'x' equation-segment.

Chapter 6.3: Generation of Material-Fabric, Pre-Genetic Boson Code

The bosons as mentioned can be thought of as force-carriers and allow all known matter particles to interact.

But when we look at bosons in more detail there are three fundamental bosons – the photon, the W and Z bosons, the gluon - and one hypothetical boson – the graviton.

This chapter looks at the on-going computation involving a process of quantization by which the very logic of bosons precipitates into the material-fabric.

As in the previous Section, the Light-Space-Time Emergence equation (3.1.3) can be used to model emergence as it proceeds to this relatively more complex form of fourfold manifestation. (3.6.5), the Potential Effect of Levels of Light on Pre-Genetic of Genetic Information equation, basically sheds light on specific types of mutation that can occur and therefore on the nature of the code-segment that is being generated by (3.1.3). As discussed before, (3.6.5) shows a particular working of (3.1.3) but is wholly contained within it.

This chapter elaborates the action of (3.1.3) and (3.6.5) in the creation of the code-segments that define the boson ecosystem logic. While this development may precede the generation of quantum particle code, for simplicity it will be assumed that this happens at the same time.

Light's Emergence as Bosons

As discussed previously the Light-Space-Time Emergence equation (3.1.3) being iterative, can be used to model emergence as it proceeds from simpler four-fold to more complex four-fold manifestations. Hence (3.1.3) has already been applied to suggest the emergence of the space-time-energy-gravity quadrumvirate, the electromagnetic spectrum, and quantum particles in

236

general. Here it will be applied to suggest the emergence of a sub-class of quantum particles, Bosons. Hence as in (6.2.1) the starting point will be assumed to be the electromagnetic spectrum.

As suggested by Equation 6.3.1, Boson Material-Fabric Pre-Genetic Code, the architecture and details of bosons can be seen to be the result of the application of the Light, Space, and Time matrices as will be elaborated:

$Light - Space - Time\ Emergence_{Bosons} =$

$$
\left\| \begin{array}{c} \left[\begin{array}{c} c_{\infty}: [Pr, Po, K, H] \\ \left(\downarrow R_{C_K} = f(R_{C_\infty}) \right) \\ c_K: [S_{Pr}, S_{Po}, S_K, S_H] \\ \left(\downarrow R_{C_N} = f(R_{C_K}) \right) \\ c_N: f(S_{Pr} \times S_{Po} \times S_K \times S_H) \\ \left(\downarrow R_{C_U} = f(R_{C_N}) \right) \\ c_U: [P, V, M, C] \\ \Uparrow \\ c_{0:[D,W,I,C]} \end{array} \right]_{Light} \\ \left[U \rightarrow \begin{array}{c} M_3 : -\infty \le t \le \infty \\ \downarrow \\ M_2 : 0 \ge t > \infty \\ \downarrow \\ M_1 : 0 > t > \infty \\ \downarrow \\ t \le E_{Cell}; TC: M_3 \rightarrow U \\ t \sim E_{Human}; TC: U \rightarrow M_3 \end{array} \right]_{Time} \end{array} \right.
\left. \begin{array}{c} \left[\begin{array}{c} M_3 \rightarrow System_X \\ (\uparrow F \rightarrow I) \\ M_2 \rightarrow S_{System_X} \\ (\uparrow Sig \rightarrow F) \\ M_1 \rightarrow Sig_X \\ (\uparrow > P_x) \\ U \rightarrow x_{EM\ Spectrum} \end{array} \right]_{Space} \\ \\ TC \rightarrow x_{Bosons} \end{array} \right\|_{\langle x_U | x_T \rangle}
$$

$\Rightarrow$

$\langle Space - Time - Energy - Gravity\ Material - Fabric\ Pre\\ - Genetic\ Code \rangle +$

$(Electromagnetic\ Spectrum\ Material - Fabric\ Pre - Genetic\ Code) +$

$FBLEEE < \cdots > + Boson\ Material - Fabric\ Pre - Genetic\ Code$

$Eq\ 6.3.1:\ Boson\ Material\text{-}Fabric\ Pre\text{-}Genetic\ Code$

Starting with the Light-Matrix, the top left-hand matrix in (6.3.1), the first line from the top, $C_\infty: [Pr, Po, K, H]$, specifies the fundamental architecture of bosons. As explored previously in this chapter, Gluons are an emergence of Light's property of Knowledge, W and Z bosons are an emergence of Light's property of Power, Photons are an emergence of Light's property of Presence, and the Graviton is an emergence of Light's property of Harmony. The fundamental architecture of these aspects hence, is an emergence of the properties of Light at ∞.

Line 3 in the Light-Matrix, $C_K: [S_{Pr}, S_{Po}, S_K, S_H]$, elaborates the sets for Presence, Power, Knowledge, and Harmony, each containing multiple elements. For example, as will be explored in the section on photons, various elements derived from the four sets define the behavior of photons and could be functions such as 'pervasiveness', 'multiple wave handler', amongst others, hence collectively describing photons' way of being. Specifically, Line 5, $C_{N:} f(S_{Pr} \times S_{Po} \times S_K \times S_H)$, suggests that unique seeds are created from a combination of such elements from all four sets, with a particular element leading, that in effect creates the distinctness possible at the level of bosons.

Line 6, $(\downarrow R_{C_U} = f(R_{C_N}))$, specifies quantization between the layer where the seeds are formed, and the physical layer we are familiar with, and as explored in Chapter 3.5 and 3.6, will result in Line 7, $C_U: [P, V, M, C]$, hence changing the material-fabric of existence. Note that as in the process describing the generation of the code-segments for space-time-energy-gravity quantization at the time of the Big Bang, the generation of the FBLEE (four-base logic-encoding ecosystem) code-segment as specified by (4.1.5) and the MF (material-fabric) code segment as specified by (4.1.6) are combined together into one equation so that the FBLEE process is apparently transparent. Note also that in (6.2.1), following the '$\Rightarrow$' the FBLEEE <...> code-segment implies that four-base logic-encoding entanglement likely with the generation of the other quantum-particle FBLEE code-segments as specified by (6.3.1).

The possibilities represented by Lines 1 through 5 hence concretize through the quantization represented by Line 6 to become the bosons with its physical (related to Presence), vital (related to Power), mental (related to Knowledge), and connection (related to Harmony) aspects now existing in material reality typified by Light moving at c. Note that just as Line 6 represents a process of quantization relating the layer of reality created by Light traveling at c with the antecedent layers, so too Lines 2 and 4 as previously discussed, also represent quantization of a more subtle kind that ultimately plays a critical part in allowing the material-fabric to express infinite diversity.

Typically it is the process as captured by the Space-Matrix that will determine if Line 6 is activated. Specifically patterns at the untransformed layer, U, will need to be overcome, as specified by the second-line from the bottom of the Space-Matrix: $(\uparrow > P_x)$. But as specified by the bottom-line of the Time-Matrix, reproduced below, it is only with the advent of the human-system that the automaticity of the action of meta-levels is reversed:

$$U \to \begin{array}{l} t \le E_{Cell}; TC: M_3 \to U \\ t \sim E_{Human}; TC: U \to M_3 \end{array}$$

Hence in the case of bosons, which in this emergence is a pre-human system, the fact that patterns do not need to be overcome means that quantization happens automatically.

Even though bosons are a pre-human system and therefore the action of meta-levels are modeled as being automatic, it is useful to review (3.6.5), **Potential Effect of Levels of Light on Pre-Genetic or Genetic Information**, to understand how the relationship with different forms of mutation has been specified:

Potential Effect of Levels of Light on Pre Genetic or Genetic Information =

$$
\begin{bmatrix}
STATIC\ \langle\|[L][S][T]TC \to x_T|_{\langle x_U|x_T\rangle}\rangle \\
\times \\
\left((Y > U: Z_Q) \vee (Y \leq U: Z_F) \vee (Y = U: Z_R)\right) \\
\ni \\
\left(\begin{array}{c} Z \in \mathbb{U}\ (Space, Time, Energy, Gravity) \\ Q:\ Quantization;\ F:\ Fragmentation;\ R:\ Random \end{array}\right)
\end{bmatrix} \to h \ni
$$

$$
h \in \left(\begin{array}{c}
[Q]:\ Constructive\ zone, \\
[Q]:\ Constructive\ zone\ \wedge\ Constructive\ mutation, \\
[F]:\ Destructive\ mutation, \\
[R]:\ Random\ mutation
\end{array}\right)
$$

Line 1 from the top in the matrix is simply a static form of (3.6.1) the Simplified Light-Space-Time Emergence equation. The static form is designated by 'STATIC', and implies that fundamental operations true of (3.6.1) are being highlighted in (3.6.5). In other word (3.6.1) already has all the operations highlighted in (3.6.5) in it, but by 'freezing' it by making it static, the essential dynamics leading to possible mutations at the genetic level can more clearly be highlighted.

Line 1 is then subjected ($\times$) to a determination of the dominant levels of light that may be active, designated by Line 2, ' $\left((Y > U: Z_Q) \vee (Y \leq U: Z_F) \vee (Y = U: Z_R)\right)$ '. Unpacking this, 'Y > U' implies meta-levels are active and as a result it is possible that Z_Q is going to take place. This also implies activation and potential change of FBLEE. The call from below, as it were, may invoke some function that already exists in the subtle-libraries 'above', so that some already existing function may influence FBLEE through ∞ -entanglement, K-entanglement, or N-entanglement. The call in this case may be a need to allow interaction between the potential range of quantum particles in process of manifesting. Some exiting 'harmony-enhancing' meta-function already

240

mathematically worked out in the subtle-libraries may fit the call and precipitate to be part of the FBLEE creation or refinement process. This may be thought of as a key-and-lock mechanism, where a deep enough visceral urge from below acting as the key, opens an entangled lock to alter FBLEE as per the visceral urge. Such a call-response system at the pre-human level is considered to be relatively automatic as compared with the emergence of human-type systems because of the complexity of many more sources of action in the case of the latter.

$'Y \leq U'$ implies that only the untransformed levels are active, and therefore also the sub-level where the speed of light is 0 is active and as a result Z_F is going to take place. $'Y = U'$ implies that all levels are active and as a result Z_R is going to take place.

Line 3 elaborates the significance of Z_Q, Z_F, and Z_R. Hence Z is the union of potential quantum-operations of space, time, energy, and gravity, designated by $'Z \in \mathbb{U}(Space, Time, Energy, Gravity)'$. But the nature of the operations is designated by $'Q: Quantization; F: Fragmentation; R: Random'$. Z_Q, then, implies that the full quantization originating from updated four-base logic-encoding ecosystems (FBLEE) can take place, and will result in lasting material change at the pre-genetic level. Z_F implies that the essential set will be fragmented and that only local libraries a the level of the local material-fabric can potentially be altered. Z_R also precludes full quantization, and that some partial local-library constructive or destructive mutation may take place.

The $' \to h \ni h \in'$ segment resolves the outcome of the operations implied by Lines 1 – 3, suggesting that the outcome will be 'h' such that ($\ni$) 'h' is an element ($\in$) of the set specified by the members '[Q]: *Constructive zone*', '[Q]: *Constructive zone AND Constructive mutation*', '[F};

241

Destructive mutation', and *'{R}: Random mutation'*. *'[Q]:*
Constructive zone' implies that the in-built buffer has been
crossed and that access to the deeper four-base logic-
encoding ecosystem (FBLEE) has been granted. '[Q]' in
this segment implies that there is the possibility that full-
quantization as specified by Line 3 of the previous matrix
can take place. Access to this zone is a prerequisite for
constructive mutation to occur, as designated by the
element   '   *[Q}: Constructive zone ∩*
Constructive mutation ', which implies that full-
quantization is going to take place and will result in
material change. The '[F]' specifies the relationship
between 'Fragmentation' in Line 3 of the previous matrix
and destructive mutation. The '[R]' specifies the
relationship between 'Random' in Line 3 of the previous
matrix and random mutation.

But as just summarized in the Time-Matrix in (6.3.1) Y is
by definition greater than U and hence quantization is
automatic. In terms of bosons such quantization implies
that wholeness becomes fully active through specific
space, time, energy, and gravity quantization to create an
holistic "ecosystem" with its own "boson logic" as it
were. The wholeness has now precipitated into the
material-fabric and is available to be consciously and
unconsciously tapped into. This "logic" or more
specifically pre-genetic code is elaborated in the
following sub-sections.

Generation of Material-Fabric Pre-Genetic Photon Code

The photon is the carrier particle of the electromagnetic
force. But in the scheme of things the electromagnetic
force pervades everything and as explored in the section
on the electromagnetic spectrum appears to be
foundational to this reality we are in. So it could be said
that it is related to Presence or is an emergence of Light's
property of Presence at the level of quantum force-
carriers.

Hence, an equation for the photon, Equation 6.3.2 could be:

$$Sig_{photon} = Xa +$$

$$\overline{Yb_{0-n}} \quad where \quad \left[\begin{array}{c} X \in [S_{System_{Pr}}] \\ Y \in [S_{System_{Pr}}, S_{System_P}, S_{System_K}, S_{System_N}] \\ a, b \text{ are integers}; a > b \end{array} \right]$$

Eq 6.3.2: Signature of Photon

While the primary element X, would continue to be the same as X for bosons, as in Equation 6.1.4, there will be additional secondary elements Y that will further qualify the attributes or functionality of photons. As an example such elements could be 'pervasiveness', 'multiple wave handler', amongst others.

Note that (6.3.2) already implies that Lines 1 – 5 in the Light Matrix (6.3.1) have been activated, and that the logic of the photon-ecosystem will automatically precipitate into the material-fabric through the action of Line 6-7 of (6.3.1). This logic is none other than the pre-genetic code as specified by (6.3.2).

The gluon is the carrier particle of what is known as the strong nuclear force and holds quarks together in their inherently composite arrangements. But we have posited that quarks are related to or are an emergence of Knowledge at the quantum-particle level. Hence it must be that the gluon is an emergence of Light's property of Knowledge at the level of quantum force-carriers.

Hence, Equation 6.3.3 is an equation for the gluon:

$$Sig_{gluon} = Xa +$$

$$\overline{Yb_{0-n}} \quad where \quad \left[\begin{array}{c} X \in [S_{System_K}] \\ Y \in [S_{System_{Pr}}, S_{System_P}, S_{System_K}, S_{System_N}] \\ a, b \ are \ integers; a > b \end{array} \right]$$

Eq 6.3.3: Signature of Gluon

The primary element X would continue to be the same as in the equation for the boson. There are likely additional secondary elements Y such as 'concentrated', 'intense connection', amongst others that would collectively determine the behavior of gluons.

Note that (6.3.3) already implies that Lines 1 – 5 in the Light Matrix (6.3.1) have been activated, and that the logic of the gluon-ecosystem will automatically precipitate into the material-fabric through the action of Line 6-7 of (6.3.1). This logic is none other than the pre-genetic code as specified by (6.3.3).

Generation of Material-Fabric Pre-Genetic W and Z Bosons Code

The W and Z bosons are the carrier particle for the weak interactions, responsible for the decay of massive quarks and leptons into lighter quarks and leptons. But this usually is accompanied by the release of energy and power and so it must be that W and Z bosons are an emergence of Light's property of Power at the level of quantum force-carriers.

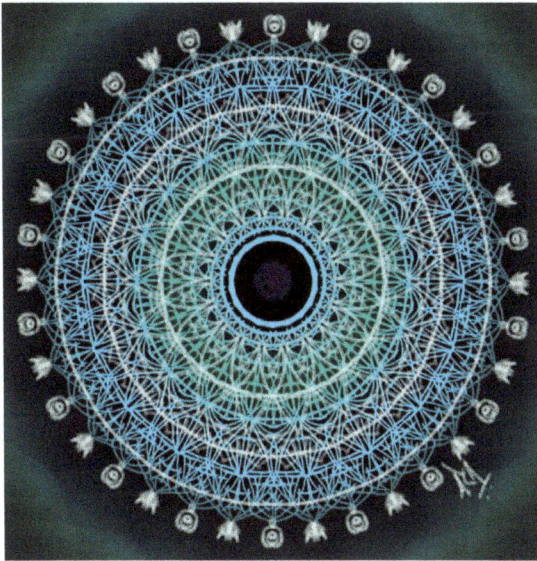

Hence, Equation 6.3.4, for W and Z bosons is as follows:

$$Sig_{W,Z\ bosons} = Xa +$$
$$\overline{Yb_{0-n}}\ where \begin{bmatrix} X \in [S_{System_P}] \\ Y \in [S_{System_{Pr}}, S_{System_P}, S_{System_K}, S_{System_N}] \\ a, b\ are\ integers; a > b \end{bmatrix}$$

Eq 6.3.4: Signature of W and Z Bosons

The primary element X would continue to be the same as in the equation for the boson. There are likely additional

247

secondary elements Y such as 'weak link', 'release of energy', amongst others that would collectively determine the behavior of W and Z bosons.

Note that (6.3.4) already implies that Lines 1 – 5 in the Light Matrix (6.3.1) have been activated, and that the logic of the W and Z-bosons-ecosystem will automatically precipitate into the material-fabric through the action of Line 6-7 of (6.3.1). This logic is none other than the pre-genetic code as specified by (6.3.4).

Generation of Material-Fabric Pre-Genetic Graviton Code

This leaves the hypothetical graviton that is thought to be the carrier particle for the force of gravity. But gravity, it is thought, is what holds astronomical objects together in relationship. Hence it must be that gravitons are an emergence of Light's property of Harmony at the level of quantum force-carriers.

Hence, Equation 6.3.5, for the hypothetical graviton is as follows:

$$Sig_{graviton} = Xa +$$
$$\overline{Yb_{0-n}} \quad where \begin{bmatrix} X \in [S_{System_N}] \\ Y \in [S_{System_{Pr}}, S_{System_P}, S_{System_K}, S_{System_N}] \\ a, b \ are \ integers; a > b \end{bmatrix}$$

Eq 6.3.5: Signature of Graviton

The primary element X would continue to be the same as in the equation for the boson. There are likely additional secondary elements Y such as 'mass curving space', 'space defining object movement', amongst others that would collectively determine the behavior of gravitons.

Note that (6.3.5) already implies that Lines 1 – 5 in the Light Matrix (6.3.1) have been activated, and that the logic of the graviton-ecosystem will automatically precipitate

into the material-fabric through the action of Line 6-7 of (6.3.1). This logic is none other than the pre-genetic code as specified by (6.3.5).

So we see that the underlying properties of light - Presence, Knowledge, Power, and Harmony - emerge as photons, gluons, W and Z bosons, and the hypothetical graviton at the quantum force-carrier level.

Summary of Material-Fabric Pre-Genetic Code at Boson Stage

Summarizing, after the boson stage iteration of the Light-Space-Time Emergence equation is complete, the following code-segments will have been generated as specified by Equation 6.3.6, Active Material-Fabric Pre-Genetic Code Segments at Boson Stage:

$$Light - Space - Time\ Emergence_{Boson} =$$

$$
\begin{Vmatrix}
\begin{bmatrix}
c_\infty : [Pr, Po, K, H] \\
\left(\downarrow R_{C_K} = f(R_{C_\infty}) \right) \\
c_K : [S_{Pr}, S_{Po}, S_K, S_H] \\
\left(\downarrow R_{C_N} = f(R_{C_K}) \right) \\
c_N : f(S_{Pr} \times S_{Po} \times S_K \times S_H) \\
\left(\downarrow R_{C_U} = f(R_{C_N}) \right) \\
c_U : [P, V, M, C] \\
\Uparrow \\
c_{0:[D,W,I,C]}
\end{bmatrix}_{Light}
&
\begin{bmatrix}
M_3 \rightarrow System_X \\
(\uparrow F \rightarrow I) \\
M_2 \rightarrow S_{System_X} \\
(\uparrow Sig \rightarrow F) \\
M_1 \rightarrow Sig_X \\
(\uparrow > P_x) \\
U \rightarrow x_{EM\ Spectrum}
\end{bmatrix}_{Space} \\
\begin{bmatrix}
U \rightarrow
\begin{matrix}
M_3 : -\infty \leq t \leq \infty \\
\downarrow \\
M_2 : 0 \geq t > \infty \\
\downarrow \\
M_1 : 0 > t > \infty \\
\downarrow \\
t \leq E_{Cell}; TC: M_3 \rightarrow U \\
t \sim E_{Human}; TC: U \rightarrow M_3
\end{matrix}
\end{bmatrix}_{Time}
&
TC \rightarrow x_{Bosons}
\end{Vmatrix}_{\langle x_U | x_T \rangle}
$$

$$\Rightarrow$$

$$\langle Space - Time - Energy - Gravity\ Material - Fabric\ Pre$$
$$- Genetic\ Code \rangle +$$

$$(ELectromagnetic\ Spectrum\ Material - Fabric\ Pre - Genetic\ Code)$$
$$+$$

$$FBLEEE < \cdots > +$$

$$\left(\begin{array}{c} Sig_{graviton} = Xa + \overline{Yb_{0-n}} \\ where \left[\begin{array}{c} X \in [S_{System_N}] \\ Y \in [S_{System_{Pr}}, S_{System_P}, S_{System_K}, S_{System_N}] \\ a, b\ are\ integers; a > b \end{array} \right] \\ Sig_{gluon} = Xa + \overline{Yb_{0-n}} \\ where \left[\begin{array}{c} X \in [S_{System_K}] \\ Y \in [S_{System_{Pr}}, S_{System_P}, S_{System_K}, S_{System_N}] \\ a, b\ are\ integers; a > b \end{array} \right] \\ Sig_{W,Z\ bosons} = Xa + \overline{Yb_{0-n}} \\ where \left[\begin{array}{c} X \in [S_{System_P}] \\ Y \in [S_{System_{Pr}}, S_{System_P}, S_{System_K}, S_{System_N}] \\ a, b\ are\ integers; a > b \end{array} \right] \\ Sig_{photon} = Xa + \overline{Yb_{0-n}} \\ where \left[\begin{array}{c} X \in [S_{System_{Pr}}] \\ Y \in [S_{System_{Pr}}, S_{System_P}, S_{System_K}, S_{System_N}] \\ a, b\ are\ integers; a > b \end{array} \right] \end{array} \right)$$

Eq. 6.3.6, *Active Material-Fabric Pre-Genetic Code Segments at Boson Stage*

250

ॐकारा गुरू तव शरणम्

Chapter 6.4: Generation of Material-Fabric, Pre-Genetic Atom Code

Quantum particles in turn create atoms, and all known atoms can be classified into one of four groups: The s-Group, p-Group, d-Group, and f-Group. But what are these groups and are they too emergences of Light as it continues its journey of crystalizing the possibility or potentiality or fourfold functionality within it?

As in the previous chapters, the Light-Space-Time Emergence equation (3.1.3) can be used to model emergence as it proceeds to this relatively more complex form of fourfold manifestation. (3.6.5), the Potential Effect of Levels of Light on Pre-Genetic of Genetic Information equation, basically sheds light on specific types of mutation that can occur and therefore on the nature of the code-segment that is being generated by (3.1.3). As discussed before, (3.6.5) shows a particular working of (3.1.3) but is wholly contained within it.

This chapter elaborates the action of (3.1.3) and (3.6.5) in the creation of the code-segments that define the logic of the atom ecosystem.

Light's Emergence as Atoms

As discussed previously the Light-Space-Time Emergence equation (3.1.3) being iterative, can be used to model emergence as it proceeds from simpler four-fold to more complex four-fold manifestations. Hence (3.1.3) has already been applied to suggest the emergence of the space-time-energy-gravity quadrumvirate, the electromagnetic spectrum, quantum particles in general, and bosons as a further instance of a particular kind of quantum particle. Here it will be applied to suggest the emergence of Atoms. But further, as implied by (3.6.5), the Potential Effect of Levels of Light on Pre-Genetic of Genetic Information equation, any process of organization, such as the architecture and cohesiveness of atoms, has the possibility of altering the material-fabric so long as the bases involved are driven primarily by a meta-level.

As suggested by Equation 6.4.1 (same as 4.2.2), Atoms Material-Fabric Pre-Genetic Code, the architecture and details of atoms can be seen to be the result of the application of the Light, Space, and Time matrices as will be elaborated:

$$Light - Space - Time\ Emergence_{Atoms} =$$

$$\left| \begin{bmatrix} c_\infty: [Pr, Po, K, H] \\ \left(\downarrow R_{C_K} = f\left(R_{C_\infty}\right) \right) \\ c_K: [S_{Pr}, S_{Po}, S_K, S_H] \\ \left(\downarrow R_{C_N} = f\left(R_{C_K}\right) \right) \\ c_N: f\left(S_{Pr} \times S_{Po} \times S_K \times S_H\right) \\ \left(\downarrow R_{C_U} = f\left(R_{C_N}\right) \right) \\ c_U: [P, V, M, C] \\ \Uparrow \\ c_{0:[D,W,I,C]} \end{bmatrix}_{Light} \begin{bmatrix} M_3 \rightarrow System_x \\ (\uparrow F \rightarrow I) \\ M_2 \rightarrow S_{System_x} \\ (\uparrow Sig \rightarrow F) \\ M_1 \rightarrow Sig_x \\ (\uparrow > P_x) \\ U \rightarrow x_{Quantum\ Particles} \end{bmatrix}_{Space} \right.$$

$$\left. \begin{bmatrix} M_3 : -\infty \le t \le \infty \\ \downarrow \\ M_2 : 0 \ge t > \infty \\ \downarrow \\ M_1 : 0 > t > \infty \\ \downarrow \\ U \rightarrow \begin{array}{l} t \le E_{Cell}; TC: M_3 \rightarrow U \\ t \sim E_{Human}; TC: U \rightarrow M_3 \end{array} \end{bmatrix}_{Time} \begin{array}{l} TC \rightarrow x_{Atoms} \end{array} \right|_{\langle x_U | x_T \rangle}$$

$$\Rightarrow$$

$$\langle Space - Time - Energy - Gravity\ Material - Fabric\ Pre - Genetic\ Code \rangle +$$

$$(Electromagnetic\ Spectrum\ Material - Fabric\ Pre - Genetic\ Code) +$$

$$(Quantum\ Particles\ Material - Fabric\ Pre - Genetic\ Code) +$$

Eq 6.4.1: Atoms Material-Fabric Pre-Genetic Code

Starting with the Light-Matrix, the top left-hand matrix in (6.4.1), the first line from the top, $C_\infty: [Pr, Po, K, H]$, specifies the fundamental architecture of atoms. As will be explored in this chapter, p-Group atoms are an emergence of Light's property of Knowledge, s-Group atoms are an emergence of Light's property of Power, d-Group atoms are an emergence of Light's property of Presence, and the f-Group atoms are an emergence of Light's property of Harmony. The fundamental architecture of these aspects hence, is an emergence of the properties of Light at ∞.

Line 3 in the Light-Matrix, $C_K: [S_{Pr}, S_{Po}, S_K, S_H]$, elaborates the sets for Presence, Power, Knowledge, and Harmony, each containing multiple elements. For example, as explored shortly in the section on the s-Group, various elements derived from the four sets define the behavior of s-Group atoms and could be functions such as 'power', 'energy', 'adventure', and 'courage', amongst others, hence collectively describing s-Groups atoms' way of being. Specifically, Line 5, $C_N: f(S_{Pr} \times S_{Po} \times S_K \times S_H)$, suggests that unique seeds are created from a combination of such elements from all four sets, with a particular element leading, that in effect creates the distinctness possible at the level of atoms.

Line 6, $(\downarrow R_{C_U} = f(R_{C_N}))$, specifies quantization between the layer where the seeds are formed, and the physical layer we are familiar with, and as explored in Chapter 3.5 and 3.6, will result in Line 7, $C_U: [P, V, M, C]$, hence changing the material-fabric of existence. Note that as in the process describing the generation of the code-segments for space-time-energy-gravity quantization at the time of the Big Bang, the generation of the FBLEE (four-base logic-encoding ecosystem) code-segment as specified by (4.1.5) and the MF (material-fabric) code segment as specified by (4.1.6) are combined together into one equation so that the FBLEE process is apparently transparent.

The possibilities represented by Lines 1 through 5 hence concretize through the quantization represented by Line 6 to become the atoms with its physical (related to Presence), vital (related to Power), mental (related to Knowledge), and connection (related to Harmony) aspects now existing in material reality typified by Light moving at c. Note that just as Line 6 represents a process of quantization relating the layer of reality created by Light traveling at c with the antecedent layers, so too Lines 2 and 4 as previously discussed, also represent quantization of a more subtle kind that ultimately plays a critical part in allowing the material-fabric to express infinite diversity.

Typically it is the process as captured by the Space-Matrix that will determine if Line 6 is activated. Specifically patterns at the untransformed layer, U, will need to be overcome, as specified by the second-line from the bottom of the Space-Matrix: $(\uparrow > P_x)$. But as specified by the bottom-line of the Time-Matrix, reproduced below, it is only with the advent of the human-system that the automaticity of the action of meta-levels is reversed:

$$U \rightarrow \begin{array}{l} t \leq E_{Cell}; \text{TC: } M_3 \rightarrow U \\ t \sim E_{Human}; \text{TC: } U \rightarrow M_3 \end{array}$$

Hence in the case of atoms, which in this emergence is a pre-human system, the fact that patterns do not need to be overcome means that quantization happens automatically.

Even though atoms are a pre-human system and therefore the action of meta-levels are modeled as being automatic, it is nonetheless useful to review (3.6.5), Potential Effect of Levels of Light on Pre-Genetic or

Genetic Information, to understand how the relationship with different forms of mutation has been specified:

Potential Effect of Levels of Light on Pre Genetic or Genetic Information =

$$
\begin{bmatrix}
STATIC\ \langle |[L][S][T]TC \to x_T|\langle x_U|x_T\rangle\rangle \\
\times \\
\left((Y > U\colon Z_Q) \lor (Y \le U\colon Z_F) \lor (Y = U\colon Z_R) \right) \\
\ni \\
\left(\begin{matrix} Z \in \mathbb{U}\ (Space, Time, Energy, Gravity) \\ Q\colon Quantization;\ F\colon Fragmentation;\ R\colon Random \end{matrix} \right)
\end{bmatrix} \to h \ni
$$

$$
h \in \left(\begin{matrix} [Q]\colon Constructive\ zone, \\ [Q]\colon Constructive\ zone\ \land\ Constructive\ mutation, \\ [F]\colon Destructive\ mutation, \\ [R]\colon Random\ mutation \end{matrix} \right)
$$

Line 1 from the top in the matrix is simply a static form of (3.6.1) the Simplified Light-Space-Time Emergence equation. The static form is designated by 'STATIC', and implies that fundamental operations true of (3.6.1) are being highlighted in (3.6.5). In other word (3.6.1) already has all the operations highlighted in (3.6.5) in it, but by 'freezing' it by making it static, the essential dynamics leading to possible mutations at the genetic level can more clearly be highlighted.

Line 1 is then subjected ($\times$) to a determination of the dominant levels of light that may be active, designated by Line 2, ' $\left((Y > U\colon Z_Q) \lor (Y \le U\colon Z_F) \lor (Y = U\colon Z_R) \right)$ '. Unpacking this, 'Y > U' implies meta-levels are active and as a result it is possible that Z_Q is going to take place. This also implies activation and potential change of FBLEE. The call from below, as it were, may invoke some function that already exists in the subtle-libraries 'above', so that some already existing function may influence FBLEE through ∞-entanglement, K-entanglement, or N-entanglement. In this case this may be the need to create a more elaborate functional basis for more of what is in Light to express itself materially. Hence the stage for the creation of 'atoms' is laid. Mathematically, such a possibility that combines the function of quantum particles together to create a larger material foundation in the form of atoms may exist at the level of some subtle-library, which then becomes the 'response' to the 'urge' from below. This may be thought of as a key-and-lock mechanism, where a deep enough visceral urge from below acting as the key, opens an entangled lock to alter FBLEE as per the visceral urge. Such a call-response system at the pre-human level is considered to be relatively automatic as compared with the emergence of human-type systems because of the complexity of many more sources of action in the case of the latter.

'$Y \leq U$' implies that only the untransformed levels are active, and therefore also the sub-level where the speed of light is 0 is active and as a result Z_F is going to take place. '$Y = U$' implies that all levels are active and as a result Z_R is going to take place.

Line 3 elaborates the significance of Z_Q, Z_F, and Z_R. Hence Z is the union of potential quantum-operations of space, time, energy, and gravity, designated by '$Z \in \mathbb{U}\,(Space, Time, Energy, Gravity)$'. But the nature of the operations is designated by 'Q: $Quantization; F$: $Fragmentation; R$: $Random$'. Z_Q, then, implies that the full quantization originating from updated four-base logic-encoding ecosystems (FBLEE) can take place, and will result in lasting material change at the pre-genetic level. Z_F implies that the essential set will be fragmented and that only local libraries a the level of the local material-fabric can potentially be altered. Z_R also precludes full quantization, and that some partial local-library constructive or destructive mutation may take place.

The '$\rightarrow h \ni h \in$' segment resolves the outcome of the operations implied by Lines 1 – 3, suggesting that the outcome will be 'h' such that ($\ni$) 'h' is an element ($\in$) of the set specified by the members '[Q]: *Constructive zone'*, '[Q]: *Constructive zone AND Constructive mutation'*, '[F]; *Destructive mutation'*, and '{R}: *Random mutation'*. '[Q]: *Constructive zone'* implies that the in-built buffer has been crossed and that access to the deeper four-base logic-encoding ecosystem (FBLEE) has been granted. '[Q]' in this segment implies that there is the possibility that full-quantization as specified by Line 3 of the previous matrix can take place. Access to this zone is a prerequisite for constructive mutation to occur, as designated by the element ' *[Q]: Constructive zone* ∩ *Constructive mutation* ', which implies that full-quantization is going to take place and will result in material change. The '[F]' specifies the relationship between 'Fragmentation' in Line 3 of the previous matrix and destructive mutation. The '[R]' specifies the relationship between 'Random' in Line 3 of the previous matrix and random mutation.

But as just summarized in the Time-Matrix in (6.4.1) Y is by definition greater than U and hence quantization is automatic. In terms of quantum particles such quantization implies that wholeness becomes fully active through specific space, time, energy, and gravity quantization to create an holistic "ecosystem" with its own "atoms logic" as it were. The wholeness has now precipitated into the material-fabric and is available to be consciously and unconsciously tapped into. This "logic" or more specifically pre-genetic code is elaborated in the following sub-sections.

Generation of Material-Fabric Pre-Genetic s-Group Code

The s-Group consists primarily of what are known as alkali metals and alkali earth metals. These alkali metals are known to easily lose electrons and form what is known as positive ions. When they lose electrons energy is gained, but when the electrons are taken up by other atoms in proximity there is a lot of energy released. Some have referred to these groups as "violent worlds" (Tweed, 2003), and it has been pointed out that stars shine because they are transmuting vast amount of hydrogen into helium, both of which are s-Group elements. So one gets the sense that the s-Group may be an emergence of Light's property of Power.

Stars and suns are also known to be the crucibles where all the different kinds of atoms are created. So these furnaces of power by virtue of their heat and high pressure are able to force electrons and protons and neutrons to come together to create all the different types of atoms known in the universe.

But philosophically what are the s-Group atoms or elements? These are atoms where there is an equal likelihood of an electron being anywhere in a symmetrical sphere around the nucleus. The other groups are all similarly defined by likelihoods of electrons being within a possible pattern around the nucleus. The patterns that distinguishes s-Group atoms is a sphere, and since all other patterns can be thought of as occurring within some sphere, in some sense this is like an imprint or precipitation or emergence that allows other kinds of emergences to surface within it. So the elements that are part of the s-Group may be thought of as the adventurers with courage who venture into a brave new world to create some foundation by which all other element-creations can follow.

The fact that hydrogen and helium are known to constitute 98% of the Universe (Heiserman, 1991) relative to other elements therefore makes sense in this view, especially since hydrogen and helium provide the fuel with which the star-furnaces manufacture all other elements.

So s-Group elements seem to embody functions such as power, energy, adventure, courage, and can be thought of as an emergence of the property of Light to do with Power.

Hence, a series of equations linked to S_{System_P} as the prime set can be suggested, starting with the s-Group mapping, as in Equation 6.4.2:

$$Element_{s-Group} = Xa + \overline{Yb_{0-n}} \quad where \left[\begin{array}{c} X \in [S_{System_P}] \\ Y \in [S_{System_{Pr}}, S_{System_P}, S_{System_K}, S_{System_N}] \\ a, b \ are \ integers; a > b \end{array} \right]$$

Eq 6.4.2: s-Group Element

Note that (6.4.2) already implies that Lines 1 – 5 in the Light Matrix (6.4.1) have been activated, and that the logic of the s-Group-ecosystem will automatically precipitate into the material-fabric through the action of Line 6-7 of (6.4.1). This logic is none other than the pre-genetic code as specified by (6.4.2).

Further, the equivalent mapping between traditional element groupings and S_{System_P} as in Equations 6.4.3 and 6.4.4 can be specified:

$$Element_{Alkali \ metal} =$$

$$Xa +$$

$$\overline{Yb_{0-n}} \quad where \left[\begin{array}{c} X \in [S_{System_P}] \\ Y \in [S_{System_{Pr}}, S_{System_P}, S_{System_K}, S_{System_N}] \\ a, b \ are \ integers; a > b \end{array} \right]$$

Eq 6.4.3: Alkali Metal Element

$$Element_{Alkali \ earth \ metal} =$$

260

$$Xa +$$

$$\overline{Yb_{0-n}} \quad where \quad \begin{bmatrix} X \in [S_{System_P}] \\ Y \in [S_{System_{Pr}}, S_{System_P}, S_{System_K}, S_{System_N}] \\ a, b \ are \ integers; a > b \end{bmatrix}$$

Eq 6.4.4: Alkali Earth Metal Element

Further, as a representative element belonging to the Alkali Metal group the equation for Lithium (Li), as in Equation 6.4.5, would be:

$$Element \ _{Lithium} = Xa +$$

$$\overline{Yb_{0-n}} \quad where \quad \begin{bmatrix} X \in [S_{System_P}] \\ Y \in [S_{System_{Pr}}, S_{System_P}, S_{System_K}, S_{System_N}] \\ a, b \ are \ integers; a > b \end{bmatrix}$$

Eq 6.4.5: Lithium

Note that the primary element X in all these cases may be a function or attribute along the lines of 'expresses power'. The secondary elements Y in all these would have some elements being the same, and then would have specific differences as one gets into sub-classes and the individual elements themselves.

Generation of Material-Fabric Pre-Genetic p-Group Code

Atoms or elements belonging to the p-Group are those with the likelihood of electrons occurring equally on either side of the nucleus, like a dumbbell of sorts.

There are some very significant elements in this group that are part of the metal, metalloid, non-metal, halogen, and noble gas sub-groupings. Carbon, Nitrogen, Oxygen, and Silicon are some of the sample elements. Looking at the types of elements present in this group it is as though all the element possibilities have been represented within it. It is perhaps that the possibility of ideas behind all elements has emerged in this group and one can hypothesize that this group may be a reflection of the property of Knowledge, forming archetypes from which all other elements are created.

Philosophically, the one spherical shell or probability cloud (s) becoming two shells like a dumbbell (p) signifies the creation of an essential polarity within a unit space. So in a sense the one becoming two is the first instance of variability in space. The two, spaced in 3-dimensions around the nucleus in a sense create six switches becoming an attractor or allowing a vaster number of different kinds of elements to surface. So there is a sense of the 'idea' of the element that comes into focus.

But further, the essential elements that allow both thinking and virtual thinking machines to come into being, are also contained within this group. Carbon is the basis of DNA and of all life. The fact that Silicon, directly below it in the periodic table and therefore sharing essential qualities, is considered the basis of all virtual thinking machines is therefore perhaps significant and may reinforce the notion that the p-Group is a precipitation of Light's property of Knowledge.

Hence, a series of equations linked to S_{System_K} as the prime set can be suggested, starting with the p-Group mapping, as in Equation 6.4.6:

$$Element_{p-Group} = Xa +$$
$$\overline{Yb_{0-n}} \ where \ \left[\begin{array}{c} X \in [S_{System_K}] \\ Y \in [S_{System_{Pr}}, S_{System_P}, S_{System_K}, S_{System_N}] \\ a, b \ are \ integers; a > b \end{array} \right]$$

Eq 6.4.6: p-Group Element

Note that (6.4.6) already implies that Lines 1 – 5 in the Light Matrix (6.4.1) have been activated, and that the logic of the p-Group-ecosystem will automatically precipitate

into the material-fabric through the action of Line 6-7 of (6.4.1). This logic is none other than the pre-genetic code as specified by (6.4.6).

Further, the equivalent mapping between traditional element groupings and S_{System_K} as in Equations 6.4.7 through 6.4.12 can also be specified:

$$Element\ _{Metal} = Xa + \overline{Yb_{0-n}}$$

$$where\ \begin{bmatrix} X \in [S_{System_K}] \\ Y \in [S_{System_{Pr}}, S_{System_P}, S_{System_K}, S_{System_N}] \\ a, b\ are\ integers; a > b \end{bmatrix}$$

Eq 6.4.7: Metal Element

$$Element\ _{Metalloid} = Xa +$$
$$\overline{Yb_{0-n}}\ \ where\ \begin{bmatrix} X \in [S_{System_K}] \\ Y \in [S_{System_{Pr}}, S_{System_P}, S_{System_K}, S_{System_N}] \\ a, b\ are\ integers; a > b \end{bmatrix}$$

Eq 6.4.8: Metalloid Element

$$Element\ _{Non-Metal} = Xa +$$
$$\overline{Yb_{0-n}}\ \ where\ \begin{bmatrix} X \in [S_{System_K}] \\ Y \in [S_{System_{Pr}}, S_{System_P}, S_{System_K}, S_{System_N}] \\ a, b\ are\ integers; a > b \end{bmatrix}$$

Eq 6.4.9: Non-Metal Element

$$Element\ _{Halogen} = Xa +$$
$$\overline{Yb_{0-n}}\ \ where\ \begin{bmatrix} X \in [S_{System_K}] \\ Y \in [S_{System_{Pr}}, S_{System_P}, S_{System_K}, S_{System_N}] \\ a, b\ are\ integers; a > b \end{bmatrix}$$

Eq 6.4.10: Halogen Element

$$Element_{\,Noble\ Gas} = Xa +$$

$$\overline{Yb_{0-n}}\ where \left[\begin{array}{c} X \in [S_{System_K}] \\ Y \in [S_{System_{Pr}}, S_{System_P}, S_{System_K}, S_{System_N}] \\ a, b\ are\ integers; a > b \end{array} \right]$$

Eq 6.4.11: Noble Gas Element

Further, as a representative element belonging to the Non-Metal group the equation for Carbon (C), as in Equation 6.4.12, would be:

$$Element_{\,Carbon} = Xa +$$

$$\overline{Yb_{0-n}}\ where \left[\begin{array}{c} X \in [S_{System_K}] \\ Y \in [S_{System_{Pr}}, S_{System_P}, S_{System_K}, S_{System_N}] \\ a, b\ are\ integers; a > b \end{array} \right]$$

Eq 6.4.12: Carbon

Note that the primary element X in all these cases may be a function or attribute along the lines of 'knowledge'. The secondary elements Y in all these would have some elements being the same, and then would have specific differences as one gets into sub-classes and the individual elements themselves.

Generation of Material-Fabric Pre-Genetic d-Group Code

The d-Group comprises the Transition Metals. These metals are generally hard and strong, exhibit corrosive resistance, and can be thought of as workhorse elements. Many industrial and well-known elements sit in this group: Titanium, Chromium, Manganese, Iron, Cobalt, Nickel, Copper, Zinc, Silver, Platinum, and Gold, amongst others.

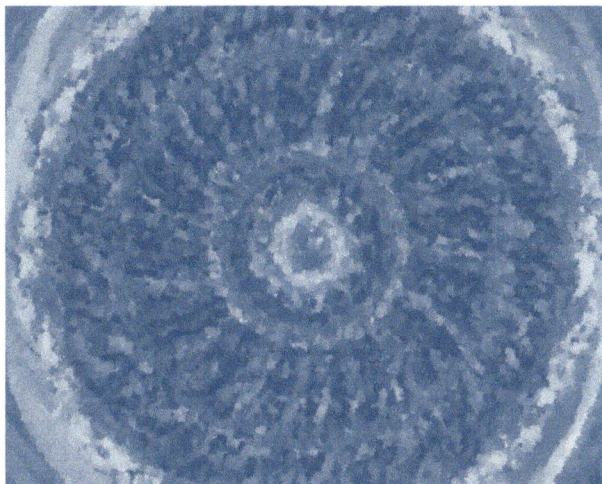

The elements and atoms in the d-Group are equally likely to show up in four possible lobes or probability-spaces around the nucleus. Four lobes occurring in multiple possible planes around the nucleus will likely create a space of stability, since there is a possibility of four lobes creating the four vertices of a tetrahedron (Fuller, 1982) that has been positioned as one of the most stable shapes in the universe.

Much of the constructed world around us is created from these elements. Further, most of the series in the group easily lose one or more electrons thereby easily combining with other atoms to form a vast array of compounds. Also, looking more broadly at the function of these elements, it can be seen that these metals exist for service, to help bring about perfection in the constructed world, to help much of the machinery in which they are used, and to assist the processes dependent on them to be completed with diligence. Hence, these transition metals appear to be an emergence of Light's property of Presence.

Therefore, a series of equations linked to $S_{System_{Pr}}$ as the prime set can be suggested, starting with the d-Group mapping, as in Equation 6.4.13:

$$Element_{d-Group} = Xa + \overline{Yb_{0-n}} \quad where \begin{bmatrix} X \in [S_{System_{Pr}}] \\ Y \in [S_{System_{Pr}}, S_{System_P}, S_{System_K}, S_{System_N}] \\ a, b \text{ are integers}; a > b \end{bmatrix}$$

Eq 6.4.13: d-Group Element

Note that (6.4.13) already implies that Lines 1 – 5 in the Light Matrix (6.4.1) have been activated, and that the logic of the d-Group-ecosystem will automatically precipitate into the material-fabric through the action of Line 6-7 of (6.4.1). This logic is none other than the pre-genetic code as specified by (6.4.13).

Further, the equivalent mapping between traditional element groupings and $S_{System_{Pr}}$, as in Equation 6.4.14, can also be specified:

$$Element_{\,Transition\,Metal} =$$

$$Xa +$$

$$\overline{Yb_{0-n}} \;\; where \; \left[\begin{array}{c} X \in [S_{System_{Pr}}] \\ Y \in [S_{System_{Pr}}, S_{System_P}, S_{System_K}, S_{System_N}] \\ a, b \; are \; integers; a > b \end{array} \right]$$

Eq 6.4.14: Transition Metal Element

Further, as a representative element belonging to the Transition-Metal group the equation for Gold (Au), as in Equation 6.4.15, would be:

$$Element_{\,Gold} = Xa +$$

$$\overline{Yb_{0-n}} \;\; where \; \left[\begin{array}{c} X \in [S_{System_{Pr}}] \\ Y \in [S_{System_{Pr}}, S_{System_P}, S_{System_K}, S_{System_N}] \\ a, b \; are \; integers; a > b \end{array} \right]$$

Eq 6.4.15: Gold

Note that the primary element X in all these cases may be a function or attribute along the lines of 'work horse'. The secondary elements Y in all these would have some elements being the same, and then would have specific differences as one gets into sub-classes and the individual elements themselves.

Generation of Material-Fabric Pre-Genetic f-Group Code

The f-Group comprises of the Lanthanides and Actinides. Philosophically, elements in the f-Group consist of 6 lobes around the nucleus within which an electron may be found. 6 lobes will exist in multiple planes around the nucleus and suggests the notion of extended relationship and collectivity: the attempt to build larger and larger bonds within a small space. Considering this it is likely that the f-Group is an emergence of Light's property of Harmony.

Thinking about Lanthanides, some interesting facts may reinforce this notion. First, the spin of electrons in a lanthanides' outer shell is aligned, creating a strong magnetic field. The notion of creating a strong magnetic field seems to be consistent with the notion of engendering a collectivity through the ordered attraction and repulsion of elements. Second, these elements curiously occur together in nature often in the same ores (Gray, 2009) and are chemically interchangeable also suggesting the notion of forming a tight intra-group collectivity.

Actinides on the other hand are inherently radioactive. This implies that these elements have inherently crossed a threshold of stability and have the urge, over their own half-lives, to decompose into other elements. This natural urge may suggest some boundary conditions on the notion of collectivity and nurturing, giving additional insight into the nature of collectivity and nurturing. Further, the entire actinide group, as opposed to the lanthanide group that is inherently stable, is unstable. It is curious that both these should be part of the f-Group, and they must provide insight into boundary conditions into the notion of collectivity in elements.

Hence, a series of equations linked to S_{System_N} as the prime set can be suggested, starting with the f-Group mapping, as in Equation 6.4.16:

$$Element_{f-Group} = Xa +$$

$$\overline{Yb_{0-n}} \quad where \quad \begin{bmatrix} X \in [S_{System_N}] \\ Y \in [S_{System_{Pr}}, S_{System_P}, S_{System_K}, S_{System_N}] \\ a, b \ are \ integers; a > b \end{bmatrix}$$

Eq 6.4.16: f-Group Element

Note that (6.4.16) already implies that Lines 1 – 5 in the Light Matrix (6.4.1) have been activated, and that the logic of the f-Group-ecosystem will automatically precipitate

into the material-fabric through the action of Line 6-7 of (6.4.1). This logic is none other than the pre-genetic code as specified by (6.4.16).

Further, the equivalent mapping between traditional element groupings and S_{System_N}, as in Equations 6.4.17 and 6.4.18, can also be specified:

$$Element_{Lanthanide} = Xa + \overline{Yb_{0-n}} \quad where \begin{bmatrix} X \in [S_{System_N}] \\ Y \in [S_{System_{Pr}}, S_{System_P}, S_{System_K}, S_{system_N}] \\ a, b \ are \ integers; a > b \end{bmatrix}$$

Eq 6.4.17: Lanthanide Element

$$Element_{Actinide} = Xa + \overline{Yb_{0-n}} \quad where \begin{bmatrix} X \in [S_{System_N}] \\ Y \in [S_{System_{Pr}}, S_{System_P}, S_{System_K}, S_{system_N}] \\ a, b \ are \ integers; a > b \end{bmatrix}$$

Eq 6.4.18: Actinide Element

Further, as a representative element belonging to the Lanthanide group the equation for Lanthanum (La), as in Equation 6.3.19, would be:

$$Element_{Lanthanum} = Xa + \overline{Yb_{0-n}}$$

$$where \begin{bmatrix} X \in [S_{System_N}] \\ Y \in [S_{System_{Pr}}, S_{System_P}, S_{System_K}, S_{system_N}] \\ a, b \ are \ integers; a > b \end{bmatrix}$$

Eq 6.4.19: Lanthanum

Note that the primary element X in all these cases may be a function or attribute along the lines of 'experiments in collectivity'. The secondary elements Y in all these would have some elements being the same, and then would have specific differences as one gets into sub-classes and the individual elements themselves.

Summary of Material-Fabric Pre-Genetic Code at Atoms Stage

Summarizing, after the atoms stage iterations of the Light-Space-Time Emergence equation are complete, the following code-segments will have been generated as specified by Equation 6.4.20, Active Material-Fabric Pre-Genetic Code Segments at Atoms Stage:

$$Light - Space - Time\ Emergence_{Atoms} =$$

$$
\left|
\begin{bmatrix}
C_\infty: [Pr, Po, K, H] \\
\left(\downarrow R_{C_K} = f(R_{C_\infty}) \right) \\
c_K: [S_{Pr}, S_{Po}, S_K, S_H] \\
\left(\downarrow R_{C_N} = f(R_{C_K}) \right) \\
c_N: f(S_{Pr} \times S_{Po} \times S_K \times S_H) \\
\left(\downarrow R_{C_U} = f(R_{C_N}) \right) \\
c_U: [P, V, M, C] \\
\Uparrow \\
C_{0:[D,W,I,C]}
\end{bmatrix}_{Light}
\begin{bmatrix}
M_3 \rightarrow System_X \\
(\uparrow F \rightarrow I) \\
M_2 \rightarrow S_{System_X} \\
(\uparrow Sig \rightarrow F) \\
M_1 \rightarrow Sig_X \\
(\uparrow > P_X) \\
U \rightarrow X_{Quantum\ Particles}
\end{bmatrix}_{Space}
\begin{bmatrix}
U \rightarrow
\begin{bmatrix}
M_3 : -\infty \leq t \leq \infty \\
\downarrow \\
M_2 : 0 \geq t > \infty \\
\downarrow \\
M_1 : 0 > t > \infty \\
\downarrow \\
t \leq E_{Cell}; TC: M_3 \rightarrow U \\
t \sim E_{Human}; TC: U \rightarrow M_3
\end{bmatrix}_{Time}
\end{bmatrix}
\begin{matrix} \\ \\ TC \rightarrow X_{Atoms} \\ \\ \end{matrix}
\right|_{\langle x_U | x_T \rangle}
$$

$$\Rightarrow$$

$$\langle Space - Time - Energy - Gravity\ Material - Fabric\ Pre - Genetic\ Code \rangle +$$

$$(Electromagnetic\ Spectrum\ Material - Fabric\ Pre - Genetic\ Code) +$$

$$(Quantum\ Particles\ Material - Fabric\ Pre - Genetic\ Code) +$$

271

$$
\left(
\begin{array}{l}
\quad \sum Element_{f-Group} = Xa + \overline{Yb_{0-n}} \\
\qquad\quad X \in [S_{System_N}] \\
where\ \left[Y \in [S_{System_{Pr}}, S_{System_P}, S_{System_K}, S_{System_N}] \right] \\
\qquad\quad a, b\ are\ integers;\ a > b \\
\quad \sum Element_{p-Group} = Xa + \overline{Yb_{0-n}} \\
\qquad\quad X \in [S_{System_K}] \\
where\ \left[Y \in [S_{System_{Pr}}, S_{System_P}, S_{System_K}, S_{System_N}] \right] \\
\qquad\quad a, b\ are\ integers;\ a > b \\
\quad \sum Element_{s-Group} = Xa + \overline{Yb_{0-n}} \\
\qquad\quad X \in [S_{System_P}] \\
where\ \left[Y \in [S_{System_{Pr}}, S_{System_P}, S_{System_K}, S_{System_N}] \right] \\
\qquad\quad a, b\ are\ integers;\ a > b \\
\quad \sum Element_{d-Group} = Xa + \overline{Yb_{0-n}} \\
\qquad\quad X \in [S_{System_{Pr}}] \\
where\ \left[Y \in [S_{System_{Pr}}, S_{System_P}, S_{System_K}, S_{System_N}] \right] \\
\qquad\quad a, b\ are\ integers;\ a > b
\end{array}
\right)
$$

Eq. 6.4.20, Active Material-Fabric Pre-Genetic Code Segments at Atoms Stage

The '$\sum x$' signifies all the possibilities for the 'x' equation-segment.

SECTION 7:   GENERATING THE GENETIC CODE FOR LIFE

This section explores the quantum-level computation and the generation of genetic code that is associated with the emergence of life through the cell, complex human attributes such as thoughts and feelings, and uniqueness of individuality.

So far the action of the Light-Space-Time Emergence has generated pre-genetic code. After the Big Bang this pre-genetic code is housed in the material-fabric, which has a universal action on all constructs arising in the universe. The emergence of matter, as discussed in Section 6, precipitates a phenomenon of material containerization. This was hypothesized as being due to an essential action of the space-time-energy-gravity quadrumvirate being contained within a smaller "space" as a result of the number of driver seeds.

In this section it is assumed that just as the space-time-energy-gravity quadrumvirate has an impact on every material emergence, so too will the action of the electromagnetic spectrum have an action on every emergence of life. Some dynamics of this possibility are explored in Chapter 7.1.

Chapter 7.2 will explore the generation of the genetic code responsible for the emergence and complexification of four-foldness through the primary molecular plans of nucleic acids, proteins, lipids, and polysaccharides.

Chapter 7.3 will relate key human attributes of sensations, urges, desires, wills, feelings, emotions, and thought to the continuing journey of fourfold complexification and suggest that genetic code is also generated to support these attributes.

Chapter 7.4 will relate truer individuality to the fourfold properties implicit in Light and suggest the generation of the genetic code tied to this.

Chapter 7.1: Effect of Electromagnetic Spectrum on Life

As explored in Section 5, while there is a significant volume of material-fabric pre-genetic code generated for the electromagnetic spectrum, this chapter will focus on some aspects of the visible-light portion of the electromagnetic spectrum that is known to have a direct impact on life.

Research indicates that in the visible light spectrum specified by light with a wavelength between 380 nanometers to 760 nanometers, each color has specific physical and psychological properties (Martel, 2018). For example, colors in the spectrum that range from red to yellow, are known as warm colors and generally are known to be stimulating, fortifying, and energizing. Colors in the spectrum that range from green to violet are known as cool colors and generally are known to be calming, sedative, and analgesic.

This suggests that there is a specific set of functions associated with each color that conceivably uses that color as a vehicle to express or concretize itself. It further suggests that there are segments of electromagnetic spectrum related code that may be fundamental to all of life and are likely deeply embedded in or influence all of life. This is evident in that colors are known to generate specific responses in human bodies. The equilibrium of the autonomic nervous system, for example, which is related to the equilibrium of the sympathetic and parasympathetic nervous system, can be influenced by the application of warm and cool colors (Spitler, 2011).

Research into generic properties associated with colors (Van Obberghen, 2007; Deppe, 2013) suggests the specific functions related to the wave-archetype or structure, the energy-profile, and the mass-potential for each of the colors.

The speed or magnetic profile of each color is assumed to be that of c as derived in Chapter 5.1. Hence as in (5.1.2) all colors will be assumed to have the magnetic properties associated with light moving at c:

$$EM\_Spectrum_{Speed} = Xa +$$

$$\overline{Yb_{0-n}} \quad where \quad \begin{bmatrix} X \in [S_{System_N}] \\ Y \in [S_{System_{Pr}}, S_{System_P}, S_{System_K}, S_{System_N}] \\ a, b \ are \ integers; a > b \end{bmatrix}$$

Red

Red is associated with the wavelength 635 to 760 nanometers.

In its association with life the energy-profile of 'red' is suggested to engender qualities to do with 'tonifying', 'warming', 'antianemia', 'antimigraine', amongst possible others. Assuming that the primary or x-element has to do with tonifying, then various y-elements in combination can create the other engendered qualities. These relationships may be depicted in Equation 7.1.1, Red Energy, in the following manner:

$$Red_{Energy} =' tonifying'.a + \overline{Yb_{0-n}}$$

$$where \begin{bmatrix} tonifying \in [S_{System_P}] \\ Y \in [S_{System_{Pr}}, S_{System_P}, S_{System_K}, S_{System_N}] \\ a, b \ are \ integers; a > b \end{bmatrix}$$

Eq 7.1.1: Red Energy

The wave-archetype or intent has to do with a primary or x-element associated with the function 'passion'. Other y-elements in combination will engender intent such as 'extroversion'. Hence, these relationships may be depicted by Equation 7.1.2, Red Intent:

$$Red_{Intent} =' passion'.a + \overline{Yb_{0-n}}$$

$$where \begin{bmatrix} passion \in [S_{System_K}] \\ Y \in [S_{System_{Pr}}, S_{System_P}, S_{System_K}, S_{System_N}] \\ a, b \ are \ integers; a > b \end{bmatrix}$$

Eq 7.1.2: Red Intent

The mass-potential or body-part most linked to red is suggested to be the 'reproductive system', the 'urogenital system', and 'system to do with blood pressure', amongst others. So there is likely a build up of mass in such a manner that facilitates the creation of these systems. An assumption can be made that there is a 'red-mass' x-element and perhaps other y-elements that yield the different systems when built up. Equation 7.1.3, Reproductive System – Red Mass Possibility, suggests how this can be depicted:

$$Reproductive \ System - Red_{Mass_{Possibility}} =' red - mass'.a + \overline{Yb_{0-n}}$$

$$where \begin{bmatrix} 'red - mass' \in [S_{System_{Pr}}] \\ Y \in [S_{System_{Pr}}, S_{System_P}, S_{System_K}, S_{System_N}] \\ a, b \ are \ integers; a > b \end{bmatrix}$$

Eq 7.1.3: Reproductive System - Red Mass Possibility

Note that (7.1.1 – 3) are proposed to be part of the material-fabric. As such they are of a pre-genetic type but also inform or even help form the cell-based genetic code.

Orange

Orange is associated with the wavelength 590 to 635 nanometers.

In its association with life the energy-profile of 'orange' is suggested to engender qualities to do with 'invigorating' and 'adrenal stimulant' amongst possible others. Assuming that the primary or x-element has to do with 'invigorating', then various y-elements in combination can create the other engendered qualities. These relationships may be depicted in Equation 7.1.4, Orange Energy, in the following manner:

$$Orange_{Energy} =' invigorating'.a + \overline{Yb_{0-n}}$$

$$where \begin{bmatrix} invigorating \in [S_{System_P}] \\ Y \in [S_{System_{Pr}}, S_{System_P}, S_{System_K}, S_{System_N}] \\ a, b \ are \ integers; a > b \end{bmatrix}$$

Eq 7.1.4: Orange Energy

The wave-archetype or intent has to do with a primary or x-element associated with the function 'joy'. Other y-elements in combination will engender intent such as 'antidepressant' and 'sociability'. Hence, these relationships may be depicted by Equation 7.1.5, Orange Intent:

$$Orange_{Intent} =' joy'.a + \overline{Yb_{0-n}}$$

$$where \begin{bmatrix} joy \in [S_{System_K}] \\ Y \in [S_{System_{Pr}}, S_{System_P}, S_{System_K}, S_{System_N}] \\ a, b\ are\ integers; a > b \end{bmatrix}$$

Eq 7.1.5: Orange Intent

The mass-potential or body-part most linked to orange is suggested to be the 'adrenal glands', 'vasomotricity', and 'musculoskeletal system', amongst possible others. So there is likely a build up of mass in such a manner that facilitates the creation of these systems. An assumption can be made that there is a 'orange-mass' x-element and perhaps other y-elements that yield the different systems when built up. Equation 7.1.6, Adrenal Gland - Orange Mass Possibility, suggests how this can be depicted:

$$Adrenal\ Glands - Orange_{Mass_{Possibility}} =' orange - mass'.a \\ + \overline{Yb_{0-n}}$$

$$where \begin{bmatrix} 'orange - mass' \in [S_{System_{Pr}}] \\ Y \in [S_{System_{Pr}}, S_{System_P}, S_{System_K}, S_{System_N}] \\ a, b\ are\ integers; a > b \end{bmatrix}$$

Eq 7.1.6: Adrenal Glands - Orange Mass Possibility

Note that (7.1.4 – 6) are proposed to be part of the material-fabric. As such they are of a pre-genetic type but also inform or even help form the cell-based genetic code.

Yellow

Yellow is associated with the wavelength 570 to 590 nanometers.

In its association with life the energy-profile of 'yellow' is suggested to engender qualities to do with 'digestive stimulant', and 'lymphatic stimulant' amongst possible others. Assuming that the primary or x-element has to do with 'stimulant', then various y-elements in combination can create the other engendered qualities.   These relationships may be depicted in Equation 7.1.7, Yellow Energy, in the following manner:

$$Yellow_{Energy} =' stimulant'.a + \overline{Yb_{0-n}}$$

$$where \begin{bmatrix} stimulant \in [S_{System_P}] \\ Y \in [S_{System_{Pr}}, S_{System_P}, S_{System_K}, S_{System_N}] \\ a, b \text{ are integers}; a > b \end{bmatrix}$$

Eq 7.1.7: Yellow Energy

The wave-archetype or intent has to do with a primary or x-element associated with the function 'optimism'. then various y-elements in combination can create the other engendered qualities. Hence, these relationships may be depicted by Equation 7.1.8, Yellow Intent:

$$Yellow_{Intent} =' optimism'.a + \overline{Yb_{0-n}}$$

$$where \begin{bmatrix} optimism \in [S_{System_K}] \\ Y \in [S_{System_{Pr}}, S_{System_P}, S_{System_K}, S_{System_N}] \\ a, b \text{ are integers}; a > b \end{bmatrix}$$

Eq 7.1.8: Yellow Intent

The mass-potential or body-part most linked to yellow is suggested to be the 'digestive system', the 'intestine', the 'spleen', the 'pancreas' amongst possible others. So there is likely a build up of mass in such a manner that facilitates the creation of these systems. An assumption can be made that there is a 'yellow-mass' x-element and perhaps other y-elements that yield the different systems when built up. Equation 7.1.9, Digestive System - Yellow Mass Possibility, suggests how this can be depicted:

$$Digestive\ System - Yellow_{Mass_{Possibility}} =' yellow - mass'.a + \overline{Yb_{0-n}}$$

$$where \begin{bmatrix} 'yellow - mass' \in [S_{System_{Pr}}] \\ Y \in [S_{System_{Pr}}, S_{System_{P}}, S_{System_{K}}, S_{System_{N}}] \\ a, b \ are \ integers; a > b \end{bmatrix}$$

Eq 7.1.9: Digestive System - Yellow Mass Possibility

Note that (7.1.7 – 9) are proposed to be part of the material-fabric. As such they are of a pre-genetic type but also inform or even help form the cell-based genetic code.

Lemon

Lemon is associated with the wavelength 550 to 570 nanometers.

In its association with life the energy-profile of 'lemon' is suggested to engender qualities to do with 'antiallergy', 'detoxifying', and 'addresses chronic conditions', amongst possible others. Assuming that the primary or x-element has to do with 'detoxifying', then various y-elements in combination can create the other engendered qualities Energy, in the following manner:

$$Red_{Energy} =' detoxifying'.a + \overline{Yb_{0-n}}$$

$$where \begin{bmatrix} detoxifying \in [S_{System_{P}}] \\ Y \in [S_{System_{Pr}}, S_{System_{P}}, S_{System_{K}}, S_{System_{N}}] \\ a, b \ are \ integers; a > b \end{bmatrix}$$

Eq 7.1.10: Lemon Energy

The wave-archetype or intent has to do with a primary or x-element associated with the function 'peace'. Other y-elements in combination will engender intent such as 'flexibility' and 'opening', amongst possible others. Hence, these relationships may be depicted by Equation 7.1.11, Lemon Intent:

$$Red_{Intent} =' peace'.a + \overline{Yb_{0-n}}$$

$$where \begin{bmatrix} peace \in [S_{System_K}] \\ Y \in [S_{System_{Pr}}, S_{System_P}, S_{System_K}, S_{System_N}] \\ a, b \ are \ integers; a > b \end{bmatrix}$$

Eq 7.1.11: Lemon Intent

The mass-potential or body-part most linked to lemon is suggested to be 'joints', 'gallbladder', 'diaphragm' and 'vision', amongst possible others. So there is likely a build up of mass in such a manner that facilitates the creation of these systems. An assumption can be made that there is a 'lemon-mass' x-element and perhaps other y-elements that yield the different systems when built up. Equation 7.1.12, Joints – Lemon Mass Possibility, suggests how this can be depicted:

$$Joints - Lemon_{Mass_{Possibility}} =' lemon - mass'.a + \overline{Yb_{0-n}}$$

$$where \begin{bmatrix} 'lemon - mass' \in [S_{System_{Pr}}] \\ Y \in [S_{System_{Pr}}, S_{System_P}, S_{System_K}, S_{System_N}] \\ a, b \ are \ integers; a > b \end{bmatrix}$$

Eq 7.1.12: Joints - Lemon Mass Possibility

Note that (7.1.10 – 12) are proposed to be part of the material-fabric. As such they are of a pre-genetic type but also inform or even help form the cell-based genetic code.

Green

Green is associated with the wavelength 520 to 550 nanometers.

In its association with life the energy-profile of 'green' is suggested to engender qualities to do with 'general balancing', 'neutralizing', and 'anti-infection' amongst possible others. Assuming that the primary or x-element has to do with 'balancing', then various y-elements in combination can create the other engendered qualities.

282

These relationships may be depicted in Equation 7.1.13, Green Energy, in the following manner:

$$Green_{Energy} =' balancing'.a + \overline{Yb_{0-n}}$$

$$where \begin{bmatrix} balancing \in [S_{System_P}] \\ Y \in [S_{System_{Pr}}, S_{System_P}, S_{System_K}, S_{System_N}] \\ a, b \ are \ integers; a > b \end{bmatrix}$$

Eq 7.1.13: Green Energy

The wave-archetype or intent has to do with a primary or x-element associated with the function 'love'. Other y-elements in combination will engender intent such as 'abundance' and 'adaptability', amongst possible others. Hence, these relationships may be depicted by Equation 7.1.14, Green Intent:

$$Green_{Intent} =' love'.a + \overline{Yb_{0-n}}$$

$$where \begin{bmatrix} love \in [S_{System_K}] \\ Y \in [S_{System_{Pr}}, S_{System_P}, S_{System_K}, S_{System_N}] \\ a, b \ are \ integers; a > b \end{bmatrix}$$

Eq 7.1.14: Green Intent

The mass-potential or body-part most linked to green is suggested to be the 'liver', 'lungs', 'thymus', and 'immune system', amongst possible others. So there is likely a build up of mass in such a manner that facilitates the creation of these systems. An assumption can be made that there is a 'green-mass' x-element and perhaps other y-elements that yield the different systems when built up. Equation 7.1.15, Liver – Green Mass Possibility, suggests how this can be depicted:

$$Liver - Green_{Mass_{Possibility}} =' green - mass'.a + \overline{Yb_{0-n}}$$

$$where \begin{bmatrix} 'green-mass' \in [S_{System_{Pr}}] \\ Y \in [S_{System_{Pr}}, S_{System_P}, S_{System_K}, S_{System_N}] \\ a, b \text{ are integers}; a > b \end{bmatrix}$$

Eq 7.1.15: Liver - Green Mass Possibility

Note that (7.1.13 – 15) are proposed to be part of the material-fabric. As such they are of a pre-genetic type but also inform or even help form the cell-based genetic code.

Cyan

Cyan is associated with the wavelength 490 to 520 nanometers.

In its association with life the energy-profile of 'cyan' is suggested to engender qualities to do with 'cleansing', 'refreshing', and 'addressing acute conditions' amongst possible others. Assuming that the primary or x-element has to do with 'cleansing', then various y-elements in combination can create the other engendered qualities. These relationships may be depicted in Equation 7.1.16, Cyan Energy, in the following manner:

$$Cyan_{Energy} =' cleansing'.a + \overline{Yb_{0-n}}$$

$$where \begin{bmatrix} cleansing \in [S_{System_P}] \\ Y \in [S_{System_{Pr}}, S_{System_P}, S_{System_K}, S_{System_N}] \\ a, b \text{ are integers}; a > b \end{bmatrix}$$

Eq 7.1.16: Cyan Energy

The wave-archetype or intent has to do with a primary or x-element associated with the function 'wholeness'. Other y-elements in combination will engender intent such as 'autonomy', and 'fulcrum of emotion vs. thought', amongst possible others. Hence, these relationships may be depicted by Equation 7.1.17, Cyan Intent:

$$Cyan_{Intent} =' wholeness'.a + \overline{Yb_{0-n}}$$

$$where \begin{bmatrix} wholeness \in [S_{System_K}] \\ Y \in [S_{System_{Pr}}, S_{System_P}, S_{System_K}, S_{System_N}] \\ a, b \ are \ integers; a > b \end{bmatrix}$$

Eq 7.1.17: Cyan Intent

The mass-potential or body-part most linked to cyan is suggested to be the 'arms and hands', 'parathyroid glands', and 'skin'. amongst others. So there is likely a build up of mass in such a manner that facilitates the creation of these systems. An assumption can be made that there is a 'cyan-mass' x-element and perhaps other y-elements that yield the different systems when built up. Equation 7.1.18, Skin – Cyan Mass Possibility, suggests how this can be depicted:

$$Skin - Cyan_{Mass_{Possibility}} =' cyan - mass'.a + \overline{Yb_{0-n}}$$

$$where \begin{bmatrix} 'cyan - mass' \in [S_{System_{Pr}}] \\ Y \in [S_{System_{Pr}}, S_{System_P}, S_{System_K}, S_{System_N}] \\ a, b \ are \ integers; a > b \end{bmatrix}$$

Eq 7.1.18: Skin - Cyan Mass Possibility

Note that (7.1.16 – 18) are proposed to be part of the material-fabric. As such they are of a pre-genetic type but also inform or even help form the cell-based genetic code.

Blue

Blue is associated with the wavelength 460 to 490 nanometers.

In its association with life the energy-profile of 'blue' is suggested to engender qualities to do with 'calming', 'anti-inflammation', and 'antipyretic', amongst possible others. Assuming that the primary or x-element has to do with 'calming', then various y-elements in combination can create the other engendered qualities. These relationships may be depicted in Equation 7.1.19, Blue Energy, in the following manner:

$$Blue_{Energy} =' calming'. a + \overline{Yb_{0-n}}$$

$$where \begin{bmatrix} calming \in [S_{System_P}] \\ Y \in [S_{System_{Pr}}, S_{System_P}, S_{System_K}, S_{System_N}] \\ a, b \ are \ integers; a > b \end{bmatrix}$$

Eq 7.1.19: Blue Energy

The wave-archetype or intent has to do with a primary or x-element associated with the function 'antistress'. Other y-elements in combination will engender intent such as 'introversion' and 'endorphins stimulant', amongst possible others. Hence, these relationships may be depicted by Equation 7.1.20, Blue Intent:

$$Blue_{Intent} =' antistress'. a + \overline{Yb_{0-n}}$$

$$where \begin{bmatrix} antistress \in [S_{System_K}] \\ Y \in [S_{System_{Pr}}, S_{System_P}, S_{System_K}, S_{System_N}] \\ a, b \ are \ integers; a > b \end{bmatrix}$$

Eq 7.1.20: Blue Intent

The mass-potential or body-part most linked to blue is suggested to be the 'nose, throat, neck', 'thyroid gland', breathing', 'ears', amongst possible others. So there is likely a build up of mass in such a manner that facilitates the creation of these systems. An assumption can be made that there is a 'blue-mass' x-element and perhaps other y-elements that yield the different systems when built up. Equation 7.1.21, Ears – Blue Mass Possibility, suggests how this can be depicted:

$$Ears - Blue_{Mass_{Possibility}} =' blue - mass'.a + \overline{Yb_{0-n}}$$

$$where \begin{bmatrix} 'blue - mass' \in [S_{System_{Pr}}] \\ Y \in [S_{System_{Pr}}, S_{System_P}, S_{System_K}, S_{System_N}] \\ a, b \ are \ integers; a > b \end{bmatrix}$$

Eq 7.1.21: Ears - Blue Mass Possibility

Note that (7.1.19 – 21) are proposed to be part of the material-fabric. As such they are of a pre-genetic type but also inform or even help form the cell-based genetic code.

Indigo

Indigo is associated with the wavelength 430 to 460 nanometers.

In its association with life the energy-profile of 'indigo' is suggested to engender qualities to do with 'sleep-inducing' and 'muscle relaxant', amongst possible others. Assuming that the primary or x-element has to do with 'relaxant', then various y-elements in combination can create the other engendered qualities. These relationships may be depicted in Equation 7.1.22, Indigo Energy, in the following manner:

$$Indigo_{Energy} =' relaxant'.a + \overline{Yb_{0-n}}$$

$$where \begin{bmatrix} relaxant \in [S_{System_P}] \\ Y \in [S_{System_{Pr}}, S_{System_P}, S_{System_K}, S_{System_N}] \\ a, b \ are \ integers; a > b \end{bmatrix}$$

Eq 7.1.22: Indigo Energy

The wave-archetype or intent has to do with a primary or x-element associated with the function 'intuition'. Other y-elements in combination will engender intent such as 'imagination' and 'anxiolytic', amongst possible others.

Hence, these relationships may be depicted by Equation 7.1.23, Indigo Intent:

$$Indigo_{Intent} =' intuition'.a + \overline{Yb_{0-n}}$$

$$where \begin{bmatrix} intuition \in [S_{System_K}] \\ Y \in [S_{System_{Pr}}, S_{System_P}, S_{System_K}, S_{System_N}] \\ a, b\ are\ integers; a > b \end{bmatrix}$$

Eq 7.1.23: Indigo Intent

The mass-potential or body-part most linked to indigo is suggested to be the 'sinuses', 'pituitary', 'forehead and occiput', and 'limbic system', amongst possible others. So there is likely a build up of mass in such a manner that facilitates the creation of these systems. An assumption can be made that there is a 'indigo-mass' x-element and perhaps other y-elements that yield the different systems when built up. Equation 7.1.24, Pituitary – Indigo Mass Possibility, suggests how this can be depicted:

$$Pituitary - Indigo_{Mass_{Possibility}} =' indigo - mass'.a + \overline{Yb_{0-n}}$$

$$where \begin{bmatrix} 'indigo - mass' \in [S_{System_{Pr}}] \\ Y \in [S_{System_{Pr}}, S_{System_P}, S_{System_K}, S_{System_N}] \\ a, b\ are\ integers; a > b \end{bmatrix}$$

Eq 7.1.24: Pituitary - Indigo Mass Possibility

Note that (7.1.22 – 24) are proposed to be part of the material-fabric. As such they are of a pre-genetic type but also inform or even help form the cell-based genetic code.

Violet

Violet is associated with the wavelength 380 to 430 nanometers.

In its association with life the energy-profile of 'violet' is suggested to engender qualities to do with 'immunostimulant', amongst possible others. Assuming that the primary or x-element has to do with 'stimulant', then various y-elements in combination can create the other engendered qualities. These relationships may be depicted in Equation 7.1.25, Violet Energy, in the following manner:

$$Violet_{Energy} =' stimulant'.a + \overline{Yb_{0-n}}$$

$$where \begin{bmatrix} stimulant \in [S_{System_P}] \\ Y \in [S_{System_{Pr}}, S_{System_P}, S_{System_K}, S_{System_N}] \\ a, b \ are \ integers; a > b \end{bmatrix}$$

Eq 7.1.25: Violet Energy

The wave-archetype or intent has to do with a primary or x-element associated with the function 'spirituality'. Other y-elements in combination will engender intent such as 'trust' and 'dignity', amongst possible others. Hence, these relationships may be depicted by Equation 7.1.26, Violet Intent:

$$Violet_{Intent} =' spirituality'.a + \overline{Yb_{0-n}}$$

$$where \begin{bmatrix} spirituality \in [S_{System_K}] \\ Y \in [S_{System_{Pr}}, S_{System_P}, S_{System_K}, S_{System_N}] \\ a, b \ are \ integers; a > b \end{bmatrix}$$

Eq 7.1.26: Violet Intent

The mass-potential or body-part most linked to violet is suggested to be the 'cerebral cortex', 'pineal gland', and 'memory, amongst possible others. So there is likely a build up of mass in such a manner that facilitates the creation of these systems. An assumption can be made that there is a 'violet-mass' x-element and perhaps other y-elements that yield the different systems when built up.

Equation 7.1.27, Cerebral Cortex– Violet Mass Possibility, suggests how this can be depicted:

$$Cerebral\ Cortex - Violet_{Mass_{Possibility}} =' violet - mass'.a + \overline{Yb_{0-n}}$$

$$where \begin{bmatrix} 'violet - mass' \in [S_{System_{Pr}}] \\ Y \in [S_{System_{Pr}}, S_{System_P}, S_{System_K}, S_{System_N}] \\ a, b\ are\ integers; a > b \end{bmatrix}$$

Eq 7.1.27: Cerebral Cortex - Violet Mass Possibility

Note that (7.1.25 – 27) are proposed to be part of the material-fabric. As such they are of a pre-genetic type but also inform or even help form the cell-based genetic code.

Chapter 7.2: Generating the Genetic Code for a Living Cell

The living cell has a universe of adaptability embedded in it. In 'The Machinery of Life', Goodsell, an Associate Professor of Molecular Biology at the Scripps Research Institute (Goodsell, 2010) suggests that every living thing on Earth uses a similar set of molecules to eat, to breathe, to move, and to reproduce. There are molecular machines that do the myriad things that distinguish living organisms that are identical in all living cells. This nanoscale machinery of cells uses four basic molecular plans with unique chemical personalities: nucleic acids, proteins, lipids, and polysaccharides.

Extrapolating from previous chapters there will likely be a vast number of iterations of the Light-Space-Time Emergence Equation to generate the four basic molecular plans. Equation 4.2.4, Cells Genetic Code, created in Chapter 4.2 that explored the post Big-Bang generation of pre-genetic code, genetic code, and post-genetic code, is reproduced here for convenience. The starting point, x_U, is 'Molecules', and the ending point, x_T, is 'Cells':

$Light - Space - Time\ Emergence_{Cells} =$

$$
\left[
\begin{array}{l}
\left[
\begin{array}{c}
c_\infty : [Pr, Po, K, H] \\
\left(\downarrow R_{C_K} = f\left(R_{C_\infty}\right) \right) \\
c_K : [S_{Pr}, S_{Po}, S_K, S_H] \\
\left(\downarrow R_{C_N} = f\left(R_{C_K}\right) \right) \\
c_N : f(S_{Pr} \times S_{Po} \times S_K \times S_H) \\
\left(\downarrow R_{C_U} = f\left(R_{C_N}\right) \right) \\
c_U : [P, V, M, C] \\
\Uparrow \\
c_{0:[D,W,I,C]}
\end{array}
\right]_{Light} \\
\\
\left[
\begin{array}{c}
M_3 : -\infty \le t \le \infty \\
\downarrow \\
M_2 : 0 \ge t > \infty \\
\downarrow \\
M_1 : 0 > t > \infty \\
\downarrow \\
U \to \quad \begin{array}{l} t \le E_{Cell}; \, TC: M_3 \to U \\ t \sim E_{Human}; \, TC: U \to M_3 \end{array}
\end{array}
\right]_{Time} \\
\Rightarrow
\end{array}
\right.
\left[
\begin{array}{c}
M_3 \to System_X \\
(\uparrow F \to I) \\
M_2 \to S_{System_X} \\
(\uparrow Sig \to F) \\
M_1 \to Sig_x \\
(\uparrow > P_x) \\
U \to x_{Molecules}
\end{array}
\right]_{Space}
$$

$$ TC \to x_{Cells} $$

$$ \langle x_U | x_T \rangle $$

$\langle Space - Time - Energy - Gravity \ Material - Fabric \ Pre$
$- Genetic \ Code \rangle \ +$

$\langle Electromagnetic \ Spectrum \ \ Material - Fabric \ Pre - Genetic \ Code \rangle$
$+$

$\langle Quantum - Particle \ \ Material - Fabric \ Pre - Genetic \ Code \rangle \ +$

$\langle Atoms \ \ Material - Fabric \ Pre - Genetic \ \ Code \rangle \ +$

$\langle Molecules \ \ Material - Fabric \ Pre - Genetic \ Code \rangle + LSTE \ \langle ... \rangle$
$+ \ Cells \ Genetic \ Code$

There are many likely iterations of this equation between molecules and cells and this in general is depicted by '*LSTE* $\langle ... \rangle$', where LSTE signifies iterations(s) of the Light-Space-Time Emergence equation. Also note that an assumption is made that all the previous material-fabric code-segments are intimately active at the level of cellular genetic code. This may be through the device of entanglement, or perhaps even by the material-fabric code segments being embedded in cellular genetic code, or perhaps by some protein-structures that exist in each cell that give the cell the capacity to read the pre-genetic code embedded in the material-fabric.

The emergence of a quadrumvirate cell-based structure specified by four molecular plans is significant in that the genetic code is now being housed in a post material-fabric structure. It is likely the collective and simultaneous action of nucleic acids, polysaccharides, lipids, and proteins that facilitates this post material-fabric housing ability. While in this chapter the code generated for each of these molecular plans appears to be separate, as an emergence of properties of Light they should be thought of as happening in such a manner so as to encapsulate the essential unity of the four properties.

Strictly speaking (4.2.4) should reflect the transition between the pre-genetic to genetic housing structure, and this is depicted as 'PGGT <...>', Pre-Genetic to Genetic Transition, in Equation 7.2.1, Cells Genetic Code – PGGT:

$$Light - Space - Time\ Emergence_{Cells} =$$

$$\left[\begin{array}{c} \left[\begin{array}{c} c_\infty : [Pr, Po, K, H] \\ \left(\downarrow R_{C_K} = f(R_{C_\infty})\right) \\ c_K : [S_{Pr}, S_{Po}, S_K, S_H] \\ \left(\downarrow R_{C_N} = f(R_{C_K})\right) \\ c_N : f(S_{Pr} \; x \; S_{Po} \; x \; S_K \; x \; S_H) \\ \left(\downarrow R_{C_U} = f(R_{C_N})\right) \\ c_U : [P, V, M, C] \\ \Uparrow \\ c_{0:[D,W,I,C]} \end{array}\right]_{Light} \quad \left[\begin{array}{c} M_3 \to System_X \\ (\uparrow F \to I) \\ M_2 \to S_{System_X} \\ (\uparrow Sig \to F) \\ M_1 \to Sig_x \\ (\uparrow > P_x) \\ U \to x_{Molecules} \end{array}\right]_{Space} \\ \left[U \to \begin{array}{c} M_3 : -\infty \le t \le \infty \\ \downarrow \\ M_2 : 0 \ge t > \infty \\ \downarrow \\ M_1 : 0 > t > \infty \\ \downarrow \\ t \le E_{Cell}; TC : M_3 \to U \\ t \sim E_{Human}; TC : U \to M_3 \end{array}\right]_{Time} \quad TC \to x_{Cells} \end{array}\right] \langle x_U | x_T \rangle$$

$$\Rightarrow$$

$\langle Space - Time - Energy - Gravity \; Material - Fabric \; Pre - Genetic \; Code \rangle +$

$\langle Electromagnetic \; Spectrum \; Material - Fabric \; Pre - Genetic \; Code \rangle +$

$\langle Quantum - Particle \; Material - Fabric \; Pre - Genetic \; Code \rangle +$

$\langle Atoms \; Material - Fabric \; Pre - Genetic \; Code \rangle +$

$\langle Molecules \; Material - Fabric \; Pre - Genetic \; Code \rangle + LSTE \; \langle ... \rangle + PGGT < \cdots > +$

Cells Genetic Code

Eq 7.2.1: Cells Genetic Code - PGGT

Light's Emergence as Living Cells

As discussed previously the Light-Space-Time Emergence equation (3.1.3) being iterative, can be used to model emergence as it proceeds from simpler four-fold to more complex four-fold manifestations. Hence (3.1.3) has

already been applied to suggest the emergence of the space-time-energy-gravity quadrumvirate, the electromagnetic spectrum, quantum particles in general, bosons as a further instance of a particular kind of quantum particle, and atoms. Here it will be applied to suggest the emergence of Living Cells as suggested by (7.2.1). But further, as implied by (3.6.5), the Potential Effect of Levels of Light on Pre-Genetic of Genetic Information equation, any process of deeper organization, such as is responsible for the architecture and cohesiveness of living cells, has the possibility of altering the material-fabric or a post material-fabric housing structure so long as the bases involved are driven primarily by a meta-level.

Starting with the Light-Matrix, the top left-hand matrix in (7.2.1), the first line from the top, $C_\infty: [Pr, Po, K, H]$, specifies the fundamental architecture of living cells. As will be explored shortly, Nucleic Acids are an emergence of Light's property of Knowledge, Polysaccharides are an emergence of Light's property of Power, Proteins are an emergence of Light's property of Presence, and Lipids are an emergence of Light's property of Harmony. The fundamental architecture of these aspects hence, is an emergence of the properties of Light at ∞.

Line 3 in the Light-Matrix, C_K: $[S_{Pr}, S_{Po}, S_K, S_H]$, elaborates the sets for Presence, Power, Knowledge, and Harmony, each containing multiple elements. For example, as will be explored in the section on Proteins, various elements derived from the four sets define the behavior of Proteins and could be functions such as 'exist for service', 'to bring about perfection at the level of the cell', 'extreme diligence and perseverance', amongst others, hence collectively describing Proteins' way of being. Specifically, Line 5, $C_{N:}$ $f(S_{Pr} \times S_{Po} \times S_K \times S_H)$, suggests that unique seeds are created from a combination of such elements from all four sets, with a particular element leading, that in effect creates the distinctness possible at the level of cells.

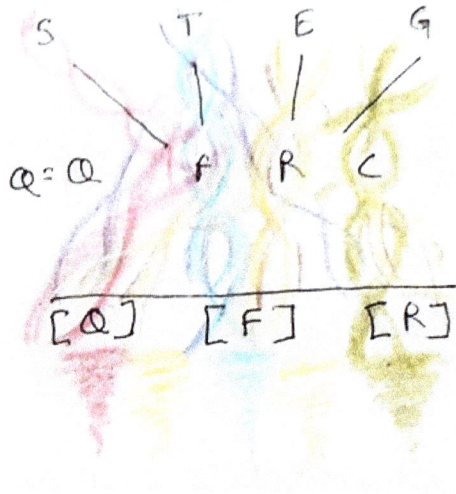

Line 6, $(\downarrow R_{C_U} = f(R_{C_N}))$, specifies quantization between the layer where the seeds are formed, and the physical layer we are familiar with, and as explored in Chapter 3.5 and 3.6, will result in Line 7, C_U: $[P, V, M, C]$, hence changing the material-fabric or post material-fabric housing structure. Note that as in the process describing the generation of the code-segments for space-time-energy-gravity quantization at the time of the Big Bang, the generation of the FBLEE (four-base logic-encoding ecosystem) code-segment as specified by (4.1.5) and the MF (material-fabric) code segment as specified by (4.1.6) are combined together into one equation so that the FBLEE process is apparently transparent.

The possibilities represented by Lines 1 through 5 hence concretize through the quantization represented by Line 6 to become or enhance living cells with its subtle physical (related to Presence), vital (related to Power), mental (related to Knowledge), and connection (related to Harmony) aspects now existing in material reality typified by Light moving at c. Note that just as Line 6 represents a process of quantization relating the layer of reality created by Light traveling at c with the antecedent layers, so too Lines 2 and 4 as previously discussed, also represent quantization of a more subtle kind that ultimately plays a critical part in allowing the material-fabric to express infinite diversity.

Typically it is the process as captured by the Space-Matrix that will determine if Line 6 is activated. Specifically patterns at the untransformed layer, U, will need to be overcome, as specified by the second-line from the bottom of the Space-Matrix: $(\uparrow > P_x)$. But as specified by the bottom-line of the Time-Matrix, reproduced below, it is only with the advent of the human-system that the automaticity of the action of meta-levels is reversed:

$$U \rightarrow \begin{array}{l} t \leq E_{Cell}; \text{TC}: M_3 \rightarrow U \\ t \sim E_{Human}; \text{TC}: U \rightarrow M_3 \end{array}$$

Hence in the case of living cells, which in this emergence is a pre-human system, the fact that patterns do not need to be overcome means that quantization more or less happens automatically.

Even though cells are a pre-human system and therefore the action of meta-levels are modeled as being automatic, it is nonetheless useful to review (3.6.5), Potential Effect of Levels of Light on Pre-Genetic or Genetic Information, to understand how the relationship with different forms of mutation has been specified:

Potential Effect of Levels of Light on Pre Genetic or Genetic Information =

$$\begin{bmatrix} STATIC \ \langle |[L][S][T]TC \rightarrow x_T|_{\langle x_U|x_T\rangle}\rangle \\ \times \\ \left((Y > U \colon Z_Q) \vee (Y \le U \colon Z_F) \vee (Y = U \colon Z_R) \right) \\ \ni \\ \left(\begin{matrix} Z \in \mathbb{U} \ (Space, Time, Energy, Gravity) \\ Q \colon Quantization; F \colon Fragmentation; R \colon Random \end{matrix} \right) \end{bmatrix} \rightarrow h \ni$$

$$h \in \left(\begin{matrix} [Q] \colon Constructive\ zone, \\ [Q] \colon Constructive\ zone \wedge Constructive\ mutation, \\ [F] \colon Destructive\ mutation, \\ [R] \colon Random\ mutation \end{matrix} \right)$$

Line 1 from the top in the matrix is simply a static form of (3.6.1) the Simplified Light-Space-Time Emergence equation. The static form is designated by 'STATIC', and implies that fundamental operations true of (3.6.1) are being highlighted in (3.6.5). In other word (3.6.1) already has all the operations highlighted in (3.6.5) in it, but by 'freezing' it by making it static, the essential dynamics leading to possible mutations at the genetic level can more clearly be highlighted.

Line 1 is then subjected ($\times$) to a determination of the dominant levels of light that may be active, designated by Line 2, ' $\left((Y > U \colon Z_Q) \vee (Y \le U \colon Z_F) \vee (Y = U \colon Z_R) \right)$ '. Unpacking this, 'Y > U' implies meta-levels are active and as a result it is possible that Z_Q is going to take place. This also implies activation and potential change of FBLEE. The call from below, as it were, may invoke some function that already exists in the subtle-libraries 'above', so that some already existing function may influence FBLEE through ∞ -entanglement, K-entanglement, or N-entanglement. In this case the range of already existing molecules in constant interaction with one another, may yearn for coming together in a more cohesive way so that much greater fourfold functionality that makes them 'alive' in a self-sustaining way comes into being. Such a call may invoke the precipitation of a cell-based ecosystem meta-function. This may be thought of as a key-and-lock mechanism, where a deep enough visceral urge from below acting as the key, opens an entangled lock to alter FBLEE as per the visceral urge. Such a call-response system at the pre-human level is considered to be relatively automatic as compared with the emergence of human-type systems because of the complexity of many more sources of action in the case of the latter.

'$Y \leq U$' implies that only the untransformed levels are active, and therefore also the sub-level where the speed of light is 0 is active and as a result Z_F is going to take place. '$Y = U$' implies that all levels are active and as a result Z_R is going to take place.

Line 3 elaborates the significance of Z_Q, Z_F, and Z_R. Hence Z is the union of potential quantum-operations of space, time, energy, and gravity, designated by '$Z \in U$ ($Space, Time, Energy, Gravity$)'. But the nature of the operations is designated by 'Q: $Quantization$; F: $Fragmentation$; R: $Random$'. Z_Q, then, implies that the full quantization originating from updated four-base logic-encoding ecosystems (FBLEE) can take place, and will result in lasting material change at the genetic level. Z_F implies that the essential set will be fragmented and that only local libraries a the level of local cellular-level DNA can potentially be altered. Z_R also precludes full quantization, and that some partial local-library constructive or destructive mutation may take place.

The '→ $h \ni h \in$' segment resolves the outcome of the operations implied by Lines 1 – 3, suggesting that the outcome will be 'h' such that ($\ni$) 'h' is an element ($\in$) of the set specified by the members '[Q]: *Constructive zone'*, '[Q}: *Constructive zone AND Constructive mutation'*, '[F}; *Destructive mutation'*, and '{R}: *Random mutation'*. '[Q]: *Constructive zone'* implies that the in-built buffer has been crossed and that access to the deeper four-base logic-encoding ecosystem (FBLEE) has been granted. '[Q}' in this segment implies that there is the possibility that full-quantization as specified by Line 3 of the previous matrix can take place. Access to this zone is a prerequisite for constructive mutation to occur, as designated by the element '[Q}: *Constructive zone* $\cap$ *Constructive mutation* ', which implies that full-quantization is going to take place and will result in material change. The '[F]' specifies the relationship between 'Fragmentation' in Line 3 of the previous matrix and destructive mutation. The '[R]' specifies the relationship between 'Random' in Line 3 of the previous matrix and random mutation.

But as just summarized in the Time-Matrix in (7.2.1) Y is by definition greater than U and hence quantization is automatic. In terms of living cells such quantization implies that wholeness becomes fully active through specific space, time, energy, and gravity quantization to create an holistic "ecosystem" with its own "living cell logic" as it were. The wholeness has now precipitated into the genetic code in the cellular structure. Note that such precipitation is captured by the action of 'PGGT <...>' that can be assumed to be triggered by Z_Q.

Generation of Cellular-Level Genetic Code for Nucleic Acids

Nucleic acids basically encode information. They store and transmit the genome, the hereditary information needed to keep the cell alive. They function as the cell's librarians and contain information on how to make proteins and when to make them.

They are hence, the keepers of a cell's knowledge, its wisdom, its ability to make laws, the vehicle to spread knowledge within cells and to the next generation of cells. Being so, one can see that there is similarity with the set for system-knowledge highlighted earlier in Chapter 2.3. Reproducing Equation 2.3.3:

$$Ssystem_K \ni [Wisdom, Law\ Making, Spread\ of\ Knowledge\ ...]$$

Nucleic acids can therefore be thought of as a precipitation of system-knowledge at the cellular level.

Hence, a nucleic acid will have a generalized signature, as in Equation 7.2.2, derived from the system-knowledge family:

$$Sig_{nucleic\ acid} = Xa +$$

$$\overline{Yb_{0-n}}\ where \left[\begin{array}{c} X \in [Ssystem_K] \\ Y \in [Ssystem_{Pr}, Ssystem_P, Ssystem_K, Ssystem_N] \\ a, b\ are\ integers; a > b \end{array} \right]$$

Eq 7.2.2: Nucleic Acid

The primary element X could be an attribute or function such as 'keeper of knowledge'. Secondary elements Y could be 'protein laws', 'generational knowledge', amongst others. The collectivity of elements as per the equation would specify the character of a nucleic acid.

Note that (7.2.2) already implies that Lines 1 – 5 in the Light Matrix (7.2.1) have been activated, and that the logic of the nucleic-acid-ecosystem will precipitate into the post material-fabric cell-based genetic structure through the action of Line 6-7 of (7.2.1). This action of precipitating into the cell-based structure as opposed to the material-fabric is designated by 'PGGT <...>' as specified in (7.2.1).

DNA and RNA, two types of nucleic acids, would hence have the equations as specified by Equation 7.2.3 and 7.2.4 respectively.

$$Sig_{DNA} = Xa +$$
$$\overline{Yb_{0-n}} \quad where \begin{bmatrix} X \in [\,S_{System_K}] \\ Y \in [S_{System_{Pr}}, S_{System_P}, S_{System_K}, S_{System_N}] \\ a, b \ are \ integers; a > b \end{bmatrix}$$

Eq 7.2.3: DNA

$$Sig_{RNA} = Xa +$$
$$\overline{Yb_{0-n}} \quad where \begin{bmatrix} X \in [\,S_{System_K}] \\ Y \in [S_{System_{Pr}}, S_{System_P}, S_{System_K}, S_{System_N}] \\ a, b \ are \ integers; a > b \end{bmatrix}$$

Eq 7.2.4: RNA

The primary element X in (7.2.3) and (7.2.4) would be the same as that for nucleic acids (7.2.2). The secondary elements Y however will be a larger and more specific set with many elements in common with (7.2.2). Note also that the genetic code as specified by the following sub-sections on proteins, polysaccharides, and lipids, would be translated into DNA-format and stored in nucleic acids as specified by (7.2.3).

Generation of Cellular-Level Genetic Code for Proteins

Proteins are the cells work-horses. Look anywhere in a cell and one will see proteins at work. Proteins are built in thousands of shapes and sizes, each performing a different function. As Goodsell describes, "some are built simply to adopt a defined shape, assembling into rods, nets, hollow spheres, and tubes. Some are molecular motors, using energy to rotate, or flex, or crawl. Many are chemical catalysts that perform chemical reactions atom-by-atom, transferring and transforming chemical groups exactly as needed." With their wide potential for diversity, proteins are constructed to perform most of the everyday tasks of the cells. In fact human cells build around 30,000 different kinds of proteins to execute on the diverse array of cellular level tasks.

Proteins hence, exist for service, to bring about perfection at the level of the cell, are characterized by extreme diligence and perseverance, and so on. Being so, one can see that there is similarity with the set for system-presence highlighted earlier in Chapter 2.3. Reproducing Equation 2.3.1:

$$S_{System_{Pr}} \ni [Service, Perfection, Diligence, Perseverance, ...]$$

Proteins can therefore be thought of as a precipitation of system-presence at the cellular level.

Hence, a protein could have a generalized signature, as in Equation 7.2.5, derived from the system-presence family:

$$Sig_{protein} = Xa +$$
$$\overline{Yb_{0-n}} \quad where \quad \begin{bmatrix} X \in [S_{System_{Pr}}] \\ Y \in [S_{System_{Pr}}, S_{System_P}, S_{System_K}, S_{System_N}] \\ a, b \ are \ integers; a > b \end{bmatrix}$$

Eq 7.2.5: Protein

This could yield a vast number of functional proteins. In fact it may be possible that the 30,000 or so known proteins created by the human cell could each be specified by a signature equation of this nature. It may be possible to map existing proteins to functionality as suggested by the four sets of molecular plans.

Note that (7.2.5) already implies that Lines 1 – 5 in the Light Matrix (7.2.1) have been activated, and that the logic of the protein-ecosystem will precipitate into the post material-fabric cell-based genetic structure through the action of Line 6-7 of (7.2.1). This action of precipitating into the cell-based structure as opposed to the material-fabric is designated by 'PGGT <...>' as specified in (7.2.1).

Consider Insulin, for example. Insulin regulates the metabolism of carbohydrates, fats and protein by promoting the absorption of, especially, glucose from the blood into fat, liver and skeletal muscle cells. Equation 7.2.6 for Insulin would hence be:

$$Sig_{insulin} = Xa +$$

$$\overline{Yb_{0-n}} \ where \ \begin{bmatrix} X \in [S_{System_{Pr}}] \\ Y \in [S_{System_{Pr}}, S_{System_P}, S_{System_K}, S_{System_N}] \\ a, b \ are \ integers; a > b \end{bmatrix}$$

Eq 7.2.6: Insulin

The primary element X could be an attribute or function such as 'workhorse. Secondary elements Y could be 'metabolic regulation, 'glucose absorption', 'blood to fat channel', amongst others. The collectivity of elements as per the equation would specify the character of insulin.

Consider Histones as another example. They are the chief protein components of chromatin, acting as spools around which DNA winds, and playing a role in gene regulation. Without histones, the unwound DNA in chromosomes would be very long (a length to width ratio of more than 10 million to 1 in human DNA). Equation 7.2.7 for Histones would hence be:

$$Sig_{histones}$$
$$= Xa$$

$$+ \overline{Yb_{0-n}} \ where \ \begin{bmatrix} X \in [S_{System_{Pr}}] \\ Y \in [S_{System_{Pr}}, S_{System_P}, S_{System_K}, S_{System_N}] \\ a, b \ are \ integers; a > b \end{bmatrix}$$

Eq 7.2.7: Histones

The secondary element Y would have elements such as 'gene regulation', 'unwound DNA management', amongst others.

Generation of Cellular-Level Genetic Code for Lipids

Lipids by themselves are tiny molecules, but when grouped together form the largest structures of the cell. When placed in water, lipid molecules aggregate to form huge waterproof sheets. These sheets easily form boundaries at multiple levels and allow concentrated interactions and work to be performed within a cell. Hence, the nucleus and the mitochondria are contained within lipid-defined compartments. Similarly, each cell itself is contained within a lipid-defined boundary.

Lipids are therefore promoters of relationship, of harmony in the cell, of nurturing the cell-level division of labor, of allowing specialization and uniqueness to emerge, hence perhaps of earlier forms of compassion and love, and so on. The notion of such early forms of compassion is consistent with the biologist's perspective that at some point a gene for compassion was developed in pre-human species (Wright, 2009). Being so, one can see that there is similarity with the set for system-nurturing highlighted earlier in Chapter 2.3. Reproducing Equation 2.3.4:

$$S_{System_N} \ni [Love, Compassion, Harmony, Relationship \ ...]$$

This function of harmonization suggests that lipids can therefore be thought of as a precipitation of system-nurturing at the cellular level.

Lipids could have a generalized signature, as in Equation 7.2.8, derived from the system-nurturing family:

$$Sig_{lipid} = Xa +$$

$$\overline{Yb_{0-n}} \quad where \quad \left[\begin{array}{c} X \in [S_{System_N}] \\ Y \in [S_{System_{Pr}}, S_{System_P}, S_{System_K}, S_{System_N}] \\ a, b \text{ are integers}; a > b \end{array} \right]$$

Eq 7.2.8: Lipid

Note that (7.2.8) already implies that Lines 1 – 5 in the Light Matrix (7.2.1) have been activated, and that the logic of the lipid-ecosystem will precipitate into the post material-fabric cell-based genetic structure through the action of Line 6-7 of (7.2.1). This action of precipitating into the cell-based structure as opposed to the material-fabric is designated by 'PGGT <...>' as specified in (7.2.1).

Specific lipids such as monoglycerides and phospholipids could have the following equations:

$$Sig_{monoglyceride} = Xa +$$

$$\overline{Yb_{0-n}} \quad where \quad \left[\begin{array}{c} X \in [S_{System_N}] \\ Y \in [S_{System_{Pr}}, S_{System_P}, S_{System_K}, S_{System_N}] \\ a, b \text{ are integers}; a > b \end{array} \right]$$

Eq 7.2.9: Monoglyceride

$$Sig_{phospholipids} = Xa +$$

$$\overline{Yb_{0-n}} \quad where \quad \left[\begin{array}{c} X \in [S_{System_N}] \\ Y \in [S_{System_{Pr}}, S_{System_P}, S_{System_K}, S_{System_N}] \\ a, b \text{ are integers}; a > b \end{array} \right]$$

Eq 7.2.10: Phospholipids

The primary element X shared by each of the lipids could be an attribute or function such as 'compartmentalization'. Secondary elements Y could be of the nature of 'work breakdown', 'intra-cell love', amongst others, and would vary with each different kind of lipid.

Generation of Cellular-Level Genetic Code for Polysaccharides

Polysaccharides are long, often branched chains of sugar molecules. Sugars are covered with hydroxyl groups, which associate to form storage containers. As a result polysaccharides function as the storehouse of cell's energy. In addition polysaccharides are also used to build some of the most durable biological structures. The stiff shell of insects, for example are made of long polysaccharides.

Polysaccharides function to create energy, power, courage, strength thereby readying the cell for adventure, and so on. Being so, one can see that there is similarity with the set for system-power highlighted previously in Chapter 2.3. Reproducing Equation 2.3.2:

$$S_{System_P} \ni [Power, Courage, Adventure, Justice, ...]$$

Providing energy and strength, polysaccharides can be thought of as a precipitation of system-power at the cellular level.

Polysaccharides could have a generalized signature, as in Equation 7.2.11, derived from the system-power family:

$$Sig_{polysaccharide} = Xa + \overline{Yb_{0-n}} \ where \begin{bmatrix} X \in [S_{System_P}] \\ Y \in [S_{System_{Pr}}, S_{System_P}, S_{System_K}, S_{System_N}] \\ a, b \ are \ integers; a > b \end{bmatrix}$$

Eq 7.2.11: Polysaccharide

Note that (7.2.11) already implies that Lines 1 – 5 in the Light Matrix (7.2.1) have been activated, and that the logic of the polysaccharide-ecosystem will precipitate into the post material-fabric cell-based genetic structure through the action of Line 6-7 of (7.2.1). This action of precipitating into the cell-based structure as opposed to the material-fabric is designated by 'PGGT <...>' as specified in (7.2.1).

Glycogen is an example of a polysaccharide. Glycogen forms an energy reserve that can be quickly mobilized to meet a sudden need for glucose, but one that is less compact and more immediately available as an energy reserve than say triglycerides. Equation 7.2.12 for Glycogen follows:

$$Sig_{glycogen} = Xa +$$

$$\overline{Yb_{0-n}} \quad where \left[\begin{array}{c} X \in [S_{System_P}] \\ Y \in [S_{System_{Pr}}, S_{System_P}, S_{System_K}, S_{System_N}] \\ a, b \text{ are integers}; a > b \end{array} \right]$$

Eq 7.1.12: Glycogen

Cellulose is another example of a polysaccharide. Cellulose is a polymer made with repeated glucose units bonded together by beta-linkages. Humans and many animals lack an enzyme to break the beta-linkages, so they do not digest cellulose. Equation 7.2.13 for Cellulose follows:

$$Sig_{cellulose} = Xa +$$

$$\overline{Yb_{0-n}} \quad where \left[\begin{array}{c} X \in [S_{System_P}] \\ Y \in [S_{System_{Pr}}, S_{System_P}, S_{System_K}, S_{System_N}] \\ a, b \text{ are integers}; a > b \end{array} \right]$$

Eq 7.2.13: Cellulose

The primary element for the preceding polysaccharides X would be along the lines of 'energy storage'. The secondary elements may vary, with a Y element for Glycogen being 'rapid energy deployment' for example, and a Y element for Cellulose being 'bonded energy', for example.

Summary of Cellular-Level Genetic Code

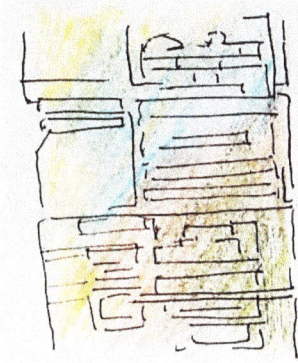

Summarizing, after the cellular stage iterations of the Light-Space-Time Emergence equation are complete, the following code-segments will have been generated as specified by Equation 7.2.14, Active Pre-Genetic and Genetic Code at the Cellular Stage:

$$Light - Space - Time\ Emergence_{Cells} =$$

$$\left\| \begin{array}{c} \left[\begin{array}{c} c_\infty : [Pr, Po, K, H] \\ \left(\downarrow R_{C_K} = f(R_{C_\infty}) \right) \\ c_K : [S_{Pr}, S_{Po}, S_K, S_H] \\ \left(\downarrow R_{C_N} = f(R_{C_K}) \right) \\ c_N : f(S_{Pr} \times S_{Po} \times S_K \times S_H) \\ \left(\downarrow R_{C_U} = f(R_{C_N}) \right) \\ c_U : [P, V, M, C] \\ \Uparrow \\ c_{0:[D,W,I,C]} \end{array} \right]_{Light} \begin{array}{c} \left[\begin{array}{c} M_3 \to System_X \\ (\uparrow F \to I) \\ M_2 \to S_{System_X} \\ (\uparrow Sig \to F) \\ M_1 \to Sig_x \\ (\uparrow > P_x) \\ U \to x_{Molecules} \end{array} \right]_{Space} \\ \\ \left[U \to \begin{array}{c} M_3 : -\infty \le t \le \infty \\ \downarrow \\ M_2 : 0 \ge t > \infty \\ \downarrow \\ M_1 : 0 > t > \infty \\ \downarrow \\ t \le E_{Cell}; TC: M_3 \to U \\ t \sim E_{Human}; TC: U \to M_3 \end{array} \right]_{Time} \begin{array}{c} TC \to x_{Cells} \end{array} \end{array} \right\| \langle x_U | x_T \rangle$$

$$\Rightarrow$$

$$\langle Space - Time - Energy - Gravity\ Material - Fabric\ Pre$$
$$- Genetic\ Code \rangle +$$

$$\langle Electromagnetic\ Spectrum\ Material - Fabric\ Pre - Genetic\ Code \rangle$$
$$+$$

$$\langle Quantum - Particle\ Material - Fabric\ Pre - Genetic\ Code\rangle +$$

$$\langle Atoms\ Material - Fabric\ Pre - Genetic\ Code\rangle +$$

$$\langle Molecules\ Material - Fabric\ Pre - Genetic\ Code\rangle + LSTE\ \langle ...\rangle$$
$$+ PGGT < \cdots > +$$

$$
\left(
\begin{array}{c}
\sum Sig_{lipid} = Xa + \overline{Yb_{0-n}} \\[4pt]
where \left[
\begin{array}{c}
X \in [S_{System_N}] \\
Y \in [S_{System_{Pr}}, S_{System_P}, S_{System_K}, S_{System_N}] \\
a, b\ are\ integers;\ a > b
\end{array}
\right] \\[16pt]
\sum Sig_{nucleic\ acid} = Xa + \overline{Yb_{0-n}} \\[4pt]
where \left[
\begin{array}{c}
X \in [S_{System_K}] \\
Y \in [S_{System_{Pr}}, S_{System_P}, S_{System_K}, S_{System_N}] \\
a, b\ are\ integers;\ a > b
\end{array}
\right] \\[16pt]
\sum Sig_{polysaccharide} = Xa + \overline{Yb_{0-n}} \\[4pt]
where \left[
\begin{array}{c}
X \in [S_{System_P}] \\
Y \in [S_{System_{Pr}}, S_{System_P}, S_{System_K}, S_{System_N}] \\
a, b\ are\ integers;\ a > b
\end{array}
\right] \\[16pt]
\sum Sig_{protein} = Xa + \overline{Yb_{0-n}} \\[4pt]
where \left[
\begin{array}{c}
X \in [S_{System_{Pr}}] \\
Y \in [S_{System_{Pr}}, S_{System_P}, S_{System_K}, S_{System_N}] \\
a, b\ are\ integers;\ a > b
\end{array}
\right]
\end{array}
\right)
$$

Eq. 7.2.14, Active Pre-Genetic and Genetic Code at the Cellular Stage

Note that the final code segment following 'PGGT <...>' contains the code for all possible ($\sum x$) lipids, nucleic acids, polysaccharides, and proteins, respectively. As mentioned previously in the sub-section on nucleic acids, this final

code segment would also be contained in the nucleic acid segment in some form.

Each equation-segment in (7.2.14) will generate a vast body of "code". As a reminder the language of this code is essentially four-fold, deriving from properties of light, and function-based.

Chapter 7.3: Generating the Genetic Code for Sensations, Urges, Feelings, and Thoughts

As human beings we experience sensations, urges and desires and wills, feelings and emotions, and thought. These are key aspects of our being and becoming and critical aspects of how choice at both the individual and collective levels may be determined. Further, there is known to be a tight relationship between these fundamental capacities of being and the effect on bodies. This implies that there is a deeply embedded relationship between these capacities and the actual functioning of bodies and cells.

This chapter suggests that this deeply embedded relationship is founded on genetic code that is created with the emergence of these capacities. This chapter hence, will go over the quantum-level computation that suggests how light emerges as these capacities, how these precipitate into the post material-fabric genetic housing structure, and what the specific equation-segment function-based code manifests as.

Light's Emergence as Fundamental Capacities of Self

As discussed previously the Light-Space-Time Emergence equation (3.1.3) being iterative, can be used to model emergence as it proceeds from simpler four-fold to more complex four-fold manifestations. Hence (3.1.3) has already been applied to suggest the emergence of the space-time-energy-gravity quadrumvirate, the electromagnetic spectrum, quantum particles in general, bosons as a further instance of a particular kind of quantum particle, atoms, and living cells.

Here, a revised form of it building off the PGGT form of the cells genetic code (7.2.1) will be applied to suggest the emergence of fundamental capacities of self, Equation 7.3.1, Fundamental Capacities of Self. The architecture and details of capacities of self can be seen to be the result of the application of the Light, Space, and Time matrices as will be elaborated. But further, as implied by (3.6.5), the Potential Effect of Levels of Light on Pre-Genetic of Genetic Information equation, any process of deeper organization, such as is responsible for the architecture and cohesiveness of capacities of self, has the possibility of altering the material-fabric so long as the bases involved are driven primarily by a meta-level.

Hence:

$$Light - Space - Time\ Emergence_{Fundamental\ Capacities\ of\ Self} =$$

$$
\left| \begin{bmatrix} \begin{array}{c} c_\infty : [Pr, Po, K, H] \\ \left(\downarrow R_{C_K} = f(R_{C_\infty}) \right) \\ c_K : [S_{Pr}, S_{Po}, S_K, S_H] \\ \left(\downarrow R_{C_N} = f(R_{C_K}) \right) \\ c_N : f(S_{Pr} \times S_{Po} \times S_K \times S_H) \\ \left(\downarrow R_{C_U} = f(R_{C_N}) \right) \\ c_U : [P, V, M, C] \\ \Uparrow \\ c_{0:[D,W,I,C]} \end{array} \end{bmatrix}_{Light} \begin{bmatrix} \begin{array}{c} M_3 \rightarrow System_X \\ (\uparrow F \rightarrow I) \\ M_2 \rightarrow S_{System_X} \\ (\uparrow Sig \rightarrow F) \\ M_1 \rightarrow Sig_x \\ (\uparrow > P_x) \\ U \rightarrow x_{Cells} \end{array} \end{bmatrix}_{Space} \\
\begin{bmatrix} \begin{array}{c} M_3 : -\infty \le t \le \infty \\ \downarrow \\ M_2 : 0 \ge t > \infty \\ \downarrow \\ M_1 : 0 > t > \infty \\ \downarrow \\ U \rightarrow \begin{array}{c} t \le E_{Cell}; TC: M_3 \rightarrow U \\ t \sim E_{Human}; TC: U \rightarrow M_3 \end{array} \end{array} \end{bmatrix}_{Time} \begin{array}{c} TC \rightarrow x_{Capacities\ of\ Self} \end{array} \right| \langle x_U | x_T \rangle
$$

$\Rightarrow$

$\langle Space - Time - Energy - Gravity\ Material - Fabric\ Pre$
$- Genetic\ Code \rangle +$

$\langle Electromagnetic\ Spectrum\ \ Material - Fabric\ Pre - Genetic\ Code \rangle$
$+$

$\langle Quantum - Particle\ Material - Fabric\ Pre - Genetic\ Code \rangle +$

313

$\langle Atoms\ Material - Fabric\ Pre - Genetic\ Code\rangle +$

$\langle Molecules\ Material - Fabric\ Pre - Genetic\ Code\rangle + LSTE\ \langle ... \rangle$
$$+ PGGT < \cdots > +$$

$Cells\ Genetic\ Code + LSTE < \cdots$
$> + Fundamental\ Capacities\ of\ Self\ Genetic\ Code$

Eq 7.3.1: Fundamental Capacities of Self Genetic Code

Starting with the Light-Matrix, the top left-hand matrix in (7.3.1), the first line from the top, C_∞: $[Pr, Po, K, H]$, specifies the architecture of the fundamental capacities of self to be introduced in this chapter. Hence, Thoughts are an emergence of Light's property of Knowledge, Urges, Desires, and Wills are an emergence of Light's property of Power, Sensations are an emergence of Light's property of Presence, and Feelings and Emotions are an emergence of Light's property of Harmony. The fundamental architecture of these aspects hence, is an emergence of the properties of Light at ∞.

Line 3 in the Light-Matrix, C_K: $[S_{Pr}, S_{Po}, S_K, S_H]$, elaborates the sets for Presence, Power, Knowledge, and Harmony, each containing multiple elements. For example, as will be explored in the section on Sensations, various elements derived from the four sets define the behavior of Sensations and could be functions such as 'tangible', 'take notice of', amongst others, hence collectively describing Sensations' way of being. Specifically, Line 5, C_N: $f(S_{Pr}\ x\ S_{Po}\ x\ S_K\ x\ S_H)$, suggests that unique seeds are created from a combination of such elements from all four sets, with a particular element leading, that in effect creates the distinctness possible at the level of fundamental capacities of self.

Line 6, ($\downarrow R_{C_U} = f(R_{C_N})$), specifies quantization between the layer where the seeds are formed, and the physical layer we are familiar with, and as explored in

Chapter 3.5 and 3.6, will result in Line 7, C_U: $[P, V, M, C]$, hence changing the material-fabric or post material-fabric housing structure. Note that as in the process describing the generation of the code-segments for space-time-energy-gravity quantization at the time of the Big Bang, the generation of the FBLEE (four-base logic-encoding ecosystem) code-segment as specified by (4.1.5) and the MF (material-fabric) code segment as specified by (4.1.6) are combined together into one equation so that the FBLEE process is apparently transparent.

The possibilities represented by Lines 1 through 5 hence concretize through the quantization represented by Line 6 to become or further enhance the capacities of self with its subtle physical (related to Presence), vital (related to Power), mental (related to Knowledge), and connection (related to Harmony) aspects now existing in material reality typified by Light moving at c.   Note that just as Line 6 represents a process of quantization relating the layer of reality created by Light traveling at c with the antecedent layers, so too Lines 2 and 4 as previously discussed, also represent quantization of a more subtle kind that ultimately plays a critical part in allowing the material-fabric to express infinite diversity.

Typically it is the process as captured by the Space-Matrix that will determine if Line 6 is activated. Specifically patterns at the untransformed layer, U, will need to be overcome, as specified by the second-line from the bottom of the Space-Matrix: $(\uparrow > P_x)$. But as specified by the bottom-line of the Time-Matrix, reproduced below, it is only with the advent of the human-system that the automaticity of the action of meta-levels is reversed:

$$U \rightarrow \begin{array}{l} t \leq E_{Cell}; \text{TC: } M_3 \rightarrow U \\ t \sim E_{Human}; \text{TC: } U \rightarrow M_3 \end{array}$$

Hence in the case of sensations, which in this emergence is largely a post-human system, the fact that patterns do need to be overcome means that quantization requires effort to happen. Given this, it is useful to review Equation 3.6.5, Potential Effect of Levels of Light on Pre-Genetic of Genetic Information:

Potential Effect of Levels of Light on Pre Genetic or Genetic Information =

$$\begin{bmatrix} STATIC \ \langle\|[L][S][T]TC \rightarrow x_T|_{\langle x_U|x_T\rangle}\rangle \\ \times \\ \Big((Y > U{:}\,Z_Q) \lor (Y \leq U{:}\,Z_F) \lor (Y = U{:}\,Z_R)\Big) \\ \ni \\ \Big(\begin{array}{c} Z \in \mathbb{U}\ (Space, Time, Energy, Gravity) \\ Q{:}\ Quantization; F{:}\ Fragmentation; R{:}\ Random \end{array}\Big) \end{bmatrix} \rightarrow h \ni$$

$$h \in \left(\begin{array}{c} [Q]{:}\ Constructive\ zone, \\ [Q]{:}\ Constructive\ zone\ \land\ Constructive\ mutation, \\ [F]{:}\ Destructive\ mutation, \\ [R]{:}\ Random\ mutation \end{array}\right)$$

Line 1 from the top in the matrix is simply a static form of (3.6.1) the Simplified Light-Space-Time Emergence equation. The static form is designated by 'STATIC', and implies that fundamental operations true of (3.6.1) are being highlighted in (3.6.5). In other word (3.6.1) already has all the operations highlighted in (3.6.5) in it, but by 'freezing' it by making it static, the essential dynamics leading to possible mutations at the genetic level can more clearly be highlighted.

Line 1 is then subjected ($\times$) to a determination of the dominant levels of light that may be active, designated by Line 2, ' $\Big((Y > U{:}\,Z_Q) \lor (Y \leq U{:}\,Z_F) \lor (Y = U{:}\,Z_R)\Big)$ '. Unpacking this, 'Y > U' implies meta-levels are active and as a result it is possible that Z_Q is going to take place. This also implies activation and potential change of FBLEE. The call from below, as it were, may invoke some function that already exists in the subtle-libraries 'above', so that some already existing function may influence FBLEE through ∞ -entanglement, K-entanglement, or N-entanglement. Collections of cells perhaps yearn for more elaborate mechanisms by which to begin to dialog with the world. The response could be functions related

316

to sensations, wills, emotions, and thoughts that have already been mathematically formed in some subtle-library. This may be thought of as a key-and-lock mechanism, where a deep enough visceral urge from below acting as the key, opens an entangled lock to alter FBLEE as per the visceral urge.

'$Y \leq U$' implies that only the untransformed levels are active, and therefore also the sub-level where the speed of light is 0 is active and as a result Z_F is going to take place. '$Y = U$' implies that all levels are active and as a result Z_R is going to take place.

Line 3 elaborates the significance of Z_Q, Z_F, and Z_R. Hence Z is the union of potential quantum-operations of space, time, energy, and gravity, designated by '$Z \in \mathbb{U}\ (Space, Time, Energy, Gravity)$'. But the nature of the operations is designated by 'Q: *Quantization; F: Fragmentation; R: Random*'. Z_Q, then, implies that the full quantization originating from updated four-base logic-encoding ecosystems (FBLEE) can take place, and will result in lasting material change at the pre-genetic or genetic level. Z_F implies that the essential set will be fragmented and that only local libraries a the level of local cellular-level DNA can potentially be altered. Z_R also precludes full quantization, and that some partial local-library constructive or destructive mutation may take place.

The '$\rightarrow h \ni h \in$' segment resolves the outcome of the operations implied by Lines 1 – 3, suggesting that the outcome will be 'h' such that ($\ni$) 'h' is an element ($\in$) of the set specified by the members '[Q]: *Constructive zone*', '[Q}: *Constructive zone AND Constructive mutation*', '[F}; *Destructive mutation*', and '{R}: *Random mutation*'. '[Q]: *Constructive zone*' implies that the in-built buffer has been crossed and that access to the deeper four-base logic-encoding ecosystem (FBLEE) has been granted. '[Q}' in this segment implies that there is the possibility that full-quantization as specified by Line 3 of the previous matrix can take place. Access to this zone is a prerequisite for constructive mutation to occur, as designated by the element   '   [Q]: *Constructive zone* ∩ *Constructive mutation* ', which implies that full-quantization is going to take place and will result in material change. The '[F]' specifies the relationship between 'Fragmentation' in Line 3 of the previous matrix and destructive mutation. The '[R]' specifies the relationship between 'Random' in Line 3 of the previous matrix and random mutation.

But as just summarized in the Time-Matrix in (7.3.1) Y is by definition not greater than U and hence quantization is not automatic. In terms of fundamental capacities of self, such quantization implies that increasing wholeness can become fully active through specific space, time, energy, and gravity quantization to create an holistic "ecosystem" with its own "fundamental capacities of self logic" as it were. Possible precipitation into the post material-fabric housing structure is captured by the action of 'PGGT <…>' that can be assumed to be triggered by Z_Q.

Generation of Cellular-Level Genetic Code for Sensations

Sensations are those things we experience with our senses. We see things, hear things, and smell things, taste things, can touch things. This ability to enter into relationship with objects through sensation is nothing other than a result of the emergence of Light's property of Presence. We become present to Presence through the device of sensation. Sensation can be thought of as the means by which this property of Light – Presence - molds or ingrains itself in us as human beings. Its potentiality, all which is contained in this aspect of Light, becomes available to us through the power of sensation. Hence an equation, Equation 7.3.2, will generally represent the family of sensations. Some elements that it would comprise of may be 'tangible', 'take notice of', amongst others.

$$Sig_{sensation} = Xa +$$

$$\overline{Yb_{0-n}} \quad where \quad \begin{bmatrix} X \in [S_{System_{Pr}}] \\ Y \in [S_{System_{Pr}}, S_{System_P}, S_{System_K}, S_{System_N}] \\ a, b \ are \ integers; a > b \end{bmatrix}$$

Eq 7.3.2: Sensation

There could also be equations for hearing, seeing, tasting, touching, and smelling.

But there is also a deeper experience of sensation that is possible. When we see things, for instance, what are we seeing? Is it just the surface rendering of the play of matter, or do we see that the fullness of Light is still there, with all its potentiality and possibility, in the smallest thing we look at? Do we see that the whole universe and more is present in all its fullness in the least thing that we easily ignore, or belittle, or loathe? When we touch things is it the seeming concreteness of the play of the particles or atoms or chains of molecules that we touch? Or is it the Love and Light and the vastness of all that IS that allows itself to be as a small corner that we touch so as to make infinity be felt by something so finite?

Such a deeper contact offered through sensation suggests a subset of (7.3.2) with secondary elements perhaps described as 'fullness of Light', 'contacting infinity', amongst others, thus also yielding an equation form (7.3.3):

319

$$Sig_{deeper-sensation} = Xa +$$

$$\overline{Yb_{0-n}} \quad where \quad \begin{bmatrix} X \in [S_{System_{Pr}}] \\ Y \in [S_{System_{Pr}}, S_{System_P}, S_{System_K}, S_{System_N}] \\ a, b \ are \ integers; a > b \end{bmatrix}$$

Eq 7.3.3: Deeper Sensation

Note that (7.3.2 and 7.3.3) already implies that Lines 1 – 5 in the Light Matrix (7.3.1) have been activated, and that the logic of the sensations- ecosystem will precipitate into the post material-fabric cell-based genetic structure through the action of Line 6-7 of (7.3.1). This action of precipitating into the cell-based structure as opposed to the material-fabric is designated by 'PGGT <...>' as specified in (7.3.1).

Generation of Cellular-Level Genetic Code for Urges, Desires & Wills

Urges and desires and wills are similarly a play of the emergence of Light's property of Power. In the mystery of focus, the vastness of Light has projected itself in us into an apparent smallness that is in reality everything that is. And this smallness is trying through urge and desire and will to connect viscerally or even intentionally to other smallnesses that similarly are nothing other than the fullness of Light projected into a small smorgasbord of selected function. So the urge or desire for food, or companionship, or of possession, or of climbing a peak, is nothing other than Light's compressed property of Power, trying to reach more of the fullness

320

that it is through a fulfillment of the urge or desire or will that it masquerades as. Hence urges can be represented as Equation 7.3.4:

$$Sig_{urges} = Xa +$$

$$\overline{Yb_{0-n}} \quad where \left[\begin{array}{c} X \in [S_{System_P}] \\ Y \in [S_{System_{Pr}}, S_{System_P}, S_{System_K}, S_{System_N}] \\ a, b \ are \ integers; a > b \end{array} \right]$$

Eq 7.3.4: Urges

Elements may be of the type of 'grasp', 'possess', 'deeply connect', amongst others.

Note that (7.3.4) already implies that Lines 1 – 5 in the Light Matrix (7.3.1) have been activated, and that the logic of the urges-ecosystem will precipitate into the post material-fabric cell-based genetic structure through the action of Line 6-7 of (7.3.1). This action of precipitating into the cell-based structure as opposed to the material-fabric is designated by 'PGGT <...>' as specified in (7.3.1).

Generation of Cellular-Level Genetic Code for Feelings & Emotions

Feelings and emotions are a play of the emergence of Light's property of Harmony or Nurturing. Its instrument is the Heart and it generates an array of emotions that are an indication or active radar of whether we are moving toward or away from a reality of harmony, whether based on our small self or some larger Self of Light. Gradually, by navigating with these emotions and feelings we can get to a state where we always feel positive emotions which basically means we have more truly entered into relationship with some larger continent of Light. An equation for feelings is as represented by Equation 7.3.5:

$$Sig_{feelings} = Xa +$$
$$\overline{Yb_{0-n}} \quad where \begin{bmatrix} X \in [S_{System_N}] \\ Y \in [S_{System_{Pr}}, S_{System_P}, S_{System_K}, S_{System_N}] \\ a, b \ are \ integers; a > b \end{bmatrix}$$

Eq 7.3.5: Feelings

Note that (7.3.5) already implies that Lines 1 – 5 in the Light Matrix (7.3.1) have been activated, and that the logic of the feelings-ecosystem will precipitate into the post material-fabric cell-based genetic structure through the action of Line 6-7 of (7.3.1). This action of precipitating into the cell-based structure as opposed to the material-fabric is designated by 'PGGT <...>' as specified in (7.3.1).

Generation of Cellular-Level Genetic Code for Thought

Thoughts are a play of the emergence of Light's property of Knowledge. Through the thought we can become greater or conceptualize things greater or begun to enter into relationship with some things other than our small self. Thought allows us to connect to more "othernesses" or even the oneness of the reality of Light. An equation for thoughts is the following:

$$Sig_{thoughts} = Xa +$$
$$\overline{Yb_{0-n}} \quad where \begin{bmatrix} X \in [S_{System_K}] \\ Y \in [S_{System_{Pr}}, S_{System_P}, S_{System_K}, S_{System_N}] \\ a, b \ are \ integers; a > b \end{bmatrix}$$

Eq 7.3.6: Thoughts

Note that (7.3.5) already implies that Lines 1 – 5 in the Light Matrix (7.3.1) have been activated, and that the logic of the thoughts-ecosystem will precipitate into the post material-fabric cell-based genetic structure through the action of Line 6-7 of (7.3.1). This action of precipitating into the cell-based structure as opposed to the material-fabric is designated by 'PGGT <...>' as specified in (7.3.1).

Summary of Cellular-Level Genetic Code with Fundamental Capacities of Self

Summarizing, after the cellular stage, iterations including the fundamental capacities of self, of the Light-Space-Time Emergence equation are complete, the following code-segments will have been generated as specified by Equation 7.3.7, Active Pre-Genetic and Genetic Code at the Cellular Stage Including Capacities of Self:

$$Light - Space - Time\ Emergence_{Fundamental\ Capacities\ of\ Self} =$$

$$\left\| \begin{bmatrix} c_\infty : [Pr, Po, K, H] \\ \left(\downarrow R_{C_K} = f\left(R_{C_\infty}\right)\right) \\ c_K : [S_{Pr}, S_{Po}, S_K, S_H] \\ \left(\downarrow R_{C_N} = f\left(R_{C_K}\right)\right) \\ c_N : f(S_{Pr} \times S_{Po} \times S_K \times S_H) \\ \left(\downarrow R_{C_U} = f\left(R_{C_N}\right)\right) \\ c_U : [P, V, M, C] \\ \Uparrow \\ c_{0:[D,W,I,C]} \end{bmatrix}_{Light} \begin{bmatrix} M_3 \rightarrow System_X \\ (\uparrow F \rightarrow I) \\ M_2 \rightarrow S_{System_X} \\ (\uparrow Sig \rightarrow F) \\ M_1 \rightarrow Sig_X \\ (\uparrow > P_X) \\ U \rightarrow x_{Cells} \end{bmatrix}_{Space} \right.$$

$$\left\| U \rightarrow \begin{bmatrix} M_3 : -\infty \leq t \leq \infty \\ \downarrow \\ M_2 : 0 \geq t > \infty \\ \downarrow \\ M_1 : 0 > t > \infty \\ \downarrow \\ t \leq E_{Cell}; TC: M_3 \rightarrow U \\ t \sim E_{Human}; TC: U \rightarrow M_3 \end{bmatrix}_{Time} \quad TC \rightarrow x_{Capacities\ of\ Self} \right\|_{\langle x_U | x_T \rangle}$$

$\Rightarrow$

$\langle Space - Time - Energy - Gravity\ Material - Fabric\ Pre$
$\qquad - Genetic\ Code \rangle +$

$\langle Electromagnetic\ Spectrum\ \ Material - Fabric\ Pre - Genetic\ Code \rangle$
$\qquad +$

$\langle Quantum - Particle\ Material - Fabric\ Pre - Genetic\ Code \rangle +$

$\langle Atoms\ Material - Fabric\ Pre - Genetic\ Code \rangle +$

$\langle Molecules\ Material - Fabric\ Pre - Genetic\ Code \rangle + LSTE\ \langle \dots \rangle$
$\qquad + PGGT < \dots > +$

$Cells\ Genetic\ Code + LSTE < \dots > +$

$$\begin{pmatrix}
\sum Sig_{feelings} = Xa + \overline{Yb_{0-n}} \\
where \begin{bmatrix} X \in [S_{System_N}] \\ Y \in [S_{System_{Pr}}, S_{System_P}, S_{System_K}, S_{System_N}] \\ a, b \ are \ integers; a > b \end{bmatrix} \\
\sum Sig_{thoughts} = Xa + \overline{Yb_{0-n}} \\
where \begin{bmatrix} X \in [S_{System_K}] \\ Y \in [S_{System_{Pr}}, S_{System_P}, S_{System_K}, S_{System_N}] \\ a, b \ are \ integers; a > b \end{bmatrix} \\
\sum Sig_{urges} = Xa + \overline{Yb_{0-n}} \\
where \begin{bmatrix} X \in [S_{System_P}] \\ Y \in [S_{System_{Pr}}, S_{System_P}, S_{System_K}, S_{System_N}] \\ a, b \ are \ integers; a > b \end{bmatrix} \\
\sum Sig_{sensation} = Xa + \overline{Yb_{0-n}} \\
where \begin{bmatrix} X \in [S_{System_{Pr}}] \\ Y \in [S_{System_{Pr}}, S_{System_P}, S_{System_K}, S_{System_N}] \\ a, b \ are \ integers; a > b \end{bmatrix}
\end{pmatrix}$$

Eq. 7.3.7, Active Pre-Genetic and Genetic Code at the Cellular Stage Including Capacities of Self

Note that the final code segment following 'LSTE <...>' contains the code for all possible ($\sum x$) sensations, urges, feelings, and thought, respectively.

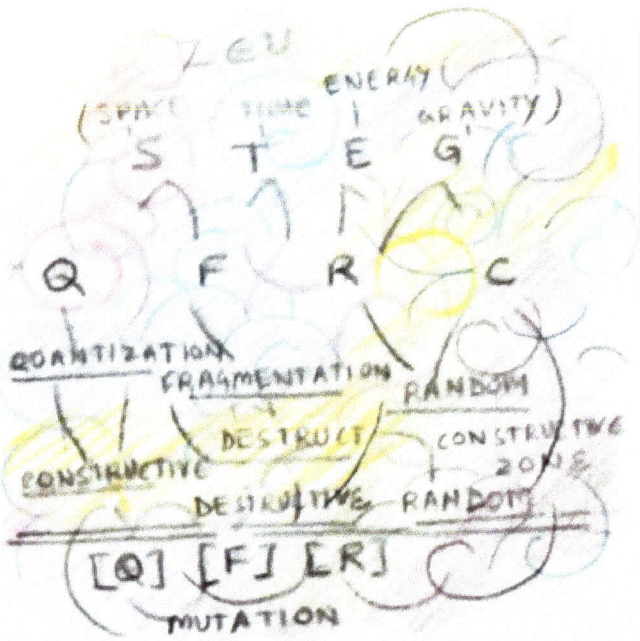

ZEU

(SPACE   TIME   ENERGY   GRAVITY)

S   T   E   G'

Q   F   R   C

QUANTIZATION
        FRAGMENTATION   RANDOM
              DESTRUCT   CONSTRUCTIVE
CONSTRUCTIVE   ZONE
        DESTRUCTIVE   RANDOM

[Q] [F] [R]
        MUTATION

Chapter 7.4: Generating the Genetic Code for Truer Individuality

At the core individuals can be thought of as projections of seeds formed in a vast continent of Light. So that core must always be there but it is often covered by surface dynamics of the physical, the vital, and the mental. These too are formed from properties of Light, but since the light has been separated from its source these movements and dynamics are incomplete in their nature.

It is easy for these dynamics and movements to completely occupy all individual processing power. This can happen to such an extent that the individual acting on the surface can become subject to these surface movements. But so long as there is the remembrance that even in separation and smallness - whether selfishness, myopia, fundamentalism, or any other ism - these movements are still light, then that can become the means by which the apparent chains around individuals can be loosened.

For in its essence these small movements are trying to extend into something other than what they are. By connecting the light in them with the larger Light this extension can yield something different than what is possible through any other means. And in the process the hold that these smaller movements have is diminished and one can begin to see or experience larger vistas of light.

Such a passage where one is always moving to larger vistas of light, more easily allows entry into a field of original seeds and it is so that each can begin to enter into conscious communion with truer individuality.

Light's Emergence as Truer Individuality

As discussed previously the Light-Space-Time Emergence equation (3.1.3) being iterative, can be used to model emergence as it proceeds from simpler four-fold to more complex four-fold manifestations. Hence (3.1.3) has already been applied to suggest the emergence of the space-time-energy-gravity quadrumvirate, the electromagnetic spectrum, quantum particles in general, bosons as a further instance of a particular kind of quantum particle, atoms, living cells, and basic capacities of the self.

Here, a revised form of it building off the truer individuality equation (4.2.6) will be applied to suggest the emergence of truer individuality, Equation 7.4.1, Truer Individuality Genetic Code. The architecture and details of truer individuality can be seen to be the result of the application of the Light, Space, and Time matrices as will be elaborated. But further, as implied by (3.6.5), the Potential Effect of Levels of Light on Pre-Genetic of Genetic Information equation, any process of deeper organization, such as is responsible for the architecture and cohesiveness of truer individuality, has the possibility of altering FBLEE and post material fabric housing structures so long as the bases involved are driven primarily by a meta-level.

$$Light - Space - Time\ Emergence_{Truer\ Individuality} =$$

$$\left\|\begin{bmatrix} c_\infty: [Pr, Po, K, H] \\ \left(\downarrow R_{C_K} = f(R_{C_\infty})\right) \\ c_K: [S_{Pr}, S_{Po}, S_K, S_H] \\ \left(\downarrow R_{C_N} = f(R_{C_K})\right) \\ c_N: f(S_{Pr} \times S_{Po} \times S_K \times S_H) \\ \left(\downarrow R_{C_U} = f(R_{C_N})\right) \\ c_U: [P, V, M, C] \\ \Uparrow \\ c_{0:[D,W,I,C]} \end{bmatrix}_{Light} \begin{bmatrix} M_3 \to System_X \\ (\uparrow F \to I) \\ M_2 \to S_{System_X} \\ (\uparrow Sig \to F) \\ M_1 \to Sig_x \\ (\uparrow > P_x) \\ U \to x_{Humans} \end{bmatrix}_{Space} \right.$$

$$\left. \begin{bmatrix} M_3: -\infty \leq t \leq \infty \\ \downarrow \\ M_2: 0 \geq t > \infty \\ \downarrow \\ M_1: 0 > t > \infty \\ \downarrow \\ U \to \begin{array}{l} t \leq E_{Cell}; TC: M_3 \to U \\ t \sim E_{Human}; TC: U \to M_3 \end{array} \end{bmatrix}_{Time} \quad TC \to x_{Truer\ Individuality} \right\|_{\langle x_U | x_T \rangle}$$

$$\Rightarrow$$

$\langle Space - Time - Energy - Gravity\ Material - Fabric\ Pre$
$\qquad\qquad - Genetic\ Code \rangle +$

$\langle Electromagnetic\ Spectrum\ Material - Fabric\ Pre - Genetic\ Code \rangle +$

$\langle Quantum - Particle\ Material - Fabric\ Pre - Genetic\ Code \rangle +$

$\langle Atoms\ Pre - Genetic\ Code \rangle$
$\qquad\qquad + \langle Molecules\ Material - Fabric\ Pre$
$\qquad\qquad - Genetic\ Code \rangle +$

$LSTE \langle \ldots \rangle + PGGT < \cdots > + \langle Cells\ Genetic\ Code \rangle + LSTE \langle \ldots \rangle +$

$(Fundamental\ Capacities\ of\ Self\ Genetic\ Code) + FBLEEE \langle \ldots \rangle +$

$(Humans\ Genetic\ Code) + LSTE \langle \ldots \rangle + FBLEEE \langle \ldots \rangle +$

$Truer\ Individuality\ Genetic\ Code$

Eq 7.4.1: *Truer Individuality Genetic Code*

Starting with the Light-Matrix, the top left-hand matrix in (7.4.1), the first line from the top, C_∞: $[Pr, Po, K, H]$, specifies the architecture of truer individuality as will be further discussed in this chapter. Hence, Knowledge-type individuals are an emergence of Light's property of Knowledge, Power-type individuals are an emergence of Light's property of Power, Service-type individuals are an emergence of Light's property of Presence, and Harmony-type individuals are an emergence of Light's property of Harmony. The fundamental architecture of these aspects hence, is an emergence of the properties of Light at ∞.

Line 3 in the Light-Matrix, C_K: $[S_{Pr}, S_{Po}, S_K, S_H]$, elaborates the sets for Presence, Power, Knowledge, and Harmony, each containing multiple elements. For example, as will be explored in the subsequent subsection, various elements derived from the four sets would define the architecture of Harmony-type individuals and could be functions such as 'driven to connect people together', amongst others, hence collectively describing Harmony-type individuals' way of being. Specifically, Line 5, $C_{N:} f(S_{Pr} \times S_{Po} \times S_K \times S_H)$, suggests that unique seeds are created from a combination of such elements from all four sets, with a particular element leading, that in effect creates the distinctness possible at the level of individuals.

Line 6, ($\downarrow$ $R_{C_U} = f(R_{C_N})$) , specifies quantization between the layer where the seeds are formed, and the physical layer we are familiar with, and as explored in Chapter 3.5 and 3.6, will result in Line 7,

C_U: $[P, V, M, C]$, hence changing the FBLEE and subsequently the post material-fabric housing structure. Note that as in the process describing the generation of the code-segments for space-time-energy-gravity quantization at the time of the Big Bang, the generation of the FBLEE (four-base logic-encoding ecosystem) code-segment as specified by (4.1.5) and the MF (material-fabric) code segment as specified by (4.1.6) are combined together into one equation so that the FBLEE process is apparently transparent.

The possibilities represented by Lines 1 through 5 hence concretize through the quantization represented by Line 6 to become or further enhance the unique individuality with its subtle physical (related to Presence), vital (related to Power), mental (related to Knowledge), and connection (related to Harmony) aspects now existing in material reality typified by Light moving at c. Note that just as Line 6 represents a process of quantization relating the layer of reality created by Light traveling at c with the antecedent layers, so too Lines 2 and 4 as previously discussed, also represent quantization of a more subtle kind that ultimately plays a critical part in allowing FBLEE and post material-fabric housing structure to express infinite diversity.

Typically it is the process as captured by the Space-Matrix that will determine if Line 6 is activated. Specifically patterns at the untransformed layer, U, will need to be overcome, as specified by the second-line from the bottom of the Space-Matrix: $(\uparrow > P_x)$. But as specified by the bottom-line of the Time-Matrix, reproduced below, it is only with the advent of the human-system that the automaticity of the action of meta-levels is reversed:

$$U \rightarrow \begin{array}{l} t \leq E_{Cell}; \text{TC: } M_3 \rightarrow U \\ t \sim E_{Human}; \text{TC: } U \rightarrow M_3 \end{array}$$

Hence in the case of truer individuality, which in this emergence is a post-human system, the fact that patterns do need to be overcome means that quantization requires effort to happen. Given this, it is useful to review Equation 3.6.5, Potential Effect of Levels of Light on Pre-Genetic of Genetic Information:

$$= \begin{bmatrix} STATIC \; \langle|[L][S][T]TC \rightarrow x_T|_{\langle x_U|x_T\rangle}\rangle \\ \times \\ \left((Y > U: Z_Q) \vee (Y \leq U: Z_F) \vee (Y = U: Z_R) \right) \\ \ni \\ \begin{pmatrix} Z \in \mathbb{U} \; (Space, Time, Energy, Gravity) \\ Q: Quantization; F: Fragmentation; R: Random \end{pmatrix} \end{bmatrix} \rightarrow h \ni$$

$$h \in \begin{pmatrix} [Q]: Constructive\ zone, \\ [Q]: Constructive\ zone \wedge Constructive\ mutation, \\ [F]: Destructive\ mutation, \\ [R]: Random\ mutation \end{pmatrix}$$

Line 1 from the top in the matrix is simply a static form of (3.6.1) the Simplified Light-Space-Time Emergence equation. The static form is designated by 'STATIC', and implies that fundamental operations true of (3.6.1) are being highlighted in (3.6.5). In other word (3.6.1) already has all the operations highlighted in (3.6.5) in it, but by 'freezing' it by making it static, the essential dynamics leading to possible mutations at the genetic level can more clearly be highlighted.

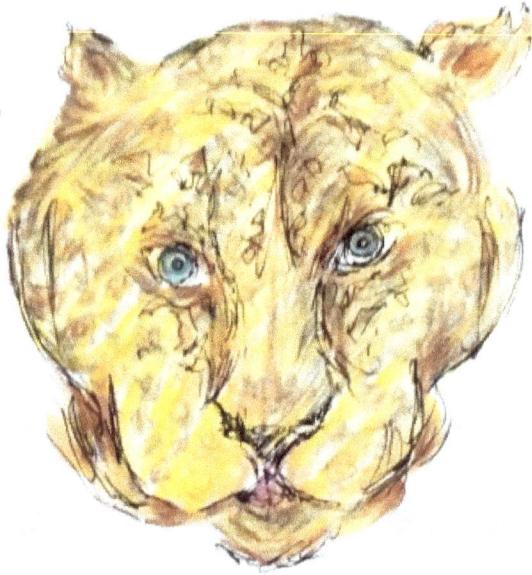

Line 1 is then subjected (×) to a determination of the dominant levels of light that may be active, designated by Line 2, ' $\left((Y > U : Z_Q) \vee (Y \leq U : Z_F) \vee (Y = U : Z_R) \right)$ '. Unpacking this, 'Y > U' implies meta-levels are active and as a result it is possible that Z_Q is going to take place. This also implies activation and potential change of FBLEE. The call from below, as it were, may invoke some function that already exists in the subtle-libraries 'above', so that some already existing function may influence FBLEE through ∞-entanglement, K-entanglement, or N-entanglement. With the emergence of material-frames that can house complex physical-vital-mental beings, the urge for more cohesive unique seeds and plays of Light manifest as truer individuality perhaps arises. A response from above initiates the precipitation of meta-functions related to a range of different truer individuality types. This may be thought of as a key-and-lock mechanism, where a deep enough visceral urge from below acting as the key, opens an entangled lock to alter FBLEE as per the visceral urge.

'$Y \leq U$' implies that only the untransformed levels are active, and therefore also the sub-level where the speed of light is 0 is active and as a result Z_F is going to take place. '$Y = U$' implies that all levels are active and as a result Z_R is going to take place.

Line 3 elaborates the significance of Z_Q, Z_F, and Z_R. Hence Z is the union of potential quantum-operations of space, time, energy, and gravity, designated by '$Z \in \mathbb{U}(Space, Time, Energy, Gravity)$'. But the nature of the operations is designated by 'Q: *Quantization*; F: *Fragmentation*; R: *Random*'. Z_Q, then, implies that the full quantization originating from updated four-base logic-encoding ecosystems (FBLEE) can take place, and will result in lasting material change at the pre-genetic or genetic level. Z_F implies that the essential set will be fragmented and that only local libraries a the level of local cellular-level DNA can potentially be altered. Z_R also precludes full quantization, and that some partial local-library constructive or destructive mutation may take place.

The '$\rightarrow h \ni h \in$' segment resolves the outcome of the operations implied by Lines 1 – 3, suggesting that the outcome will be 'h' such that ($\ni$) 'h' is an element ($\in$) of the set specified by the members '[Q]: *Constructive zone*', '[Q]: *Constructive zone AND Constructive mutation*', '[F]: *Destructive mutation*', and '{R}: *Random mutation*'. '[Q]: *Constructive zone*' implies that the in-built buffer has been crossed and that access to the deeper four-base logic-encoding ecosystem (FBLEE) has been granted. '[Q]' in this segment implies that there is the possibility that full-quantization as specified by Line 3 of the previous matrix can take place. Access to this zone is a prerequisite for constructive mutation to occur, as designated by the element ' [*Q*}: *Constructive zone* ∩ *Constructive mutation* ', which implies that full-quantization is going to take place and will result in material change. The '[F]' specifies the relationship between 'Fragmentation' in Line 3 of the previous matrix and destructive mutation. The '[R]' specifies the relationship between 'Random' in Line 3 of the previous matrix and random mutation.

But as just summarized in the Time-Matrix in (7.4.1) Y is by definition not greater than U and hence quantization is not automatic. In terms of truer individuality such quantization implies that increasing wholeness can become fully active through specific space, time, energy, and gravity quantization to create an holistic "ecosystem" with its own "truer individuality logic" as it were. Possible precipitation into the post material-fabric housing structure is captured by the action of 'PGGT <...>' that can be assumed to be triggered by Z_Q.

Generation of Cellular-Level Genetic-Code for Truer Individuality

The truer individuality itself is some mathematical function of the four properties of Light - Harmony, Power, Knowledge, and Service, which in turn, are practically infinite sets of qualities related to one of the main properties of Light.

So it may be that one individual is primarily driven to connect people together, itself a variation of the property of Harmony. The individual may further want to do this by deeply understanding what makes these people tick, itself a variation of the property of Knowledge. So the seed or the truer individuality of this person can be thought of as a mathematical function consisting of some element or property from the set of Harmony and some element or quality from the set of Knowledge, combined together with possibly different elements from the same

336

or different sets, all with possibly different weights, but with the first weight of the need to connect people, being the strongest.

Such an individual may have an essential form as expressed in Equation 7.4.2, Harmony-Type Individual:

$$Sig_{Harmony-type} = Xa +$$

$$\overline{Yb_{0-n}} \;\; where \left[\begin{array}{c} X \in [S_{System_N}] \\ Y \in [S_{System_{Pr}}, S_{System_P}, S_{System_K}, S_{System_N}] \\ a, b \; are \; integers; a > b \end{array} \right]$$

Eq 7.4.2: Harmony-Type Individual

But equally, and consistent with the emergent properties of Light there could be essential Service-Type, Power-Type, and Knowledge-Type individuals as well, with infinite variation in the precise nature of the core seed, as captured in Equations 7.4.3-5:

$$Sig_{Service-type} = Xa + \overline{Yb_{0-n}}$$

$$where \left[\begin{array}{c} X \in [S_{System_{Pr}}] \\ Y \in [S_{System_{Pr}}, S_{System_P}, S_{System_K}, S_{System_N}] \\ a, b \; are \; integers; a > b \end{array} \right]$$

Eq 7.4.3: Service-Type Individual

$$Sig_{Power-type} = Xa + \overline{Yb_{0-n}}$$

$$where \left[\begin{array}{c} X \in [S_{System_P}] \\ Y \in [S_{System_{Pr}}, S_{System_P}, S_{System_K}, S_{System_N}] \\ a, b \; are \; integers; a > b \end{array} \right]$$

Eq 7.4.4: Power-Type Individual

$$Sig_{Knowledge-type} = Xa + \overline{Yb_{0-n}}$$

$$where \begin{bmatrix} X \in [S_{System_K}] \\ Y \in [S_{System_{Pr}}, S_{System_P}, S_{System_K}, S_{System_N}] \\ a, b \ are \ integers; a > b \end{bmatrix}$$

Eq 7.4.5: Knowledge-Type Individual

Note that (7.4.2 - 5) already implies that Lines 1 – 5 in the Light Matrix (37.4.1) have been activated, and that the logic of the truer-individuality- ecosystem will precipitate into the post material-fabric cell-based genetic structure through the action of Line 6-7 of (7.4.1). This action of precipitating into the cell-based structure as opposed to the material-fabric is designated by 'PGGT <...>' as specified in (7.4.1).

Summary of Cellular-Level Genetic Code with Truer Individuality

Summarizing, after the cellular stage iterations - including the truer individuality - of the Light-Space-Time Emergence equation are complete, the following code-segments will have been generated as specified by Equation 7.4.6, Active Pre-Genetic and Genetic Code at the Cellular Stage with Truer Individuality:

$Light - Space - Time \ Emergence_{Truer \ Individuality} =$

$$\left| \begin{array}{l} \left[\begin{array}{c} c_\infty : [Pr, Po, K, H] \\ \left(\downarrow R_{C_K} = f(R_{C_\infty}) \right) \\ c_K : [S_{Pr}, S_{Po}, S_K, S_H] \\ \left(\downarrow R_{C_N} = f(R_{C_K}) \right) \\ c_N : f(S_{Pr} \; x \; S_{Po} \; x \; S_K \; x \; S_H) \\ \left(\downarrow R_{C_U} = f(R_{C_N}) \right) \\ c_U : [P, V, M, C] \\ \Uparrow \\ c_{0 : [D, W, I, C]} \end{array} \right]_{Light} \quad \left[\begin{array}{c} M_3 \rightarrow System_X \\ (\uparrow F \rightarrow I) \\ M_2 \rightarrow S_{System_X} \\ (\uparrow Sig \rightarrow F) \\ M_1 \rightarrow Sig_x \\ (\uparrow > P_x) \\ U \rightarrow x_{Human} \end{array} \right]_{Space} \\[2em] \left[\begin{array}{c} M_3 : -\infty \leq t \leq \infty \\ \downarrow \\ M_2 : 0 \geq t > \infty \\ \downarrow \\ M_1 : 0 > t > \infty \\ \downarrow \\ U \rightarrow \begin{array}{c} t \leq E_{Cell}; \text{TC}: M_3 \rightarrow U \\ t \sim E_{Human}; \text{TC}: U \rightarrow M_3 \end{array} \end{array} \right]_{Time} \quad TC \rightarrow x_{Truer \; Individuality} \end{array} \right| \langle x_U | x_T \rangle$$

$\Rightarrow$

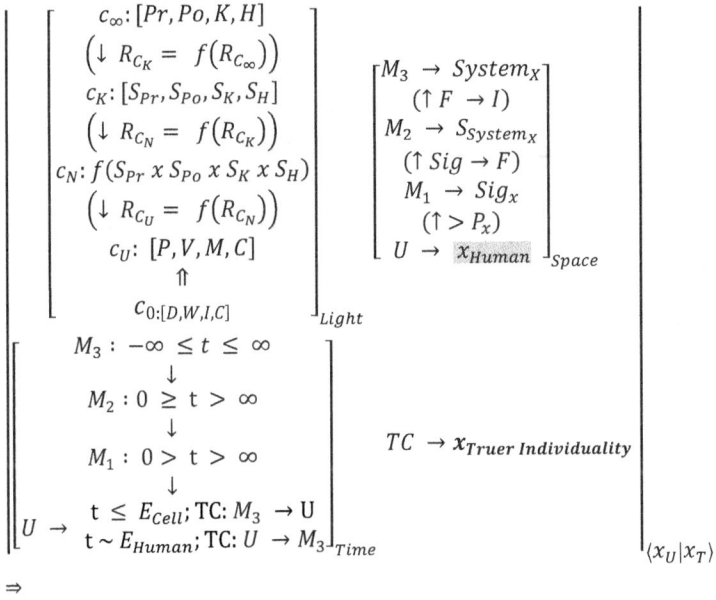

$\langle Space - Time - Energy - Gravity \; Material - Fabric \; Pre \\ - Genetic \; Code \rangle +$

$\langle Electromagnetic \; Spectrum \; Material - Fabric \; Pre - Genetic \; Code \rangle +$

$\langle Quantum - Particle \; Material - Fabric \; Pre - Genetic \; Code \rangle +$

$\langle Atoms \; Pre - Genetic \; Code \rangle +$

$\langle Molecules \; Material - Fabric \; Pre - Genetic \; Code \rangle + LSTE \; \langle ... \rangle +$

$PGGT < \cdots > + \langle Cells \; Genetic \; Code \rangle \; + LSTE \; \langle ... \rangle +$

$(Fundamental \; Capacities \; of \; Self \; Genetic \; Code) +$

$FBLEEE \; \langle ... \rangle \; + (Humans \; Genetic \; Code) + LSTE \; \langle ... \rangle + FBLEEE \; \langle ... \rangle +$

$$\left(\begin{array}{l} \sum Sig_{Harmony-type} = Xa + \overline{Yb_{0-n}} \\ \text{where } \left[\begin{array}{c} X \in [S_{System_N}] \\ Y \in [S_{System_{Pr}}, S_{System_P}, S_{System_K}, S_{System_N}] \\ a, b \text{ are integers; } a > b \end{array} \right] \\ \sum Sig_{Knowledge-type} = Xa + \overline{Yb_{0-n}} \\ \text{where } \left[\begin{array}{c} X \in [S_{System_K}] \\ Y \in [S_{System_{Pr}}, S_{System_P}, S_{System_K}, S_{System_N}] \\ a, b \text{ are integers; } a > b \end{array} \right] \\ \sum Sig_{Power-type} = Xa + \overline{Yb_{0-n}} \\ \text{where } \left[\begin{array}{c} X \in [S_{System_P}] \\ Y \in [S_{System_{Pr}}, S_{System_P}, S_{System_K}, S_{System_N}] \\ a, b \text{ are integers; } a > b \end{array} \right] \\ \sum Sig_{Service-type} = Xa + \overline{Yb_{0-n}} \\ \text{where } \left[\begin{array}{c} X \in [S_{System_{Pr}}] \\ Y \in [S_{System_{Pr}}, S_{System_P}, S_{System_K}, S_{System_N}] \\ a, b \text{ are integers; } a > b \end{array} \right] \end{array} \right)$$

Eq. 7.4.6, Active Pre-Genetic and Genetic Code at the Cellular Stage with Truer Individuality

Note that the final code segment following 'FBLEEE <...>' contains the code for all possible ($\sum x$) human-types.

SECTION 8:   GENERATING GENETIC-CODE FOR COMPLEX ORGANIZATIONS

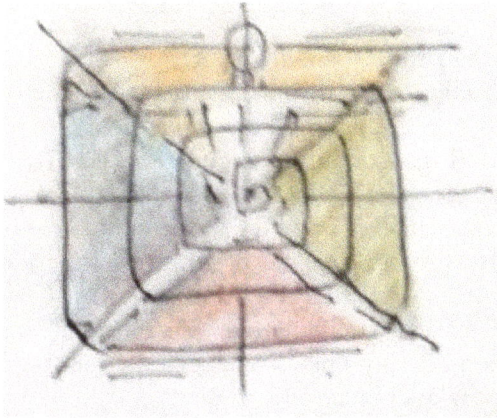

Having traced the computations resulting in the emergence of space-time-energy-gravity, the electromagnetic spectrum, matter, life, and even human becoming as manifest in sensation, urges, feelings, and thoughts, and truer individuality, we now turn our attention to study the computation resulting in the emergence of complex organizations and possible genetic code related to that.

In the previous sections we have seen that there is a gradual addition to pre-genetic and then genetic code based on the emergence of more complex fourfold functionality. The condition for constructive mutation is that some meta-level is active. When this happens there is a possibility that the existing four-base logic-encoding ecosystems (FBLEE) are enhanced or that new ones are created. Note that it is such FBLEE action that alters the quantum-level interface between the material and antecedent layers of light in effect changing the basis by which matter materializes. In other words FBLEE changes what matter can be. This will be explored in greater detail in a subsequent section on matter. Further, through the action of the light-space-time emergence equation (LSTEE) FBLEE can materialize in the material-fabric, or in the post material-fabric cellular structure following action of pre-genetic to genetic transition (PGGT).

But at its heart it is good to get clear that it is the complexification of four-foldness that is the driver of changes to FBLEE and subsequent possible changes to matter and the genetic substance of living cells. This is so because such complexification essentially necessitates that existing habits, of any kind, are continually broken and in the process can admit of greater and more sweeping actions of light from the meta-levels. It is such action that makes clear how the emergence of complex organization actually links to changes such as to matter and to the make-up of genetic content.

Chapter 8.1 examines the emergence of mega-organizations and the cellular-level genetic code it can generate. Similarly Chapter 8.2 examines the emergence of sustainable global civilization and its impact on cellular-level genetic code.

We have traced the emergence of the four properties of Light through the primary fourfold emergence of space-time-energy-gravity, the electromagnetic spectrum, through matter, through life, and even through human becoming as manifest in sensation, urges, feelings, and thoughts, and truer individuality. At each of these levels it seems it is a balanced combination of all four of the properties that creates stability and makes that structure or organism or being, sustainable. And this has to be since the four properties occur simultaneously in light and must be impelled to hold to that relationship even when projected from it. Hence it should be possible to deductively arrive at an equation for organizational sustainability given the four-fold emergence of properties of light through subsequent layers of manifestation.

To triangulate, though, we briefly trace the four-fold emergence, reinforce the significance of this and inductively arrive at an equation for sustainability that we would have arrived at deductively.

Starting with the primary emergence of space-time-energy-gravity we see that space is defined as that in which seeds of knowledge are planted – hence, existing as a means to express Light's property of knowledge. Time is defined as the inevitability of the seeds of knowledge maturing into fullness, in spite of any opposition. Hence time is seen as an emergence of Light's property of power. Energy is seen as that which accumulates to create matter. Hence it is an emergence of Light's property of Presence. Gravity is that by which relationship between objects comes into being – hence an emergence of Light's property of harmony.

With the electromagnetic spectrum we see that the spectrum as a whole is defined or architected by the speed of light c, as an emergence of nurturing, the wavelength λ, as an emergence of knowledge, the frequency f, as an emergence of power, and mass-potential as an emergence of presence. These occur simultaneously and constitute the wholeness of the electromagnetic spectrum.

Similarly, at the level of the quantum particle it can be seen that the integrity and functioning of the atom also depends on the integration of all four properties of light. Hence quarks, an emergence of knowledge, leptons, an emergence of power, bosons, an emergence of harmony or nurturing, and the Higgs-boson, an emergence of presence, act together to create the structure and functionality of every single atom.

What this reinforces is that every single thing must have been created through some integration and balance of the four properties of light.

Going up the organizational scale to the next level of complexity, to the level of the atom, the same pattern is found again: every single atom belongs to one of four groups, and in unison these four groups orchestrate the set of combinations of known compounds.

Hence, Alkali Metals and Alkali Earth Metals configured by the s-Group, from the family of system-power, Metals, Metalloid, Non-Metals, Halogens, and Noble Gases configured by the p-Group, from the family of system-knowledge, Transition Metals configured by the d-Group, from the family of system-presence, and Lanthanides and Actinides configured by the f-Group, from the family of system-nurturing, act together to create the complex array of compounds that form the entire material from which the physical and subsequently even the organic world around us is constructed.

Going up the organizational scale to another level of complexity, to the level of the cell, the same pattern is found again: basically every single cell of every single living creature that has ever been studied by humankind also has a similar balance and integration of these four sets of forces acting together.

Hence, proteins, from the family of presence, nucleic acids, from the family of knowledge, polysaccharides, from the family of power, and lipids, from the family of nurturing, work together to create the balanced functioning of every living cell.

The hypothesis hence, is that even at larger scale, in considering the most innovative organization at the level of teams, corporations, or markets, a similar integration and balance of the four sets of properties or forces may yield the best results.   Recent developments in sustainability investment models ranging from Socially Responsible Investing (Logue, 2008), the Global Reporting Initiative (Willis, 2003), the Dow Jones Sustainability Index (Hope & Fowler, 2007), the Principles of Responsible Investing (Harvard-Edu, 2014), all reinforce the concept of investment criteria becoming broader to be based not only on economic, but on environmental, social, and governance factors as well. Further evidence suggests that financial returns on such broad-based investment models continually beat financial returns on regular funds such as the S&P 500, for instance (Openshaw, 2015).   In their study of major transitions in evolution Smith and Szathmary (Smith & Szathmary, 1995) chronicle the development of life toward increasing complexity as an application of successful collaboration and even co-evolution by which species evolve by changing together as system pressures increase.   This hypothesis can be summarized by the following graph:

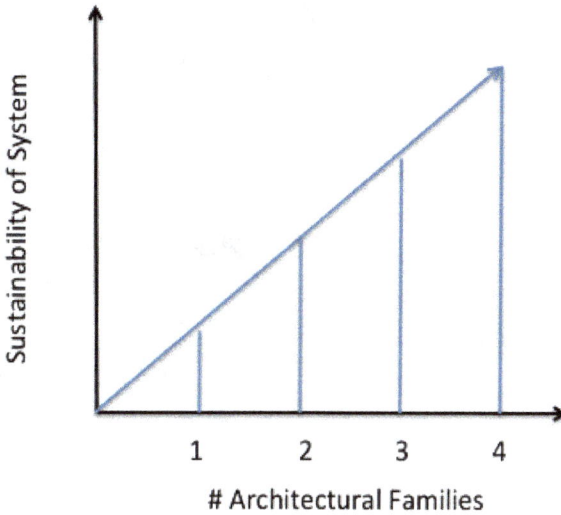

Figure 8.1.1 Sustainability of CAS

An equation for the sustainability of systems, *Sustainability$_{Systems}$*, where the interaction between the four families of forces is instrumental can be created. Hence, as in Equation 8.1.1:

$$Sustainability_{Systems}$$
$$\propto Interaction\ (S_{System_{Pr}}, S_{System_P}, S_{System_K}, S_{System_N})$$

Eq 8.1.1: Sustainability of Systems

This notion of the balance of the fourfold properties of Light as fundamental in creating sustainable mega-organization will be borne out in the next subsections of this chapter.

Light's Emergence as Stable Mega-Organization

As discussed previously the Light-Space-Time Emergence equation (3.1.3) being iterative, can be used to model emergence as it proceeds from simpler four-fold to more complex four-fold manifestations. Hence (3.1.3) has already been applied to suggest the emergence of the space-time-energy-gravity quadrumvirate, the electromagnetic spectrum, quantum particles in general, bosons as a further instance of a particular kind of quantum particle, atoms, living cells, basic capacities of the self, and truer individuality.

SUSTAINABILITY OF SYSTEM CAS :

ARCHITECTURAL FAMILIES

Here, a revised form of it building off the truer individuality equation (7.4.1) will be applied to suggest the emergence of truer individuality, Equation 8.1.2, Stable Mega-Organization Genetic Code. The architecture and details of a sustainable mega-organization can be seen to be the result of the application of the Light, Space, and Time matrices as will be elaborated. But further, as implied by (3.6.5), the Potential Effect of Levels of Light on Pre-Genetic of Genetic Information equation, any process of deeper organization, such as is responsible for the architecture and cohesiveness of sustainable mega-organizations, has the possibility of altering FBLEE and post material fabric housing structures so long as the bases involved are driven primarily by a meta-level.

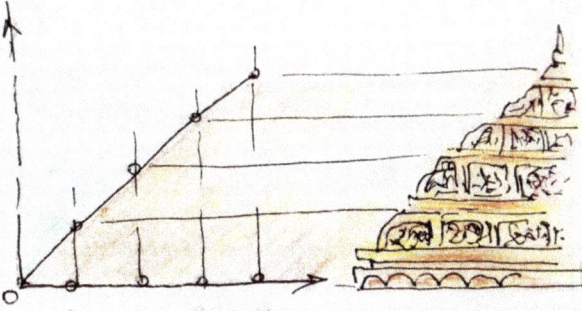

$$Light - Space - Time\ Emergence_{Stable\ Mega-Orgnization} =$$

$$\left|\begin{matrix} \begin{bmatrix} c_\infty: [Pr, Po, K, H] \\ \left(\downarrow R_{C_K} = f\left(R_{C_\infty}\right)\right) \\ c_K: [S_{Pr}, S_{Po}, S_K, S_H] \\ \left(\downarrow R_{C_N} = f\left(R_{C_K}\right)\right) \\ c_N: f(S_{Pr}\ x\ S_{Po}\ x\ S_K\ x\ S_H) \\ \left(\downarrow R_{C_U} = f\left(R_{C_N}\right)\right) \\ c_U: [P, V, M, C] \\ \Uparrow \\ c_{0:[D,W,I,C]} \end{bmatrix}_{Light} \begin{bmatrix} M_3 \rightarrow System_X \\ (\uparrow F \rightarrow I) \\ M_2 \rightarrow S_{System_X} \\ (\uparrow Sig \rightarrow F) \\ M_1 \rightarrow Sig_x \\ (\uparrow > P_x) \\ U \rightarrow x_{Humans} \end{bmatrix}_{Space} \\ \begin{bmatrix} M_3: -\infty \leq t \leq \infty \\ \downarrow \\ M_2: 0 \geq t > \infty \\ \downarrow \\ M_1: 0 > t > \infty \\ \downarrow \\ U \rightarrow \begin{array}{l} t \leq E_{Cell}; TC: M_3 \rightarrow U \\ t \sim E_{Human}; TC: U \rightarrow M_3 \end{array} \end{bmatrix}_{Time} \quad TC \rightarrow x_{Stable\ Mega-Organization} \end{matrix}\right| \langle x_U | x_T \rangle$$

$$\Rightarrow$$

$$\langle Space - Time - Energy - Gravity\ Material - Fabric\ Pre \\ - Genetic\ Code \rangle +$$

$$\langle Electromagnetic\ Spectrum\ Material - Fabric\ Pre - Genetic\ Code \rangle +$$

349

$\langle Quantum - Particle\ Material - Fabric\ Pre - Genetic\ Code \rangle +$

$\langle Atoms\ Pre - Genetic\ Code \rangle +$

$\langle Molecules\ Material - Fabric\ Pre - Genetic\ Code \rangle + LSTE\ \langle ... \rangle +$

$PGGT < \cdots > + \langle Cells\ Genetic\ Code \rangle + LSTE\ \langle ... \rangle +$

$(Fundamental\ Capacities\ of\ Self\ Genetic\ Code) + FBLEEE\ \langle ... \rangle +$

$(Humans\ Genetic\ Code) + LSTE\ \langle ... \rangle + FBLEEE\ \langle ... \rangle +$

$Stable\ Mega - Organization\ Genetic\ Code$

Eq 8.1.2: *Stable Mega-Organization Genetic Code*

Starting with the Light-Matrix, the top left-hand matrix in (8.1.2), the first line from the top, $C_\infty : [Pr, Po, K, H]$, specifies the architecture of mega-organization as will be discussed shortly. Hence, Academic organizations are an emergence of Light's property of Knowledge, Political organizations are an emergence of Light's property of Power, Social organizations are an emergence of Light's property of Presence, and Commercial organizations can be an emergence of Light's property of Harmony. The fundamental architecture of these aspects hence, is an emergence of the properties of Light at ∞.

Line 3 in the Light-Matrix, C_K: $[S_{Pr}, S_{Po}, S_K, S_H]$, elaborates the sets for Presence, Power, Knowledge, and Harmony, each containing multiple elements. For example, as will be explored in the following subsection, various elements derived from the four sets would define the architecture of a Silicon Valley type mega-commercial organization and could be functions such as 'commercial immunity', 'power of aesthetics', 'ease of informal meetings', amongst others, hence collectively describing mega-commercial organization's way of being. Specifically, Line 5, C_N: $f(S_{Pr} \times S_{Po} \times S_K \times S_H)$, suggests that unique seeds are created from a combination of such elements from all four sets, with a particular element leading, that in effect creates the distinctness possible at the level of mega-organization.

Line 6, $(\downarrow R_{C_U} = f(R_{C_N}))$, specifies quantization between the layer where the seeds are formed, and the physical layer we are familiar with, and as explored in Chapter 3.5 and 3.6, will result in Line 7, C_U: $[P, V, M, C]$, hence changing the FBLEE and subsequently the post material-fabric housing structure. Note that as in the process describing the generation of the code-segments for space-time-energy-gravity quantization at the time of the Big Bang, the generation of the FBLEE (four-base logic-encoding ecosystem) code-segment as specified by (4.1.5) and the MF (material-fabric) code-segment as specified by (4.1.6) are combined together into one equation so that the FBLEE process is apparently transparent.

The possibilities represented by Lines 1 through 5 hence concretize through the quantization represented by Line 6 to become or further enhance the mega-organization with its subtle physical (related to Presence), vital (related to Power), mental (related to Knowledge), and connection (related to Harmony) aspects now existing in material reality typified by Light moving at c. Note that just as

Line 6 represents a process of quantization relating the layer of reality created by Light traveling at c with the antecedent layers, so too Lines 2 and 4 as previously discussed, also represent quantization of a more subtle kind that ultimately plays a critical part in allowing FBLEE and post material-fabric housing structure to express infinite diversity.

Typically it is the process as captured by the Space-Matrix that will determine if Line 6 is activated. Specifically patterns at the untransformed layer, U, will need to be overcome, as specified by the second-line from the bottom of the Space-Matrix: $(\uparrow > P_x)$. But as specified by the bottom-line of the Time-Matrix, reproduced below, it is only with the advent of the human-system that the automaticity of the action of meta-levels is reversed:

$$U \rightarrow \begin{array}{l} t \le E_{Cell}; TC: M_3 \rightarrow U \\ t \sim E_{Human}; TC: U \rightarrow M_3 \end{array}$$

Hence in the case of mega-organizations, which in this emergence is a post-human system, the fact that patterns do need to be overcome means that quantization requires effort to happen. Given this, it is useful to review Equation 3.6.5, Potential Effect of Levels of Light on Pre-Genetic of Genetic Information:

$$Potential\ Effect\ of\ Levels\ of\ Light\ on\ Pre$$
$$- Genetic\ or\ Genetic\ Information =$$
$$\begin{bmatrix} STATIC\ \langle |[L][S][T]TC \rightarrow x_T|_{\langle x_U|x_T\rangle}\rangle \\ \times \\ \left((Y > U: Z_Q) \vee (Y \le U: Z_F) \vee (Y = U: Z_R) \right) \\ \ni \\ \left(\begin{array}{c} Z \in \mathbb{U}\ (Space, Time, Energy, Gravity) \\ Q: Quantization; F: Fragmentation; R: Random \end{array} \right) \end{bmatrix} \rightarrow h \ni$$

$$h \in \left(\begin{array}{c} [Q]: Constructive\ zone, \\ [Q]: Constructive\ zone \wedge Constructive\ mutation, \\ [F]: Destructive\ mutation, \\ [R]: Random\ mutation \end{array} \right)$$

Line 1 from the top in the matrix is simply a static form of (3.6.1) the Simplified Light-Space-Time Emergence equation. The static form is designated by 'STATIC', and implies that fundamental operations true of (3.6.1) are being highlighted in (3.6.5). In other word (3.6.1) already has all the operations highlighted in (3.6.5) in it, but by 'freezing' it by making it static, the essential dynamics leading to possible mutations at the genetic level can more clearly be highlighted.

Line 1 is then subjected ($\times$) to a determination of the dominant levels of light that may be active, designated by Line 2, ' $\left((Y > U: Z_Q) \vee (Y \leq U: Z_F) \vee (Y = U: Z_R) \right)$ '. Unpacking this, '$Y > U$' implies meta-levels are active and as a result it is possible that Z_Q is going to take place. This also implies activation and potential change of FBLEE. The call from below, as it were, may invoke some function that already exists in the subtle-libraries 'above', so that some already existing function may influence FBLEE through ∞ -entanglement, K-entanglement, or N-entanglement. In this case the call may have to do with organizing larger and larger collectivities of humans so allow an even richer materialization of four-foldness. The response may cause the precipitation of some mathematically possible mega-organization meta-function already existing in a subtle-library to influence FBLEE. This may be thought of as a key-and-lock mechanism, where a deep enough visceral urge from below acting as the key, opens an entangled lock to alter FBLEE as per the visceral urge.

'$Y \leq U$' implies that only the untransformed levels are active, and therefore also the sub-level where the speed of light is 0 is active and as a result Z_F is going to take place. '$Y = U$' implies that all levels are active and as a result Z_R is going to take place.

Line 3 elaborates the significance of Z_Q, Z_F, and Z_R. Hence Z is the union of potential quantum-operations of space, time, energy, and gravity, designated by ' $Z \in \mathbb{U}\ (Space, Time, Energy, Gravity)$ '. But the nature of the operations is designated by ' Q: $Quantization$; F: $Fragmentation$; R: $Random$ '. Z_Q, then, implies that the full quantization originating from updated four-base logic-encoding ecosystems (FBLEE) can take place, and will result in lasting material change at the genetic level. Z_F implies that the essential set will be fragmented and that only local libraries a the level of local cellular-level DNA can potentially be altered. Z_R also precludes full quantization, and that some partial local-library constructive or destructive mutation may take place.

The ' $\rightarrow h \ni h \in$ ' segment resolves the outcome of the operations implied by Lines 1 – 3, suggesting that the outcome will be 'h' such that ($\ni$) 'h' is an element ($\in$) of the set specified by the members '[Q]: $Constructive\ zone$', '[Q}: $Constructive\ zone\ AND\ Constructive\ mutation$', '[F}; $Destructive\ mutation$', and '{R}: $Random\ mutation$'. '[Q]: $Constructive\ zone$' implies that the in-built buffer has been crossed and that access to the deeper four-base logic-encoding ecosystem (FBLEE) has been granted. '[Q}' in this segment implies that there is the possibility that full-quantization as specified by Line 3 of the previous matrix can take place. Access to this zone is a prerequisite for constructive mutation to occur, as designated by the element ' [Q}: $Constructive\ zone \cap Constructive\ mutation$ ', which implies that full-quantization is going to take place and will result in material change. The '[F]' specifies the relationship between 'Fragmentation' in Line 3 of the previous matrix and destructive mutation. The '[R]' specifies the relationship between 'Random' in Line 3 of the previous matrix and random mutation.

But as just summarized in the Time-Matrix in (8.1.2) Y is by definition not greater than U and hence quantization is not automatic. In terms of mega-organizations such quantization implies that increasing wholeness can become fully active through specific space, time, energy, and gravity quantization to create an holistic "ecosystem" with its own "mega-organization logic" as it were. Possible precipitation into the post material-fabric housing structure is captured by the action of 'PGGT <...>' that can be assumed to be triggered by Z_Q.

Generation of Cellular-Level Genetic Code for Stable Mega-Organizations

As suggested by inductively derived (8.1.1), a sustainable organization is the result of two dynamics: maturity along each of the architectural force dimensions, and a robust interaction between all for sets of forces. These dynamics are implicit in (8.1.2) as borne out in the previous subsection.

Consider the example of Silicon Valley. While Silicon Valley may be primarily a commercial organization, yet we begin to glimpse something of what may become

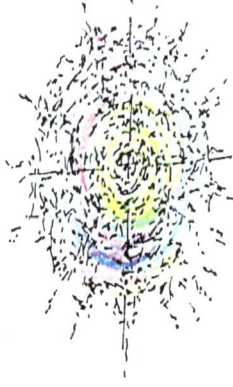

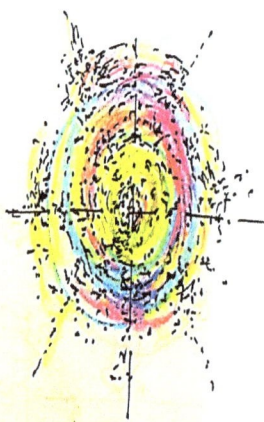

possible when the four emergent forces of light can begin to interact with one another in free fashion. That is, without the motive of doing everything for the sake of generating money only. In the case of Silicon Valley we see that the climate and beauty of the Bay Area, thereby likely representing the impetus of nurturing and harmony, began to attract many talented people into the vicinity. Over time the talent pool became progressively diversified. Educational institutes, such as Stanford University and University of Berkeley cropped up and became centers of cutting-edge research, representing the impetus of knowledge. The Armed Forces were attracted to the area for the same reason, and came with their huge requirements for research for research's sake, and their funds in support of the area, representing the impetus of power. Graduates from the universities started companies that began to in turn support the universities with funds. Talent moved around from company to company like people from department to department. Thus we see that even when individual companies failed, Silicon Valley functioned as a larger organization and was able to retain the talent in the area, was able further, to buffer the shocks to some extent, hence preparing the ground for future waves of innovation. This reality of becoming a container for dynamic experimentation perhaps represents the impetus of service in that the container is serving the dynamic components existing within it.

Pragmatically speaking this dynamic of functioning as a larger mega-organization meant that some level of commercial-immunity began to develop, so that people losing their jobs was not necessarily considered as that stressful an event. There was a freeing, thus, from the purely commercial element. If the various powers of beauty, aesthetics, knowledge, power in the form of money, plus the myriad streams of talent – from engineers, scientists, managers, lawyers etc., did not exist in such close proximity, and further, if the various professionals could not thus support each other through their continued informal meetings, through the urge to create the next wave of innovation, through the urge to pursue progress for the sake of progress, then they would have left for other areas and the phenomenon of Silicon Valley would never have been.

Usually modern day organizations cannot provide these various buffers and opportunities for interaction that Silicon Valley provides, and hence when hit with adversity, more often than not, simply crumble. This is perhaps due to the fact that it is always only one-dimension that drives them, and when the health of that is threatened or falls, the organization resorts to tactics that will ensure that that dimension looks good at any cost, in the bargain sacrificing the development of the other dimensions, and therefore its longer term health.

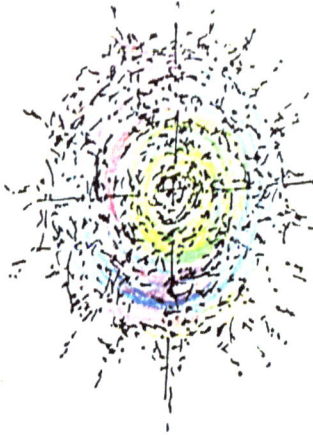

The four-forces led organization will be something like a community. Having attained freedom from an exaggerated commercial impetus, people will 'live' their 'jobs' because it is what fulfills them. Such a freedom is what will allow the four primary powers to manifest to greater and greater degree within them and their environment. Seeing thus, their ability to become centers of knowledge and wisdom, or mutuality and harmony, or power and leadership and energy, or perfection and service, or some unique combination of these primary forces, so increase, a sense of satisfaction with life will more easily accompany all that they continue to do. Under the freer flow of these powers their uniqueness will be refined and flourish, and correspondingly, so too will the uniqueness of their respective organizations.

358

An organization may be political and hence be primarily driven by the power of leadership and courage, or it may be social and hence be primarily driven by the power of service and perfection, or it may be commercial and hence be primarily driven by the power of mutuality and harmony, or it may be research-oriented and academic and hence be primarily driven by the powers of knowledge and wisdom, but always each of the other powers will also be behind it, fulfilling and completing its primary urge.   Thus the dynamic of unique primary powers being supported by an increasing plethora of secondary powers is represented by the following four organizational equations, (8.1.3 – 6) and illustrates how the four powers of light emerge even in complex human-based collectivities:

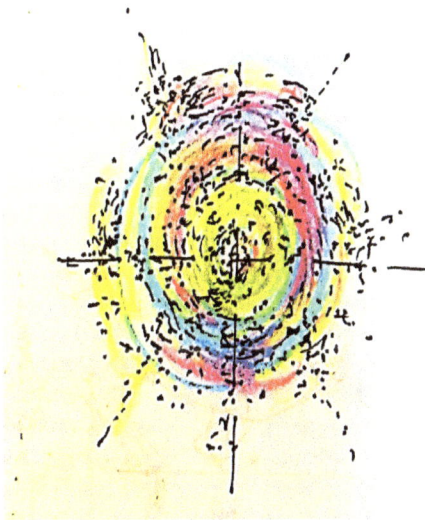

$$Sig_{Political} = Xa + \overline{Yb_{0-n}}$$

$$where \begin{bmatrix} X \in [S_{System_P}] \\ Y \in [S_{System_{Pr}}, S_{System_P}, S_{System_K}, S_{System_N}] \\ a, b \ are \ integers; a > b \end{bmatrix}$$

Eq 8.1.3: Political Organization

$$Sig_{Social} = Xa + \overline{Yb_{0-n}}$$

359

$$where \begin{bmatrix} X \in [S_{System_{Pr}}] \\ Y \in [S_{System_{Pr}}, S_{System_P}, S_{System_K}, S_{System_N}] \\ a, b \ are \ integers; a > b \end{bmatrix}$$

Eq 8.1.4: Social Organization

$$Sig_{Commercial} = Xa + \overline{Yb_{0-n}}$$

$$where \begin{bmatrix} X \in [S_{System_N}] \\ Y \in [S_{System_{Pr}}, S_{System_P}, S_{System_K}, S_{System_N}] \\ a, b \ are \ integers; a > b \end{bmatrix}$$

Eq 8.1.5: Commercial Organization

$$Sig_{Academic} = Xa + \overline{Yb_{0-n}}$$

$$where \begin{bmatrix} X \in [S_{System_K}] \\ Y \in [S_{System_{Pr}}, S_{System_P}, S_{System_K}, S_{System_N}] \\ a, b \ are \ integers; a > b \end{bmatrix}$$

Eq 8.1.6: Academic Organization

Note that (8.1.3 - 6) already implies that Lines 1 – 5 in the Light Matrix (3.1.3) have been activated, and that the logic of the mega-organization- ecosystem will precipitate into the post material-fabric cell-based genetic structure through the action of Line 6-7 of (8.1.2). This action of precipitating into the cell-based structure as opposed to the material-fabric is designated by 'PGGT <...>' as specified in (8.1.2).

Summary of Cellular-Level Genetic Code with Stable Mega-Organization

Summarizing, after the cellular stage iterations - including the sustainable mega-organization - of the Light-Space-Time Emergence equation are complete, the following code-segments will have been generated as specified by Equation 8.1.7, Active Pre-Genetic and Genetic Code at the Cellular Stage with Stable Mega-Organization:

$$Light - Space - Time\ Emergence_{Stable\ Mega-Orgnization} =$$

$$\left| \begin{bmatrix} c_\infty: [Pr, Po, K, H] \\ \left(\downarrow R_{C_K} = f(R_{C_\infty}) \right) \\ c_K: [S_{Pr}, S_{Po}, S_K, S_H] \\ \left(\downarrow R_{C_N} = f(R_{C_K}) \right) \\ c_N: f(S_{Pr} \times S_{Po} \times S_K \times S_H) \\ \left(\downarrow R_{C_U} = f(R_{C_N}) \right) \\ c_U: [P, V, M, C] \\ \Uparrow \\ c_{0:[D,W,I,C]} \end{bmatrix}_{Light} \begin{bmatrix} M_3 \rightarrow System_X \\ (\uparrow F \rightarrow I) \\ M_2 \rightarrow S_{System_X} \\ (\uparrow Sig \rightarrow F) \\ M_1 \rightarrow Sig_x \\ (\uparrow > P_x) \\ U \rightarrow x_{Humans} \end{bmatrix}_{Space} \\ \begin{bmatrix} M_3 : -\infty \leq t \leq \infty \\ \downarrow \\ M_2 : 0 \geq t > \infty \\ \downarrow \\ M_1 : 0 > t > \infty \\ \downarrow \\ U \rightarrow \ \ t \leq E_{Cell}; TC: M_3 \rightarrow U \\ t \sim E_{Human}; TC: U \rightarrow M_3 \end{bmatrix}_{Time} \quad TC \rightarrow x_{Stable\ Mega-Organization} \right| \langle x_U | x_T \rangle$$

$$\Rightarrow$$

$$\langle Space - Time - Energy - Gravity\ Material - Fabric\ Pre - Genetic\ Code \rangle +$$

$$\langle Electromagnetic\ Spectrum\ Material - Fabric\ Pre - Genetic\ Code \rangle +$$

$$\langle Quantum - Particle\ Material - Fabric\ Pre - Genetic\ Code \rangle +$$

$$\langle Atoms\ Pre - Genetic\ Code \rangle +$$

$$\langle Molecules\ Material - Fabric\ Pre - Genetic\ Code \rangle + LSTE \langle ... \rangle +$$

$$PGGT < \cdots > + \langle Cells\ Genetic\ Code \rangle + LSTE \langle ... \rangle +$$

$$\langle Fundamental\ Capacities\ of\ Self\ Genetic\ Code \rangle + FBLEEE \langle ... \rangle +$$

$(Humans\ Genetic\ Code) + LSTE\ \langle ... \rangle + FBLEEE\ \langle ... \rangle\ +$

$$
\left(
\begin{array}{c}
\sum Sig_{Commercial} = Xa + \overline{Yb_{0-n}} \\[4pt]
where \left[
\begin{array}{c}
X \in \left[S_{System_N} \right] \\
Y \in \left[S_{System_{Pr}}, S_{System_P}, S_{System_K}, S_{System_N} \right] \\
a, b\ are\ integers;\ a > b
\end{array}
\right] \\[14pt]
\sum Sig_{Academic} = Xa + \overline{Yb_{0-n}} \\[4pt]
where \left[
\begin{array}{c}
X \in \left[S_{System_K} \right] \\
Y \in \left[S_{System_{Pr}}, S_{System_P}, S_{System_K}, S_{System_N} \right] \\
a, b\ are\ integers;\ a > b
\end{array}
\right] \\[14pt]
\sum Sig_{Political} = Xa + \overline{Yb_{0-n}} \\[4pt]
where \left[
\begin{array}{c}
X \in \left[S_{System_P} \right] \\
Y \in \left[S_{System_{Pr}}, S_{System_P}, S_{System_K}, S_{System_N} \right] \\
a, b\ are\ integers;\ a > b
\end{array}
\right] \\[14pt]
\sum Sig_{Social} = Xa + \overline{Yb_{0-n}} \\[4pt]
where \left[
\begin{array}{c}
X \in \left[S_{System_{Pr}} \right] \\
Y \in \left[S_{System_{Pr}}, S_{System_P}, S_{System_K}, S_{System_N} \right] \\
a, b\ are\ integers;\ a > b
\end{array}
\right]
\end{array}
\right)
$$

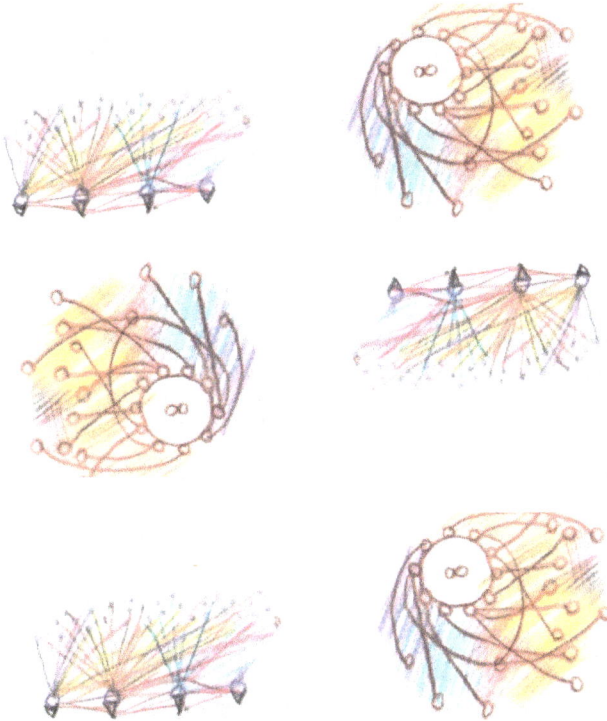

Note that the final code segment following 'FBLEEE <...>' contains the code for all possible ($\sum x$) mega-organization types.

Chapter 8.2: Generating the Cellular-Level Genetic Code for Sustainable Global Civilization

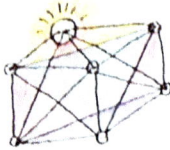

This chapter will consider the quantum-level computation necessary in the creation of a sustainable global civilization and the genetic code segments that accompany that.

Light's Emergence as Sustainable Global Civilization

As discussed in previous chapters the Light-Space-Time Emergence equation (3.1.3) being iterative, can be used to model emergence as it proceeds from simpler four-fold to more complex four-fold manifestations. Hence (3.1.3) has already been applied to suggest the emergence of the space-time-energy-gravity quadrumvirate, the electromagnetic spectrum, quantum particles in general, bosons as a further instance of a particular kind of quantum particle, atoms, living cells, basic capacities of the self, truer individuality, and mega-organizations.

Here it will be applied to suggest the emergence of sustainable global civilization. A revised form of the stable mega-organization equation (8.1.2) will be applied to suggest the emergence of sustainable global civilization, Equation 8.2.1, Sustainable Global Civilization Genetic Code. The architecture and details of a sustainable global civilization can be seen to be the result of the application of the Light, Space, and Time matrices as will be elaborated. But further, as implied by (3.6.5), the Potential Effect of Levels of Light on Pre-Genetic of Genetic Information equation, any process of deeper organization, such as is responsible for the architecture and cohesiveness of sustainable global civilization, has the possibility of altering FBLEE and post material fabric housing structures so long as the bases involved are driven primarily by a meta-level.

$$Light - Space - Time\ Emergence_{Sustainable\ Global\ Civilization} =$$

$$\left| \begin{bmatrix} \begin{matrix} c_{\infty}: [Pr, Po, K, H] \\ \left(\downarrow R_{C_K} = f(R_{C_{\infty}}) \right) \\ c_K: [S_{Pr}, S_{Po}, S_K, S_H] \\ \left(\downarrow R_{C_N} = f(R_{C_K}) \right) \\ c_N: f(S_{Pr}\ x\ S_{Po}\ x\ S_K\ x\ S_H) \\ \left(\downarrow R_{C_U} = f(R_{C_N}) \right) \\ c_U: [P, V, M, C] \\ \Uparrow \\ c_{0:[D,W,I,C]} \end{matrix} \end{bmatrix}_{Light} \begin{bmatrix} M_3 \rightarrow System_X \\ (\uparrow F \rightarrow I) \\ M_2 \rightarrow S_{System_X} \\ (\uparrow Sig \rightarrow F) \\ M_1 \rightarrow Sig_x \\ (\uparrow > P_x) \\ U \rightarrow x_{Stable\ Mega-Organization} \end{bmatrix}_{Space} \right.$$

$$\left. \begin{bmatrix} M_3 : -\infty \le t \le \infty \\ \downarrow \\ M_2 : 0 \ge t > \infty \\ \downarrow \\ M_1 : 0 > t > \infty \\ \downarrow \\ U \rightarrow \begin{matrix} t \le E_{Cell}; TC: M_3 \rightarrow U \\ t \sim E_{Human}; TC: U \rightarrow M_3 \end{matrix} \end{bmatrix}_{Time} \quad TC \rightarrow x_{S Sustainbale\ Global\ Civilization} \right| \langle x_U | x_T \rangle$$

$$\Rightarrow$$

$$\langle Space - Time - Energy - Gravity\ Material - Fabric\ Pre - Genetic\ Code \rangle +$$

$$\langle Electromagnetic\ Spectrum\ Material - Fabric\ Pre - Genetic\ Code \rangle +$$

$$\langle Quantum - Particle\ Material - Fabric\ Pre - Genetic\ Code \rangle +$$

$$\langle Atoms\ Pre - Genetic\ Code \rangle +$$

$$\langle Molecules\ Material - Fabric\ Pre - Genetic\ Code \rangle + LSTE \langle ... \rangle +$$

$$PGGT < \cdots > + \langle Cells\ Genetic\ Code \rangle + LSTE \langle ... \rangle +$$

$$(Fundamental\ Capacities\ of\ Self\ Genetic\ Code) + FBLEEE \langle ... \rangle +$$

$$(Humans\ Genetic\ Code) + LSTE \langle ... \rangle + FBLEEE \langle ... \rangle +$$

$(Stable\ Mega - Organization\ Genetic\ Code) + LSTE < \cdots > +$

Sustainable Global Civilization Genetic Code

Eq 8.2.1: Sustainable Global Civilization Genetic Code

Starting with the Light-Matrix, the top left-hand matrix in (3.1.3), the first line from the top, $C_\infty : [Pr, Po, K, H]$, specifies the architecture of sustainable global civilization as will be further discussed in this chapter. A key to such a civilization is also pointed to by the previously derived equation (8.1.1), reproduced here for convenience:

$$Sustainability_{Systems}$$
$$\propto Interaction\ (S_{System_{Pr}}, S_{System_P}, S_{System_K}, S_{System_N})$$

This requires maturity by large organized parts of the world, whether nations or regional blocs. The maturity is such that the fourfold properties of Light are adequately expressed. Hence in the example of nations to be illustrated shortly, a country like India is in its deeper essence an emergence of Light's property of Knowledge, a country like Japan is in its deeper essence an emergence of Light's property of Power, a country like Thailand is in its deeper essence an emergence of Light's property of Presence, and a country like UK is in in its deeper essence an emergence of Light's property of Harmony. The fundamental architecture of the combination of such emergences, which as per (8.1.1) is required for sustainability, is hence an emergence of the properties of Light at ∞.

Line 3 in the Light-Matrix, C_K: $[S_{Pr}, S_{Po}, S_K, S_H]$, elaborates the sets for Presence, Power, Knowledge, and Harmony, each containing multiple elements. For example, as will be explored, various elements derived from the four sets define the architecture of a knowledge-essence country like India and could be functions such as 'exceptional capacity for penetrating behind the surface', 'meaningfully synthesizing many streams of development', amongst others, hence collectively describing different aspects of a knowledge-essence country's way of being. Specifically, Line 5, C_N: $f(S_{Pr} \times S_{Po} \times S_K \times S_H)$, suggests that unique seeds are created from a combination of such elements from all four sets, with a particular element leading, that in effect creates the distinctness possible at the level of sustainable global civilization.

Line 6, ($\downarrow R_{C_U} = f(R_{C_N})$), specifies quantization between the layer where the seeds are formed, and the physical layer we are familiar with, and as explored in Chapter 3.5 and 3.6, will result in Line 7, C_U: $[P, V, M, C]$, hence changing the FBLEE and subsequently the post material-fabric housing structure. Note that as in the process describing the generation of the code-segments for space-time-energy-gravity quantization at the time of the Big Bang, the generation of the FBLEE (four-base logic-encoding ecosystem) code-segment as specified by (4.1.5) and the MF (material-fabric) code-segment as specified by (4.1.6) are combined together into one equation so that the FBLEE process is apparently transparent.

The possibilities represented by Lines 1 through 5 hence concretize through the quantization represented by Line 6 to enhance sustainable global civilization with subtle physical (related to Presence), vital (related to Power), mental (related to Knowledge), and connection (related to Harmony) aspects now existing in material reality typified by Light moving at c.   Note that just as Line 6 represents a process of quantization relating the layer of reality created by Light traveling at c with the antecedent layers, so too Lines 2 and 4 as previously discussed, also represent quantization of a more subtle kind that ultimately plays a critical part in allowing FBLEE and post material-fabric housing structure to express infinite diversity.

Typically it is the process as captured by the Space-Matrix that will determine if Line 6 is activated. Specifically patterns at the untransformed layer, U, will need to be overcome, as specified by the second-line from the bottom of the Space-Matrix: $(\uparrow > P_x)$. But as specified by the bottom-line of the Time-Matrix, reproduced below, it is only with the advent of the human-system that the automaticity of the action of meta-levels is reversed:

$$U \rightarrow \begin{array}{l} t \le E_{Cell}; TC: M_3 \rightarrow U \\ t \sim E_{Human}; TC: U \rightarrow M_3 \end{array}$$

Hence in the case of sustainable global civilization, which in this emergence is a post-human system, the fact that patterns do need to be overcome means that quantization requires effort to happen. Given this, it is useful to review Equation 3.6.5, Potential Effect of Levels of Light on Pre-Genetic of Genetic Information:

Potential Effect of Levels of Light on Pre Genetic or Genetic Information =

$$\begin{bmatrix} STATIC \; \langle |[L][S][T]TC \to x_T|_{\langle x_U|x_T\rangle}\rangle \\ \times \\ \Big((Y > U : Z_Q) \vee (Y \leq U : Z_F) \vee (Y = U : Z_R) \Big) \\ \ni \\ \begin{pmatrix} Z \in \mathbb{U} \; (\textit{Space}, \textit{Time}, \textit{Energy}, \textit{Gravity}) \\ Q: \textit{Quantization}; F: \textit{Fragmentation}; R: \textit{Random} \end{pmatrix} \end{bmatrix} \to h \ni$$

$$h \in \begin{pmatrix} [Q]: \textit{Constructive zone}, \\ [Q]: \textit{Constructive zone} \wedge \textit{Constructive mutation}, \\ [F]: \textit{Destructive mutation}, \\ [R]: \textit{Random mutation} \end{pmatrix}$$

Line 1 from the top in the matrix is simply a static form of (3.6.1) the Simplified Light-Space-Time Emergence equation. The static form is designated by 'STATIC', and implies that fundamental operations true of (3.6.1) are being highlighted in (3.6.5). In other word (3.6.1) already has all the operations highlighted in (3.6.5) in it, but by 'freezing' it by making it static, the essential dynamics leading to possible mutations at the genetic level can more clearly be highlighted.

Line 1 is then subjected ($\times$) to a determination of the dominant levels of light that may be active, designated by Line 2, ' $\Big((Y > U : Z_Q) \vee (Y \leq U : Z_F) \vee (Y = U : Z_R) \Big)$ '. Unpacking this, '$Y > U$' implies meta-levels are active and as a result it is possible that Z_Q is going to take place. This also implies activation and potential change of FBLEE. The call from below, as it were, may invoke some function that already exists in the subtle-libraries 'above', so that some already existing function may influence FBLEE through ∞ -entanglement, K-entanglement, or N-entanglement. With the existence of stable mega-organizations, the yearning to increase the possibility of a sustainable global civilization may awaken from the cumulative Light-constructs already materialized. Such a call may invoke the precipitation of some sustainable-global-civilization meta-function that exists as a mathematical possibility 'above' in some subtle-library. This may be thought of as a key-and-lock mechanism, where a deep enough visceral urge from below acting as the key, opens an entangled lock to alter FBLEE as per the visceral urge.

'$Y \leq U$' implies that only the untransformed levels are active, and therefore also the sub-level where the speed of light is 0 is active and as a result Z_F is going to take place. '$Y = U$' implies that all levels are active and as a result Z_R is going to take place.

Line 3 elaborates the significance of Z_Q, Z_F, and Z_R. Hence Z is the union of potential quantum-operations of space, time, energy, and gravity, designated by '$Z \in \mathbb{U}\,(Space, Time, Energy, Gravity)$'. But the nature of the operations is designated by '$Q: Quantization; F: Fragmentation; R: Random$'. Z_Q, then, implies that the full quantization originating from updated four-base logic-encoding ecosystems (FBLEE) can take place, and will result in lasting material change at the genetic level. Z_F implies that the essential set will be fragmented

and that only local libraries at the level of local cellular-level DNA can potentially be altered. Z_R also precludes full quantization, and that some partial local-library constructive or destructive mutation may take place.

The '$\rightarrow h \ni h \in$' segment resolves the outcome of the operations implied by Lines 1 – 3, suggesting that the outcome will be 'h' such that ($\ni$) 'h' is an element ($\in$) of the set specified by the members '[Q]: *Constructive zone*', '[Q]: *Constructive zone AND Constructive mutation*', '[F]: *Destructive mutation*', and '{R}: *Random mutation*'. '[Q]: *Constructive zone*' implies that the in-built buffer has been crossed and that access to the deeper four-base logic-encoding ecosystem (FBLEE) has been granted. '[Q}' in this segment implies that there is the possibility that full-quantization as specified by Line 3 of the previous matrix can take place. Access to this zone is a prerequisite for constructive mutation to occur, as designated by the element ' [Q}: *Constructive zone* ∩ *Constructive mutation* ', which implies that full-quantization is going to take place and will result in material change. The '[F]' specifies the relationship between 'Fragmentation' in Line 3 of the previous matrix and destructive mutation. The '[R]' specifies the relationship between 'Random' in Line 3 of the previous matrix and random mutation.

But as just summarized in the Time-Matrix in (8.2.1) Y is by definition not greater than U and hence quantization is not automatic. In terms of sustainable global civilization such quantization implies that increasing wholeness can become fully active through specific space, time, energy, and gravity quantization to create an holistic "ecosystem" with its own "sustainable global civilization logic" as it were. Possible precipitation into the post material-fabric housing structure is captured by

371

the action of 'PGGT <...>' that can be assumed to be triggered by Z_Q.

Generation of Cellular-Level Genetic Code for Global Sustainable Civilization

A brief look at history will reinforce the idea that it is typically maturity along multiple and distinct dimensions, represented by the four properties of light, and their rich interaction that allows civilizations to endure.

Thus, those civilizations that have endured typically have a balance of all four families (Sri Aurobindo, 1971). Civilizations that have become extinct typically have had a focus on few drivers of innovation. Jared Diamond proposes five interconnected causes of collapse that may reinforce each other: non-sustainable exploitation of resources, climate changes, diminishing support from

372

friendly societies, hostile neighbors, and inappropriate attitudes for change (Diamond, 2005). But these five sources may also be thought of as symptoms that arise due to the failure to adopt the catholicity of the sources of innovation emanating from each of the four sets or families. Further, the historian Toynbee suggested that societies decay because of their over-reliance on structures that helped them solve old problems (Toynbee, 1961). It can be interpreted that being thus biased they are unable to adopt the catholicity of the sources of innovation emanating from each of the four sets of families.

Approaching Civilization from a big-picture, global basis though, the sustainability of humankind will be ensured by a balance of development amongst the four sets of forces. This means that countries and global regions must be unique and in such a way that their primary emergence is distributed amongst all four sets of forces. Further, and based on this uniqueness, there must be an open and healthy interaction amongst these centers of uniqueness.

Hence, India, Japan, Thailand, and UK will be considered as representative examples of the four distinct and emergent properties that must be balanced in the whole.

Historically at least, India, for example, in its essence, may be thought of as having an exceptional capacity for penetrating behind the surface, and further of meaningfully synthesizing many streams of development. Its primary power may thus be from the family of knowledge, with a strong secondary driver being its ability to create living harmonies. The equation for India will then likely be represented by Equation 8.2.2:

$$Sig_{India} = Xa +$$

$$\overline{Yb_{0-n}} \quad where \quad \left[\begin{array}{c} X \in [S_{System_K}] \\ Y \in [S_{System_{Pr}}, S_{System_P}, S_{System_K}, S_{System_N}] \\ a, b \ are \ integers; a > b \end{array} \right]$$

Eq 8.2.2: India (Knowledge Family)

Japan, in its essence, may be thought of as having a strong and noble warrior nature, along with a refined sense of aesthetics, amongst other qualities. Its equation, Equation 8.2.3 would be of the form:

$$Sig_{Japon} = Xa +$$

$$\overline{Yb_{0-n}} \quad where \quad \left[\begin{array}{c} X \in [S_{System_P}] \\ Y \in [S_{System_{Pr}}, S_{System_P}, S_{System_K}, S_{System_N}] \\ a, b \ are \ integers; a > b \end{array} \right]$$

Eq 8.2.3: Japan (Power Family)

UK, in its essence, may be thought of as having a strong ability to create practical, materialistic harmonies, resulting in such things as working parliaments and advanced democracy, for example. Its equation, Equation 8.2.4, would be of the form:

$$Sig_{UK} = Xa +$$

$$\overline{Yb_{0-n}} \quad where \quad \left[\begin{array}{c} X \in [S_{System_N}] \\ Y \in [S_{System_{Pr}}, S_{System_P}, S_{System_K}, S_{System_N}] \\ a, b \ are \ integers; a > b \end{array} \right]$$

Eq 8.2.4: UK (Nurturing Family)

Thailand, in its essence, may be characterized by an exceptional sense of hospitality and sweet service, with an attention to detail in the practical arrangement of things. Its equation, Equation 8.2.5, would be of the form:

$$Sig_{Thailand} = Xa +$$

$$\overline{Yb_{0-n}} \quad where \quad \begin{bmatrix} X \in [S_{System_N}] \\ Y \in [S_{System_{Pr}}, S_{System_P}, S_{System_K}, S_{System_N}] \\ a, b \ are \ integers; a > b \end{bmatrix}$$

Eq 8.2.5: Thailand (Service Family)

Note that (8.2.2 - 5) already implies that Lines 1 – 5 in the Light Matrix (3.1.3) have been activated, and that the logic of the sustainable-global-civilization- ecosystem will precipitate into the post material-fabric cell-based genetic structure through the action of Line 6-7 of (8.2.1). This action of precipitating into the cell-based structure as opposed to the material-fabric is designated by 'PGGT <...>' as specified in (8.2.1).

Similarly every nation on earth will have a uniqueness that can be represented by equations belonging to one of the four families.

As discussed in Chapter 2.5, A Possible Mutational-Sequence and the Upward-Strand, for uniqueness to emerge and mature is a process. The process is represented by Equation 2.5.1 reproduced here for convenience:

$$Sig_E = X \begin{vmatrix} C: Sig * mod \left(\int = 1 \right) \\ F: Sig \, mod \, (c) \\ I: Sig \, mod \left(\int \overline{G, e, \pi} \right) \\ M: Sig * mod \, (G) \\ V: Sig * mod \, (e) \\ P: Sig * mod \, (\pi) \end{vmatrix}$$

Equations 8.2.2 – 5 represent example of the essence of uniqueness. For these to become living practicalities requires work at many different levels within a nation. If a nation has not really done the work to animate itself with its uniqueness then it may exist in the physical or P-state, or vital of V-state, or mental or M-state, which by definition lack sufficient maturity, and interaction with other nations is then going to be compromised. Moving to the integral or I-state, will allow a nation to at least not be locked in to a point of view. A consistent I-state practiced by each nation then is the minimum requirement for a sustainable global civilization. But a consistent I-state also means that restricting patterns are always being broken and that the quantization effect of organization on the material-fabric is actively in play. In other words, separations of light are continually uniting with the larger continent of Light and changing the very nature of possibility on larger and larger scale. The next sub-section examines such a quantization effect in more detail.

Summary of Cellular-Level Genetic Code with Sustainable Global Civilization

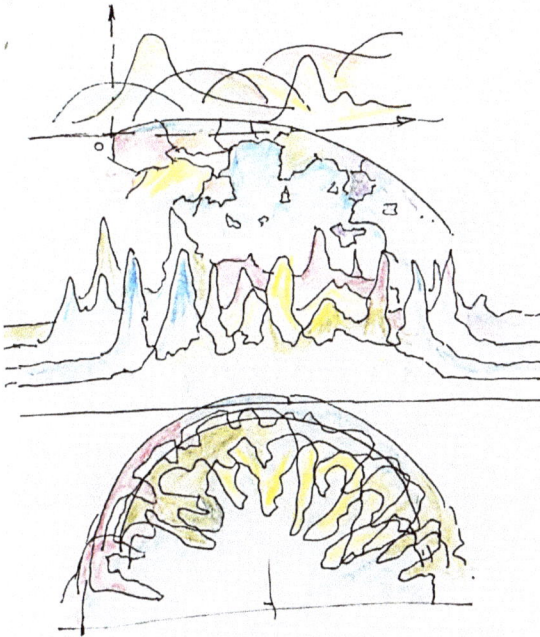

Summarizing, after the cellular stage iterations - including the sustainable global civilization - of the Light-Space-Time Emergence equation are complete, the following code-segments will have been generated as specified by Equation 8.2.6, Active Pre-Genetic and Genetic Code at the Cellular Stage with Sustainable Global Civilization:

$$Light - Space - Time\ Emergence_{Sustainable\ Global\ Civilization} =$$

377

$$\begin{Vmatrix} \begin{bmatrix} \begin{bmatrix} c_\infty : [Pr, Po, K, H] \\ \left(\downarrow R_{C_K} = f\left(R_{C_\infty}\right) \right) \\ c_K : [S_{Pr}, S_{Po}, S_K, S_H] \\ \left(\downarrow R_{C_N} = f\left(R_{C_K}\right) \right) \\ c_N : f(S_{Pr} \; x \; S_{Po} \; x \; S_K \; x \; S_H) \\ \left(\downarrow R_{C_U} = f\left(R_{C_N}\right) \right) \\ c_U : [P, V, M, C] \\ \Uparrow \\ c_{0:[D,W,I,C]} \end{bmatrix}_{Light} \\ U \to \begin{bmatrix} M_3 : -\infty \le t \le \infty \\ \downarrow \\ M_2 : 0 \ge t > \infty \\ \downarrow \\ M_1 : 0 > t > \infty \\ \downarrow \\ t \le E_{Cell}; \text{TC}: M_3 \to U \\ t \sim E_{Human}; \text{TC}: U \to M_3 \end{bmatrix}_{Time} \end{bmatrix} & \begin{bmatrix} \begin{bmatrix} M_3 \to System_X \\ (\uparrow F \to I) \\ M_2 \to S_{System_X} \\ (\uparrow Sig \to F) \\ M_1 \to Sig_X \\ (\uparrow > P_X) \end{bmatrix} \\ U \to x_{Stable\;Mega-Organization} \end{bmatrix}_{Space} \\ TC \to x_{Sust.Global\;Civilization} \end{Vmatrix} \langle x_U | x_T \rangle$$

$\Rightarrow$

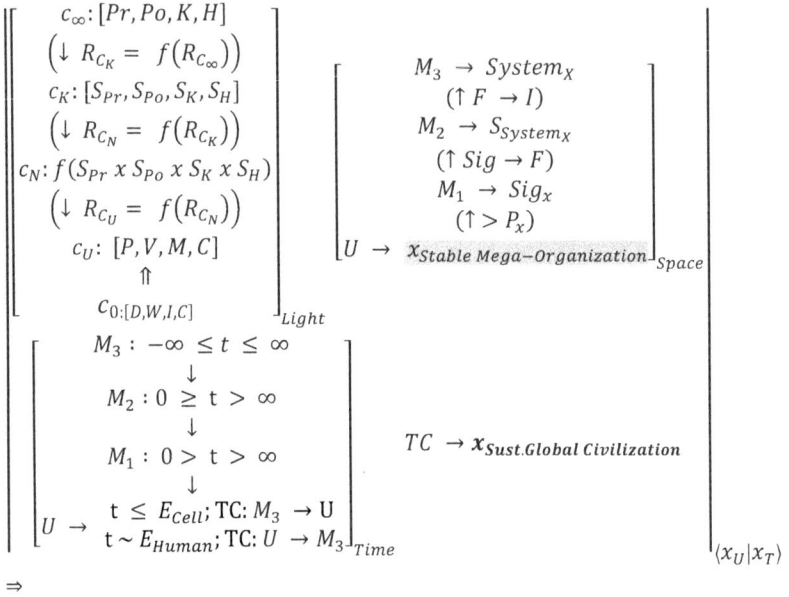

$\langle Space - Time - Energy - Gravity\ Material - Fabric\ Pre$
$- Genetic\ Code \rangle +$

$\langle Electromagnetic\ Spectrum\ Material - Fabric\ Pre - Genetic\ Code \rangle +$

$\langle Quantum - Particle\ Material - Fabric\ Pre - Genetic\ Code \rangle +$

$\langle Atoms\ Pre - Genetic\ Code \rangle +$

$\langle Molecules\ Material - Fabric\ Pre - Genetic\ Code \rangle + LSTE \langle \dots \rangle +$

$PGGT < \dots > + \langle Cells\ Genetic\ Code \rangle + LSTE \langle \dots \rangle +$

$(Fundamental\ Capacities\ of\ Self\ Genetic\ Code) + FBLEEE \langle \dots \rangle +$

$(Humans\ Genetic\ Code) + LSTE \langle \dots \rangle + FBLEEE \langle \dots \rangle +$

$(Stable\ Mega - Organization\ Genetic\ Code) + LSTE < \dots > +$

$$\left(\begin{array}{l} \displaystyle\sum Sig_{Nurturing-Nation} = Xa + \overline{Yb_{0-n}} \\[2mm] \quad X \in [S_{System_N}] \\ where \left[\begin{array}{l} Y \in [S_{System_{Pr}}, S_{System_P}, S_{System_K}, S_{System_N}] \\ a, b \ are \ integers; a > b \end{array} \right] \\[4mm] \displaystyle\sum Sig_{Knowlegde-Nation} = Xa + \overline{Yb_{0-n}} \\[2mm] \quad X \in [S_{System_K}] \\ where \left[\begin{array}{l} Y \in [S_{System_{Pr}}, S_{System_P}, S_{System_K}, S_{System_N}] \\ a, b \ are \ integers; a > b \end{array} \right] \\[4mm] \displaystyle\sum Sig_{Power-Nation} = Xa + \overline{Yb_{0-n}} \\[2mm] \quad X \in [S_{System_P}] \\ where \left[\begin{array}{l} Y \in [S_{System_{Pr}}, S_{System_P}, S_{System_K}, S_{System_N}] \\ a, b \ are \ integers; a > b \end{array} \right] \\[4mm] \displaystyle\sum Sig_{Service-Nation} = Xa + \overline{Yb_{0-n}} \\[2mm] \quad X \in [S_{System_{Pr}}] \\ where \left[\begin{array}{l} Y \in [S_{System_{Pr}}, S_{System_P}, S_{System_K}, S_{System_N}] \\ a, b \ are \ integers; a > b \end{array} \right] \end{array} \right)$$

Eq. 8.2.6, Active Pre-Genetic and Genetic Code at the Cellular Stage with Sustainable Global Civilization

Note that the final code segment following 'FBLEEE <...>' contains the code for all possible $(\sum x)$ nation types.

SECTION 9: A PRACTICAL LOOK AT GENETIC MUTATION

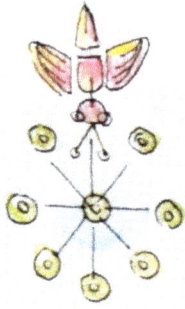

This Section will look at a couple of different examples discussed previously in the book to elaborate how genetic mutation may work in a light-based model of reality.

Hence, Chapter 9.1 will leverage the discussion on cells in Chapter 7.2 to look at how genetic mutation may

occur in a case of an adapting insect. Chapter 9.2 will leverage the discussion in Chapter 8.2 to look at how genetic mutation may occur as we strive towards a sustainable global civilization order.

Chapter 9.1:   Genetic Mutation in the Case of an Adapting Insect

Leveraging the discussion on the emergence of Light in a living cell from Chapter 7.2, Generating the Genetic Code for a Living Cell, this chapter will consider how mutation may occur in a light-based model, in the case of the adaptation of an insect.

Recall that the 'quantum' is a window into layers of reality behind the surface layer U. This insight was captured by Schrodinger's Equation (discussed in Chapter 3.3) and by Heisenberg's Uncertainty Principle (discussed in Chapter 3.4), the former suggesting that matter, or what is going to appear materially, is a function of some incredible number of superposed states containing infinite possibility, and the latter suggesting that quantum fluctuation due to an always buzzing pregnant-infinity, lies behind everything even when seemingly still.

The Light-Space-Time Emergence equation (3.1.3) derived in Chapter 3.1 may provide some insight into the dynamics of the pregnant-infinity and of quantum-certainty in the adaptabilities of the cell. Recall that the basis of cell adaptability was discussed in Chapter 7.2, and (7.2.1) modeling pre-genetic and genetic code available with the emergence of the cell is reproduced here:

$Light - Space - Time\ Emergence_{Cells} =$

$$
\left\| \begin{bmatrix} \begin{array}{c} c_\infty: [Pr, Po, K, H] \\ \left(\downarrow R_{C_K} = f(R_{C_\infty}) \right) \\ c_K: [S_{Pr}, S_{PO}, S_K, S_H] \\ \left(\downarrow R_{C_N} = f(R_{C_K}) \right) \\ c_N: f(S_{Pr} \times S_{PO} \times S_K \times S_H) \\ \left(\downarrow R_{C_U} = f(R_{C_N}) \right) \\ c_U: [P, V, M, C] \\ \Uparrow \\ c_{0:[D,W,I,C]} \end{array} \end{bmatrix}_{Light}
\begin{bmatrix} M_3 \to System_X \\ (\uparrow F \to I) \\ M_2 \to S_{System_X} \\ (\uparrow Sig \to F) \\ M_1 \to Sig_x \\ (\uparrow > P_x) \\ U \to x_{Molecules} \end{bmatrix}_{Space}
\right.
$$

$$
\left. \begin{bmatrix} M_3 : -\infty \le t \le \infty \\ \downarrow \\ M_2 : 0 \ge t > \infty \\ \downarrow \\ M_1 : 0 > t > \infty \\ \downarrow \\ U \to \begin{array}{l} t \le E_{Cell}; TC: M_3 \to U \\ t \sim E_{Human}; TC: U \to M_3 \end{array} \end{bmatrix}_{Time}
\quad TC \to x_{Cells}
\right\|_{\langle x_U | x_T \rangle}
$$

$\Rightarrow$

$\langle Space - Time - Energy - Gravity\ Material - Fabric\ Pre - Genetic\ Code \rangle +$

$\langle Electromagnetic\ Spectrum\ Material - Fabric\ Pre - Genetic\ Code \rangle +$

$\langle Quantum - Particle\ Material - Fabric\ Pre - Genetic\ Code \rangle +$

$\langle Atoms\ Material - Fabric\ Pre - Genetic\ Code \rangle +$

$\langle Molecules\ Material - Fabric\ Pre - Genetic\ Code \rangle + LSTE\ \langle ... \rangle + PGGT < \cdots > +$

$Cells\ Genetic\ Code$

The Light-Matrix, the top-left matrix in (3.1.3), suggests the possibilities in the pregnant-infinity. These possibilities, recall, have been set up by realities so created by light traveling at different speeds. While prevalent dynamics at the surface layer, U, are always influenced by dynamics from the deeper layers, there is a state that can be created that will allow a more focused and intentional activation of the meta-functions resident in the deeper layers. Such activation will potentially trigger a series of processes culminating in a state of quantum-certainty, introduced in Chapter 3.5 when taking a deeper look at S-T-E-G quantization, and as will be explored through the rest of this chapter.

The Space-Matrix, the top-right matrix in (3.1.3), suggests the dynamics involved in creating such an activation-state. Essentially this involves the overcoming of patterns at U.

Insect Adaptation Case

Consider the case of a hypothetical insect, for example, that is always prey to other predators. At some point a visceral urge to overcome some of these predators may arise in the insect. This visceral urge is a breaking of habitual patterns common to that species of insect. If it is deep enough and pervasive enough, it can be thought of as an activation-state and will allow the opening of a quantum-window so that at every micro level there is now a connection with the quantum worlds, Q, behind.

It is proposed that for effective adaptability to come about it is only such a pervasive state that will open a quantum-window to effectively stimulate the interaction between layers that will ultimately allow adaptability to occur. This activation-state can be further specified by leveraging (3.7.4) – The Generalized Equation for Direction of Mutation, that elaborates some Space-Matrix dynamics and was derived in Chapter 3.7, on Qualified Determinism.

Reproducing Equation 3.7.4:

$$Org\_Dir = DI \left(\begin{bmatrix} M_3 \rightarrow System_X \\ (\uparrow F \rightarrow I) \\ M_2 \rightarrow S_{System_X} \\ (\uparrow Sig \rightarrow F) \\ M_1 \rightarrow Sig_x \\ (\uparrow > P_P) \\ U \rightarrow x_U \end{bmatrix}^{x=p,v,m,i} \right) \rightarrow$$

$x\_matrix_{strongest} @ level_{strongest}$

The activation-state can be thought of as being invoked when $level_{strongest}$ is at least M_1. Hence, as in Equation 9.1.1:

$$Activation - State_{condition} \geq M_1$$

Eq 9.1.1: Condition for Activation-State

If the activation-state is invoked, that implies access to the four-base logic-encoding ecosystem (FBLEE).

Note that the Time-Matrix, in the lower-left side of (7.2.1), specifies that approximately, adaptability may tend to be more automatic up to the emergence of cellular organisms. This is specified by the direction $M_3 \rightarrow U$, which indicates that it is the meta-level that organizes activity at U. At approximately the human level the direction is flipped as specified by $U \rightarrow M_3$, and suggests that will, intention, feeling and the like are more important in stimulating the process of adaptability.

The visceral urge, then, may stimulate interaction with a collective-intelligence or 'specific-species-intelligence meta-function' or FBLEE, which allows the generation of a new and specific 'predator-overcoming meta-function' so that the hypothetical insect in question can now go through an adaptation to survive at least some kinds of predator attacks. This may take the form of this species of insect creating a hard-shell around it. Mathematically this new meta-function may take the following form as suggested by Equation 9.1.2:

$$Sig_{hard-protective-shell} = Xa + \overline{Yb_{0-n}}$$

$$where \begin{bmatrix} X \in [S_{System_P}] \\ Y \in [S_{System_{Pr}}, S_{System_P}, S_{System_K}, S_{System_N}] \\ a, b \ are \ integers; a > b \end{bmatrix}$$

Eq 9.1.2: Hard Protective Shell Meta-Function

Specifically, S_{System_P} relates to the set of Polysaccharides, as introduced in Chapter 7.2, and suggests that the chains of sugar molecules will adapt to become a shell to protect the insect. The primary element X, is therefore an element of the set or Polysaccharides. $S_{System_{Pr}}$ refers to the set of Proteins. S_{System_K} refers to the set of Nucleic Acids. S_{System_N} refers to the set of Lipids. Y as the secondary element will invoke the action of some proteins, some existing polysaccharides, some nucleic acids, and some lipids in bringing about the adaptation as specified by $Sig_{hard-protective-shell}$. Perhaps the nucleic acids will code or coordinate how the proteins will work with the existing polysaccharides and lipids to create a new arrangement of polysaccharides that becomes the protective layer for the insect. This new arrangement of polysaccharides then becomes the purpose of a particular new type of specialized cell that exists to protect the insect against certain types of predators.

So while the breaking of patterns as specified by the Space-Matrix was a first step in the process, the subsequent creation of a new meta-function can be thought of as a second step. This new meta-function can be thought of as happening due to some combined functioning of the Light and Space Matrices. Specifically in the Space-Matrix, the breaking of patterns, $(\uparrow > P_x)$, allows the dynamics of M_1 to become active: $M_1 \rightarrow Sig_x$. This allows the ever-present layers of light, and specifically $C_{N:} f(S_{Pr} \ x \ S_{Po} \ x \ S_K \ x \ S_H)$ in the Light-Matrix to become more consciously active so that new meta-function that may add to the insect-specific FBLEE is generated.

The third step is suggested by ($\downarrow R_{C_U} = f(R_{C_N})$) that allows specific quantization to occur. It is proposed that the hard-protective-shell adaptability becomes real through a series of tightly coordinated space, time, energy, and gravity quantization as specified by (3.5.2-7) derived in Chapter 3.5. A relevant subset of equations is reproduced here for convenience and suggests the mechanics for the adaptability along this specific line of development.

Equation 3.5.2 models space-quantization:

$$Space_{quantization} = h_{UK}\left(Xa + \overline{Yb_{0-n}}\right)$$

$$where: \begin{bmatrix} X \in [S_{System_K}] \\ Y \in [S_{System_{Pr}}, S_{System_P}, S_{System_K}, S_{System_N}] \\ a, b \text{ are integers}; a > b \end{bmatrix}$$

Recall that space-quantization further structures space so that seeds of knowledge or knowledge-potential to do with the new meta-function are now resident in the material ecosystem associated with the meta-function. In this case, the material ecosystem is the cell-based genetic structure, as indicated by the action of PGGT in (7.2.1). In such a manner the species that set this meta-function into action will more and more tap into the possibilities to bring about the new protective mechanism it seeks.

Equation 3.5.4 models time-quantization:

$$Time_{quantization} = h_{UP}\left(Xa + \overline{Yb_{0-n}}\right)$$

$$where: \begin{bmatrix} X \in [S_{System_P}] \\ Y \in [S_{System_{Pr}}, S_{System_P}, S_{System_K}, S_{System_N}] \\ a, b \ are \ integers; a > b \end{bmatrix}$$

Recall that time-quantization alters the structure of time so that phases of maturity associated with seeds in space are now encoded in time. Slow and faster periods of change are in this way connected to the species seeking to protect itself against predators.

Equation 3.5.6 models energy-quantization:

$$Energy_{quantization} = h_{UPr}(Xa + \overline{Yb_{0-n}})$$

$$where: \begin{bmatrix} X \in [S_{System_{Pr}}] \\ Y \in [S_{System_{Pr}}, S_{System_P}, S_{System_K}, S_{System_N}] \\ a, b \ are \ integers; a > b \end{bmatrix}$$

Recall that energy-quantization allows matter to be formed, and in this case, will have to do with the incremental material changes that the species will go through in forming a hard protective shell.

Equation 3.5.7 models gravity-quantization:

$$Gravity_{quantization} = h_{UN}(Xa + \overline{Yb_{0-n}})$$

$$where: \begin{bmatrix} X \in [S_{System_N}] \\ Y \in [S_{System_{Pr}}, S_{System_P}, S_{System_K}, S_{System_N}] \\ a, b \ are \ integers; a > b \end{bmatrix}$$

Recall that gravity-quantization has to do with inter-relation between the species and surrounding objects and will change the very nature of gravity to allow a subtle new balance in the species interaction with its surrounding so that the deep urge of the species can more easily be fulfilled.

Note that the previously derived equation (3.6.5) - Potential Effect of Levels of Light on Pre-Genetic or Genetic Information – may also be leveraged to understand the possible quantization effects of organization on the post material-fabric housing structure. Reproducing (3.6.5):

Potential Effect of Levels of Light on Pre Genetic or Genetic Information
=

$$
\begin{bmatrix}
STATIC \ \langle |[L][S][T]TC \rightarrow x_T |_{\langle x_U | x_T \rangle} \rangle \\
\times \\
\left((Y > U: Z_Q) \vee (Y \leq U: Z_F) \vee (Y = U: Z_R) \right) \\
\ni \\
\left(\begin{array}{c} Z \in \mathbb{U} \ (Space, Time, Energy, Gravity) \\ Q: Quantization; F: Fragmentation; R: Random \end{array} \right)
\end{bmatrix} \rightarrow h \ni
$$

$$
h \in \left(\begin{array}{c} [Q]: Constructive\ zone, \\ [Q]: Constructive\ zone \wedge Constructive\ mutation, \\ [F]: Destructive\ mutation, \\ [R]: Random\ mutation \end{array} \right)
$$

Line 1 from the top in the matrix is simply a static form of (3.6.1) the Simplified Light-Space-Time Emergence equation. The static form is designated by 'STATIC', and implies that fundamental operations true of (3.6.1) are being highlighted in (3.6.5). In other word (3.6.1) already has all the operations highlighted in (3.6.5) in it, but by 'freezing' it by making it static, the essential dynamics leading to possible mutations at the genetic level can more clearly be highlighted.

Line 1 is then subjected (×) to a determination of the dominant levels of light that may be active, designated by

Line 2, '$\left((Y > U: Z_Q) \vee (Y \le U: Z_F) \vee (Y = U: Z_R) \right)$ '. Unpacking this, '$Y > U$' implies meta-levels are active and as a result it is possible that Z_Q is going to take place. This implies activation and potential change of FBLEE. The call from below, as it were, may invoke some function that already exists in the subtle-libraries 'above', so that some already existing function may influence FBLEE through ∞ -entanglement, K-entanglement, or N-entanglement. This may be thought of as a key-and-lock mechanism, where a deep enough visceral urge from below acting as the key, opens an entangled lock to alter FBLEE as per the visceral urge. Note that in the case of the SRY gene the radical jump from chimpanzee to human (Ridley, 1999) was likely due to such a deep visceral urge that invoked higher meta-levels of light. '$Y \le U$' implies that only the untransformed levels are active, and therefore also the sub-level where the speed of light is 0 is active and as a result Z_F is going to take place. 'Y = U' implies that all levels are active and as a result Z_R is going to take place.

Line 3 elaborates the significance of Z_Q, Z_F, and Z_R. Hence Z is the union of potential quantum-operations of space, time, energy, and gravity, designated by ' $Z \in$ $\mathbb{U}$ (*Space, Time, Energy, Gravity*)'. But the nature of the operations is designated by ' *Q: Quantization; F: Fragmentation; R: Random* '. Z_Q , then, implies that the full quantization originating from updated four-base logic-encoding ecosystems (FBLEE) can take place, and will result in lasting material change at the genetic level as discussed above in invoking the subset of (3.5.2–7). Z_F implies that the essential set will be fragmented and that only local libraries can potentially be altered. Z_R also precludes full quantization, and that some partial local-library constructive or destructive mutation may take place.

The $' \to h \ni h \in '$ segment resolves the outcome of the operations implied by Lines 1 – 3, suggesting that the outcome will be 'h' such that ($\ni$) 'h' is an element ($\in$) of the set specified by the members '[Q]: *Constructive zone*', '[Q}: *Constructive zone AND Constructive mutation*', '[F}; *Destructive mutation*', and '{R}: *Random mutation*'. '[Q]: *Constructive zone*' implies that the in-built buffer has been crossed and that access to the deeper four-base logic-encoding ecosystem (FBLEE) has been granted. '[Q}' in this segment implies that there is the possibility of full-quantization as specified by Line 3 of the previous matrix can take place. Access to this zone is a prerequisite for constructive mutation to occur, as designated by the element '[Q}: *Constructive zone* ∩ *Constructive mutation* ', which implies that full-quantization is going to take place and will result in material change. The '[F]' specifies the relationship between 'Fragmentation' in Line 3 of the previous matrix and destructive mutation. The '[R]' specifies the relationship between 'Random' in Line 3 of the previous matrix and random mutation.

394

Chapter 9.2: Genetic Mutation in the Case of Creating a Sustainable Global Civilization

Leveraging the discussion on the emergence of Light in a sustainable global civilization from Chapter 8.2, Generating the Genetic Code for a Sustainable Global Civilization, this chapter will consider how mutation may occur in a light-based model, in the case of the development of a sustainable global civilization. The Light-Space-Time Emergence equation (8.2.1) is reproduced here for convenience:

$$Light - Space - Time\ Emergence_{Sustainable\ Global\ Civilization} =$$

$$\left| \left| \begin{bmatrix} c_\infty: [Pr, Po, K, H] \\ \left(\downarrow R_{C_K} = f(R_{C_\infty}) \right) \\ c_K: [S_{Pr}, S_{Po}, S_K, S_H] \\ \left(\downarrow R_{C_N} = f(R_{C_K}) \right) \\ c_N: f(S_{Pr} \times S_{Po} \times S_K \times S_H) \\ \left(\downarrow R_{C_U} = f(R_{C_N}) \right) \\ c_U: [P, V, M, C] \\ \Uparrow \\ c_{0:[D,W,I,C]} \end{bmatrix}_{Light} \left[U \to \begin{array}{c} M_3 \to System_x \\ (\uparrow F \to I) \\ M_2 \to S_{System_x} \\ (\uparrow Sig \to F) \\ M_1 \to Sig_x \\ (\uparrow > P_x) \\ x_{Stable\ Mega-Organization} \end{array} \right]_{Space} \right.$$

$$\left. \left| U \to \begin{bmatrix} M_3: -\infty \le t \le \infty \\ \downarrow \\ M_2: 0 \ge t > \infty \\ \downarrow \\ M_1: 0 > t > \infty \\ \downarrow \\ t \le E_{Cell}; TC: M_3 \to U \\ t \sim E_{Human}; TC: U \to M_3 \end{bmatrix}_{Time} \right. \quad TC \to x_{SSustainbale\ Global\ Civilization} \right| \langle x_U | x_T \rangle$$

$$\Rightarrow$$

$\langle Space - Time - Energy - Gravity\ Material - Fabric\ Pre - Genetic\ Code \rangle +$
$\langle Electromagnetic\ Spectrum\ Material - Fabric\ Pre - Genetic\ Code \rangle +$
$\langle Quantum - Particle\ Material - Fabric\ Pre - Genetic\ Code \rangle +$
$\langle Atoms\ Pre - Genetic\ Code \rangle$
$\qquad + \langle Molecules\ Material - Fabric\ Pre - Genetic\ Code \rangle +$
$LSTE \langle ... \rangle + PGGT < \cdots > + \langle Cells\ Genetic\ Code \rangle + LSTE \langle ... \rangle +$

$(Fundamental\ Capacities\ of\ Self\ Genetic\ Code) + FBLEEE \langle … \rangle +$
$(Humans\ Genetic\ Code) + LSTE \langle … \rangle + FBLEEE \langle … \rangle +$
$(Stable\ Mega - Organization\ Genetic\ Code) + LSTE < … > +$
$Sustainable\ Global\ Civilization\ Genetic\ Code$

The Light-Matrix, the top-left matrix in (8.2.1), recall, has been set up by realities so created by light traveling at different speeds. While prevalent dynamics at the surface layer, U, are always influenced by dynamics from the deeper layers, there is an activation-state that can be created that will allow a more focused and intentional activation or creation of the meta-functions in the deeper layers.

The Space-Matrix, the top-right matrix in (3.1.3), suggests the dynamics involved in creating such an activation-state. Essentially this involves the overcoming of patterns at U. Note that by definition such an activation-state necessitates moving in the direction where the speed of light increases. Note that speed of light increasing implies that the sense of separation is diminishing, and entities can have a more complete access to the fullness that they are. This can only happen when patterns get elevated. If patterns are in the U realm, being therefore untransformed, that will by definition not allow a quantum-window to open. Therefore the impact to history is termed as 'sustainable human history'.

This activation-state can be further specified by leveraging (3.7.4) – The Generalized Equation for Direction of Mutation, derived in Chapter 3.7, on Qualified Determinism.

Reproducing Equation 3.7.4:

$$Org\_Dir = DI \left(\begin{bmatrix} M_3 \rightarrow System_X \\ (\uparrow F \rightarrow I) \\ M_2 \rightarrow S_{System_X} \\ (\uparrow Sig \rightarrow F) \\ M_1 \rightarrow Sig_x \\ (\uparrow > P_P) \\ U \rightarrow x_U \end{bmatrix}^{x=p,v,m,i} \right) \rightarrow$$

$$x\_matrix_{strongest} @ level_{strongest}$$

Activation-state can be thought of as being invoked when $level_{strongest}$ is at least M_1. Hence, as suggested by the previously derived (9.1.1), reproduced here for convenience:

$$Activation - state_{condition} \geq M_1$$

If the activation-state is invoked, that implies access to the four-base logic-encoding ecosystem (FBLEE).

Deeper Nation Essence Case

As per the discussion in the previous chapter, assume that there is a deep wanting felt by a threshold number of people. Perhaps this wanting, a step towards a global sustainable civilization, is for each nation to operate more from their true essence, or I-state, as opposed to fundamentalist form. In other words a meta-function, $Sig_{Nation-essence}$, must be created that will then organize its own reality, and as indicated by Equation 9.2.1.

$$Sig_{Nation-essence} = Xa + \overline{Yb_{0-n}}$$

$$where \begin{bmatrix} X \in [S_{System_K}] \\ Y \in [S_{System_{Pr}}, S_{System_P}, S_{System_K}, S_{System_N}] \\ a, b \ are \ integers; a > b \end{bmatrix}$$

Eq 9.2.1: Nation-Essence Meta-Function

While the key aspect, or primary X-element in the nation-essence meta-function may be an element from the set of Knowledge, and specifically 'what the nation-essence means', there will be many parts, represented by the Y-elements, such as the structure, processes, institutions, culture, that will also need to be created. An important denominator in breaking patterns involves the creation of 'cohesion' at the individual and collective levels as in Equation 9.2.2, Nation-Essence-Cohesion with the primary element belonging to the set of Harmony or Nurturing:

$$Sig_{Nation-essence-cohesion} = Xa + \overline{Yb_{0-n}}$$

$$where \begin{bmatrix} X \in [S_{System_N}] \\ Y \in [S_{System_{Pr}}, S_{System_P}, S_{System_K}, S_{System_N}] \\ a, b \ are \ integers; a > b \end{bmatrix}$$

Eq 9.2.2: Nation-Essence-Cohesion Meta-Function

Hence, S_{System_K} relates to the set of Thoughts required to contextualize and frame nation-essence. $S_{System_{Pr}}$ refers to the set of Sensations required to assess and interpret the constantly arising signs of the new development – the sensory cues as it were that will allow any individual or collectivity to sense that they were on the right path. S_{System_P} refers to the set of Urges and Wills constantly required to ensure that the goal of cohesion was attained. S_{System_N} refers to the set of Feelings and Emotions that will need to be generated to attain cohesion.

As such cohesion becomes a reality, older patterns will break down and there will be easier and longer activation-state periods, potentially allowing 'miracle' to become the law.

As suggested in the previous chapter, while the breaking of patterns as specified by the Space-Matrix was a first step in the process, the subsequent creation of a new meta-function can be thought of as a second step. This new meta-function can be thought of as happening due to some combined functioning of the Light and Space Matrices. Specifically in the Space-Matrix, the breaking of patterns, $(\uparrow > P_x)$, allows the dynamics of M_1 to become active: $M_1 \rightarrow Sig_x$. This allows the ever-present layers of light, and specifically $C_{N:} f(S_{Pr} \times S_{Po} \times S_K \times S_H)$ in the Light-Matrix to become more consciously active so that new meta-function that may add to the insect-specific FBLEE is generated.

The third step is suggested by $(\downarrow R_{CU} = f(R_{CN}))$ that allows specific quantization to occur. Quantization is of fundamental importance because it is what allows the fabric of experienced reality to change. The new thought to do with nation-essence is quantized as seeds of knowledge in space, fundamentally changing the structure of space. Hence, reproducing Equation 3.5.2, space-quantization is modeled as:

$$Space_{quantization} = h_{UK}\left(Xa + \overline{Yb_{0-n}}\right)$$

$$where: \begin{bmatrix} X \in [S_{System_K}] \\ Y \in [S_{System_{Pr}}, S_{System_P}, S_{System_K}, S_{System_N}] \\ a, b \ are \ integers; a > b \end{bmatrix}$$

The phases of maturity that the seeds of knowledge will go through give time a new reality or structure. In other words time too is re-oriented to promote the outcome of the new meta-function in effect. This quantization of time, essentially giving time in this new fabric a different meaning is modeled by Equation 3.5.4, reproduced here for convenience:

$$Time_{quantization} = h_{UP}\left(Xa + \overline{Yb_{0-n}}\right)$$

$$where: \begin{bmatrix} X \in [S_{System_P}] \\ Y \in [S_{System_{Pr}}, S_{System_P}, S_{System_K}, S_{System_N}] \\ a, b \ are \ integers; a > b \end{bmatrix}$$

The new meta-function changes the very nature of matter, which is quantized to allow the material fabric of existence to promote the new meta-function in unimaginable ways. This quantization of energy, resulting in a subtly different matter is modeled by Equation 3.5.6, reproduced here for convenience:

$$Energy_{quantization} = h_{UPr}(Xa + \overline{Yb_{0-n}})$$

$$where: \begin{bmatrix} X \in [S_{System_{Pr}}] \\ Y \in [S_{System_{Pr}}, S_{System_P}, S_{System_K}, S_{System_N}] \\ a, b \ are \ integers; a > b \end{bmatrix}$$

As a result the very force of gravity is altered locally as well, so that its quantization promotes subtle interaction in the related ecosystem that will be predisposed to making the meta-function a reality. This quantization of gravity is modeled by Equation 3.5.7, reproduced here for convenience:

$$Gravity_{quantization} = h_{UN}(Xa + \overline{Yb_{0-n}})$$

$$where: \begin{bmatrix} X \in [S_{System_N}] \\ Y \in [S_{System_{Pr}}, S_{System_P}, S_{System_K}, S_{System_N}] \\ a, b \ are \ integers; a > b \end{bmatrix}$$

Note that the previously derived equation (3.6.5) - Potential Effect of Levels of Light on Pre-Genetic or Genetic Information – may also be leveraged to understand a possible quantization effect of organization on the material fabric. Reproducing (3.6.5):

Potential Effect of Levels of Light on Pre Genetic or Genetic Information =

$$\begin{bmatrix} STATIC\ \langle|[L][S][T]TC \rightarrow x_T|_{\langle x_U|x_T\rangle}\rangle \\ \times \\ \left((Y > U: Z_Q) \vee (Y \leq U: Z_F) \vee (Y = U: Z_R)\right) \\ \ni \\ \begin{pmatrix} Z \in \mathbb{U}\ (Space, Time, Energy, Gravity) \\ Q:\ Quantization;\ F:\ Fragmentation;\ R:\ Random \end{pmatrix} \end{bmatrix} \rightarrow h \ni$$

$$h \in \begin{pmatrix} [Q]:\ Constructive\ zone, \\ [Q]:\ Constructive\ zone \wedge Constructive\ mutation, \\ [F]:\ Destructive\ mutation, \\ [R]:\ Random\ mutation \end{pmatrix}$$

Line 1 from the top in the matrix is simply a static form of (3.6.1) the Simplified Light-Space-Time Emergence equation. The static form is designated by 'STATIC', and implies that fundamental operations true of (3.6.1) are being highlighted in (3.6.5). In other word (3.6.1) already has all the operations highlighted in (3.6.5) in it, but by 'freezing' it by making it static, the essential dynamics leading to possible mutations at the genetic level can more clearly be highlighted. Line 1 is then subjected ($\times$) to a determination of the dominant levels of light that may be active, designated by Line 2, $'\left((Y > U: Z_Q) \vee (Y \leq U: Z_F) \vee (Y = U: Z_R)\right)'$. Unpacking this, '$Y > U$' implies meta-levels

401

are active and as a result it is possible that Z_Q is going to take place. This implies activation and potential change of FBLEE. The call from below, as it were, may invoke some function that already exists in the subtle-libraries 'above', so that some already existing function may influence FBLEE through ∞ -entanglement, K-entanglement, or N-entanglement. This may be thought of as a key-and-lock mechanism, where a deep enough visceral urge from below acting as the key, opens an entangled lock to alter FBLEE as per the visceral urge. 'Y $\leq$ U' implies that only the untransformed levels are active, and therefore also the sub-level where the speed of light is 0 is active and as a result Z_F is going to take place. 'Y = U' implies that all levels are active and as a result Z_R is going to take place.

Line 3 elaborates the significance of Z_Q, Z_F, and Z_R. Hence Z is the union of potential quantum-operations of space, time, energy, and gravity, designated by ' $Z \in$ $\mathbb{U}$ ($Space, Time, Energy, Gravity$)'. But the nature of the operations is designated by ' Q: $Quantization; F$: $Fragmentation; R$: $Random$ '. Z_Q , then, implies that the full quantization originating from updated four-base logic-encoding ecosystems (FBLEE) can take place, and will result in lasting material change at the genetic level as discussed above in invoking the subset of (3.5.2–7). Z_F implies that the essential set will be fragmented and that only local libraries can potentially be altered. Z_R also precludes full quantization, and that some partial local-library constructive or destructive mutation may take place.

The ' $\rightarrow h \ni h \in$ ' segment resolves the outcome of the operations implied by Lines 1 – 3, suggesting that the outcome will be 'h' such that ($\ni$) 'h' is an element ($\in$) of the set specified by the members '[Q]: *Constructive zone*', '[Q]: *Constructive zone AND Constructive mutation*', '[F]; *Destructive mutation*', and '{R}: *Random mutation*'. '[Q]:

Constructive zone' implies that the in-built buffer has been crossed and that access to the deeper four-base logic-encoding ecosystem (FBLEE) has been granted. '[Q}' in this segment implies that there is the possibility of full-quantization as specified by Line 3 of the previous matrix can take place. Access to this zone is a prerequisite for constructive mutation to occur, as designated by the element '   [Q}: *Constructive zone* ∩ *Constructive mutation* ', which implies that full-quantization is going to take place and will result in material change. The '[F]' specifies the relationship between 'Fragmentation' in Line 3 of the previous matrix and destructive mutation. The '[R]' specifies the relationship between 'Random' in Line 3 of the previous matrix and random mutation.

SECTION 10: SUPER-MATTER AND A POSSIBLE FUTURE OF GENETICS

This Section summarizes the developments of the previous sections to draw insight into the nature of matter and genetics itself.

Chapter 10.1 discusses the creation of super-matter and a post-genetic housing structure from a macro perspective. Super-matter is suggested to be that type of matter created through the intervention of conscious will or cohesive want. While complexification of matter as summarized in previous sections is often an "automatic" process of Nature, super-matter is a foundation based on will or cohesive want and sets the stage for a potentially unending willed development in which functional-richness existing in Light at Infinity can manifest in this material universe. From a macro-level, meta-level bases are what will ensure the development of super-matter.

Chapter 10.2 further focuses on the nature of matter by drawing on observations from the field of astrophysics. In light of further insight into matter, equated as emerging light, and of the correlation between type of matter and structures to house genetic-type information, the chapter culminates in a general equation for future genetic-type code, and a reflection on genetics itself.

Chapter 10.1: Transitioning to Super-Matter and a Post-Genetic Structure

Super-Matter is positioned as being fundamentally different from matter. While matter can be thought of as the result of Nature's automatic working, super-matter can be thought of as the result of a conscious will and cohesive wanting that causes a deliberate process of quantization by which the very fabric of matter is changed. Quantization occurs in any case even when Nature's automatic working drives change. But with conscious will this process can be accelerated, broadened, and heightened, so that a wider and higher possibility of

function may consciously materialize. As suggested in the previous sections such quantization is related to enhancing the functional-richness of matter. Imagine through a cohesive-wanting matter being able to become more flexible, more durable, more stretchable, light at will, heavy at will, change color at will, expand in size, contract in size, amongst infinite other functions and possibilities contained in Light at infinity.

The previous sections explored numerous examples of the process by which matter becomes functionally rich. In cases getting closer to, and even beyond the human-system, this involves a visceral urge that can open the entity to the influence of meta-levels of light. This also gives access to FBLEE and as explored in the previous section, Section 9 on a practical look at genetic mutation, an enhanced function can then materialize through a process of quantization. This increasing possibility of FBLEE means that matter can transition to richer versions of itself.

Further, in considering the vast information embedded in the ubiquitous-point-instant pre-genetic ∞-entanglement library, the architectural forces pre-genetic K-entanglement library, and the organizational-uniqueness pre-genetic N-entanglement library, discussed in Chapters 2.2, 2.3, and 2.4 respectively, it is clear that many possible functions already exist in layers of light. The call from below, whether a pattern breaking will, urge, or desire, likely causes some meta-function or precipitation into FBLEE, from where materialization becomes possible through the process of quantization.

Another way to look at this is summarized in the previously derived (4.2.8) which models post-genetic code for matter:

$$Light - Space - Time\ Emergence_{Suoer-Matter} =$$

$$\left|\begin{array}{l} \begin{bmatrix} c_\infty: [Pr, Po, K, H] \\ \left(\downarrow R_{C_K} = f(R_{C_\infty})\right) \\ c_K: [S_{Pr}, S_{Po}, S_K, S_H] \\ \left(\downarrow R_{C_N} = f(R_{C_K})\right) \\ c_N: f(S_{Pr} \times S_{Po} \times S_K \times S_H) \\ \left(\downarrow R_{C_U} = f(R_{C_N})\right) \\ c_U: [P, V, M, C] \\ \Uparrow \\ c_{0:[D,W,I,C]} \end{bmatrix}_{Light} \\ \begin{bmatrix} M_3 : -\infty \le t \le \infty \\ \downarrow \\ M_2 : 0 \ge t > \infty \\ \downarrow \\ M_1 : 0 > t > \infty \\ \downarrow \\ U \to \begin{array}{l} t \le E_{Cell}; TC: M_3 \to U \\ t \sim E_{Human}; TC: U \to M_3 \end{array} \end{bmatrix}_{Time} \end{array} \right. \left. \begin{array}{l} \begin{bmatrix} M_3 \to System_X \\ (\uparrow F \to I) \\ M_2 \to S_{System_X} \\ (\uparrow Sig \to F) \\ M_1 \to Sig_x \\ (\uparrow > P_x) \\ U \to x_{Molecules} \end{bmatrix}_{Space} \\ \\ TC \to x_{Super-Matter} \end{array} \right| \langle x_U | x_T \rangle$$

$\Rightarrow$

$\langle Space - Time - Energy - Gravity\ Material - Fabric\ Pre - Genetic\ Code \rangle +$

$\langle Electromagnetic\ Spectrum\ \ Material - Fabric\ Pre - Genetic\ Code \rangle +$

$\langle Quantum - Particle\ Material - Fabric\ Pre - Genetic\ Code \rangle +$

$\langle Atoms\ Material - Fabric\ Pre - Genetic\ Code \rangle +$

$\langle Molecules\ Material - Fabric\ Pre - Genetic\ Code \rangle + LSTE \langle \dots \rangle + FBLEEE \langle \dots \rangle +$

$LSTE \langle \dots \rangle + Super - Matter\ Post - Genetic\ Code$

Equation 4.2.8 suggests computational realities yet to emerge. The essential action of space-time-energy-gravity quantization can cause the materialization of potentially infinite four-base logic-encoding ecosystems. It is conceivable that such a variation of space-time-energy-gravity quantization coupled with an increasing ability to easily move back and forth between the material and antecedent realms, perhaps the outcome of an enhanced function driven by an initiating will itself, makes the need to house genetic information in a form such as DNA alone, burdensome. It is conceivable that four-base logic-encoding ecosystems, or even the ∞-entanglement, K-entanglement, and N-entanglement libraries may be able to more directly act at the material level in some composite subtle-material post-genetic form that also includes DNA. This possibility is referred to as post-genetic code.

This chapter takes a more macro-level mathematical view based on a cosmology of light, to summarize the aggregate conditions required for the creation of functionally rich super-matter. In the process the distinction between three distinct phases of matter emerge: that of established-matter, matter-in-transition, and super-matter. The development of super-matter will naturally also lead to a post-genetic housing structure for genetic-type information.

It makes sense for the structure of genetic-type information-storage to change in each of these cases since it is something different in each case. In such a view the material-fabric is the genetic-type information storage medium for established matter. Code for the electromagnetic spectrum, quantum particles, and atoms, is stored in the material-fabric as reviewed in Sections 5 and 6.

Matter-in-transition would be associated with the emergence of life and as has been argued in this book, matter becomes more dynamic and adaptable as a result of this. Cellular-based genetic structure is the vehicle for storage for code associated with the cell. But further, as discussed, with the emergence of basic human capabilities, and truer individuality, the genetic information is further enhanced. These were reviewed in detail in Section 7. It has also been argued in Section 8, that the emergence of complex organization, which incudes mega-organization and a sustainable global civilization, will likely make further changes to the cellular-based genetic code.

When we arrive at super-matter, which is inevitable with the increase in functional-richness, then it stands to reason that there will likely come into existence some post-genetic structure that will house essential genetic-type information in a very different way.

The Light-Space-Time Emergence equation (3.1.3) derived in Chapter 3.1, and reproduced here for convenience, will be leveraged to derive summary equations to distinguish between the different phases of matter. The following derivations run parallel to equations created in Chapter 2.6 that focused on systematization of mutational-sequences along the upward-strand. Reproducing (3.1.3):

$Emergence_{light-space-time} =$

$$\left| \begin{array}{l} \left[\begin{array}{l} c_\infty: [Pr, Po, K, H] \\ (\downarrow R_{C_K} = f(R_{C_\infty})) \\ c_K: [S_{Pr}, S_{Po}, S_K, S_H] \\ (\downarrow R_{C_N} = f(R_{C_K})) \\ c_N: f(S_{Pr} \times S_{Po} \times S_K \times S_H) \\ (\downarrow R_{C_U} = f(R_{C_N})) \\ c_U: [P, V, M, C] \\ \Uparrow \\ \quad c_{0:[D,W,I,C]} \end{array} \right]_{Light} \left[\begin{array}{l} M_3 \rightarrow System_X \\ (\uparrow F \rightarrow I) \\ M_2 \rightarrow S_{System_X} \\ (\uparrow Sig \rightarrow F) \\ M_1 \rightarrow Sig_x \\ (\uparrow > P_{x)} \\ U \rightarrow x_U \end{array} \right]_{Space} \\ \left[U \rightarrow \begin{array}{l} M_3 : -\infty \leq t \leq \infty \\ \downarrow \\ M_2 : 0 \geq t > \infty \\ \downarrow \\ M_1 : 0 > t > \infty \\ \downarrow \\ t \sim E_{Cell}; TC: M_3 \rightarrow U \\ t \sim E_{Human}; TC: U \rightarrow M_3 \end{array} \right]_{Time} \quad TC \rightarrow x_T \end{array} \right| \langle x_U | x_T \rangle$$

Simplifying each of the main matrices yielded (3.6.1), the simplified version of the Light-Space-Time Emergence equation, reproduced here for convenience:

$$Emergence_{light-space-time} = |[L][S][T]TC \rightarrow x_T|_{\langle x_U | x_T \rangle}$$

As explained by Neil Turok in his book The Universe Within (Turok, 2012), Euler's formula reproduced below, can be used to model many naturally occurring phenomena because of its sinusoidal oscillation between narrow bounds as x increases. Reproducing Euler's formula:

$$e^{ix} = \cos x + i \sin x$$

The sum of the squares of the ordinary and complex parts, on the right side of the equation, is one. In quantum theory this ensures that the probabilities for all possible outcomes add up to one. Hence this formula is useful when considering the macro-level effects of quantization originating from FBLEE.

Further, in modeling matter a modified notation of the Schrodinger wavefunction as interpreted by Feynman is leveraged. This version features the integral sign, $\int$, meaning that all terms to the right of it have to be summed up for all space and time till the moment when the wavefunction is required to be known.

Summary Equations for Established-Matter

Hence, combining (3.6.1) with Feynman's interpretation of the Schrodinger wavefunction, with the Euler formula yields the following equations, 10.1.1-6 for matter:

$$\psi_{established-matter} = \left| \int e^{i \int |[L][S][T]|} \right|_U$$

Eq. 10.1.1: Wavefunction for Established-Matter

(10.1.1) suggests that so long as the basis of matter is untransformed, specified by the U following the vertical-brackets, the outcome is not going to be any different from established matter, specified by $\psi_{established-matter}$. U implies that there is not any action of the conscious will yet. This makes sense since conscious will arises with close-to human-systems. Note also that the presumption here is that matter as we perceive it – comprising of quantum particle, atoms, molecules – is perhaps only at the start of its journey. This observation is reinforced through the possibility of infinite information existing in meta-layers of Light that has yet to precipitate materially.

Further, as specified by Equation 10.1.2, the Probability-View of Established-Matter, the very basis of matter is going to be either the untransformed physical (P_U), the untransformed vital (V_U), the untransformed mental (M_U), or the untransformed integral (I_U):

$$|\psi_{established-matter}|^2 = P_U^2 + V_U^2 + M_U^2 + I_U^2 = 1$$

Eq. 10.1.2: Probability-View for Established-Matter

As specified by (10.1.2) the probability that any of these bases will be leveraged in the creation of matter adds up to one.

Another way of viewing established matter is by leveraging the non-wavefunction form as in (3.6.2) and adapting it to create Equation 10.1.3, Simplified Light-Space-Time Matrix Form for Established-Matter:

$$Established - Matter = \left\| |[L][S][T]TC \rightarrow x_T|\langle x_U|x_T\rangle \right\|_U$$

Eq. 10.1.3: Simplified Light-Space-Time Matrix Form for Established-Matter

Summary Equations for Matter-in-Transition

Equation 10.1.4, Wavefunction for Matter-in-Transition, suggests the mixed bases for matter, as specified by dynamics of both the untransformed (U) and the meta-levels (M_x), which therefore results in matter-in-transition.

$$\Psi_{matter-in-transition} = \left| \int e^{i \int |[L][S][T]|} \right|_{U\&M_x}$$

Eq. 10.1.4: Wavefunction for Matter-in-Transition

The influence of meta-level, M_x, implies that action of visceral urge or desire or will. This is mixed with Nature's existing results, specified by U. As already discussed, such matter would be the result of the dynamics of life and its emergences.

As specified by Equation 10.1.5, Probability-View for Matter-in-Transition, the probability that any of the untransformed (U) and transformed (T) bases will be leveraged in the creation of matter adds up to one.

$$|\Psi_{matter-in-transition}|^2$$
$$= P_U^2 + P_T^2 + V_U^2 + V_T^2 + M_U^2 + M_T^2 + I_U^2$$
$$+ I_T^2 = 1$$

Eq. 10.1.5: Probability View for Matter-in-Transition

Note also that another way of viewing matter in transition is by leveraging the non-wavefunction form as in (3.6.3) and adapting it to create Equation 10.1.6, Simplified Light-Space-Time Matrix Form for Matter-in-Transition:

$$Matter - in - transition$$
$$= \left| |[L][S][T]TC \rightarrow x_T| \langle x_U | x_T \rangle \right|_{U\&M_x}$$

Eq. 10.1.6: Simplified Light-Space-Time Matrix Form for Matter-in-Transition

Summary Equations for Super-Matter

Equation 10.1.7, Wavefunction for Super-Matter, suggests some transformed bases for matter, as specified by dynamics of the meta-levels (M_x), which therefore results in super-matter. It is essentially more consciousness, whether of visceral urge, desire, or will, that is driving change.

$$\psi_{super-matter} = \left\| \int e^{i \int |[L][S][T]|} \right\|_{M_x}$$

Eq. 10.1.7: Wavefunction for Super-Matter

As specified by Equation 10.1.8, Probability-View for Super-Matter, the probability that the transformed (T) bases will be leveraged in the creation of matter adds up to one.

$$\left| \psi_{super-matter} \right|^2 = P_T^2 + V_T^2 + M_T^2 + I_T^2 = 1$$

Eq. 10.1.8: Probability View for Super-Matter

Note also that another way of viewing super-matter is by leveraging the non-wavefunction form for as in (3.6.4) and adapting it to create Equation 10.1.9, Simplified Light-Space-Time Matrix Form for Super-Matter:

$$Super - Matter = \left\| |[L][S][T]TC \rightarrow x_T|_{\langle x_U | x_T \rangle} \right\|_{M_x}$$

Eq. 10.1.9: Simplified Light-Space-Time Matrix Form for Super-Matter

Chapter 10.2: The Nature of Matter and a Possible Future of "Genetics"

The previous chapter further discussed the effect of functional richness in the creation of super-matter. The question then, is what is matter? Further, the previous chapter also suggested a tight relationship or correlation between the type of matter and the type of information-storage required for genetic-type information. This chapter will take a deeper look at the nature of matter by drawing on some observations from the field of astrophysics. This deeper look will also suggest an equation for future genetic-type code.

From discussion in the previous chapter it is clear that if the cumulative effect of overcoming habitual patterns increases the likelihood of super-matter being formed, this implies the increase too of four-fold quantization. Space, as the seeding ground of knowledge, may therefore also require expansion to continue to allow seeding to take place. But if space needs to expand in such a scenario, then the fact that scientists have observed that the universe is expanding may be related to this. This chapter hence also explores a possible link between the expansion of the universe and an increase in functional-richness that implies the expansion of space. Such a link provides insight into the nature of matter.

The entertainment of such an idea perhaps can be strengthened through consideration of the cosmological principle that suggests that the spatial distribution of matter in the universe is homogenous and isotropic when viewed on a threshold scale of 250 million light years. The nine-year cosmic microwave background image created with the Wilkinson Microwave Anisotropy Probe (WMAP) depicts 13.77 Billion year old temperature differences that correspond to the seeds that became the galaxies (NASA-WMAP, 2014). This homogeneity in temperature difference in the early universe suggests too a process of homogenous four-fold quantization that set into motion the development of galaxies.

Complexity of Matter, Dark Matter, & Dark Energy

According to NASA only 5% of the matter in the universe is visible, while as much as 27% is 'dark matter'. It is the 5% of 'visible' matter though that forms the familiar stars, galaxies, or the atoms (NASA-darkmatter, 2016). But since, as explored in Chapter 3.7 on Qualified Determinism, the architectural sets at M_2 are infinite the question is why would all functions need to manifest in the same way?

As stated in an article 'Dark Energy: The Biggest Mystery in the Universe' (Panek, 2010), "Sight itself has blinded us to the Universe". It should be possible for other matter-based constructions to come into being based on the functionality they exist for. For instance there could be other fields that create other kinds of elementary particles, that create other kinds of atoms and that results in other kinds of structures not visible with our current instrumentation.

In reference to the four sets at M_2 (as discussed in Chapter 2.3 on Architectural Forces K-Entanglement Pre-Genetic Library) it may also be that as the basis of matter complexifies as it journeys through the electromagnetic spectrum, quantum particles, the atom, cellular life, complex individual and further organizational development, the sets which have been positioned to each contain infinite elements, concretely manifest more of their function-elements.

This suggestion has been the basis of the equations derived in previous chapters. The complexity of matter may therefore be related to the number of manifested function-elements of the set of four sets. This idea is consistent with the notion of the "adjacent possible" suggested by Kaufmann (Kaufmann, 2003) in which innovation is positioned as a recombination of existing parts to create new value – or of existing sets to combine parts of themselves to create new elements based on new circumstance. If MS signifies manifested-set, so that the cardinality or number of elements in the combined set is the union of the four manifested-sets, this may be summarized by the following equation, Equation 10.2.1, Complexity of Matter in Terms of Manifested Set:

$$Matter_{Complexity} \propto$$
$$\left| MS_{System_{Pr}} \cup MS_{System_P} \cup MS_{System_K} \cup MS_{System_N} \right|$$

Eq 10.2.1: Complexity of Matter in Terms of Manifested Set

It is also possible to hypothesize an Equation, 10.2.2, relating complexity of matter in terms of visible and dark matter:

$$Matter_{Complexity} \propto [Visible\ Matter + Dark\ Matter]$$

Eq. 10.2.2: Complexity of Matter in Terms of Visible and Dark Matter

Further, as suggested by the previous discussions on the quantization of space, space is not empty but is a seeding ground for a variety of emergences. The notion of space having 'amazing' properties was, according to a report on 'dark energy' by the Harvard-Smithsonian Center for Astrophysics, first suggested by Einstein (Harvard-Smithsonian Center for Astrophysics, 2004). Quoting: "Einstein was the first person to realize that empty space is not nothingness. Space has amazing properties, many of which are just beginning to be understood." Einstein had suggested the existence of a 'dark energy' about 100 years ago (NASA-Supernova, 2001) as a property of space that caused the expansion of the universe. Dark Energy is estimated to comprise as much as 68% of the universe.

Reinterpretation of Cosmological Expansion-Contraction Dynamics

Keeping in mind the complexification of matter as it journeys from the field-level through the quantum-, atomic-, and cellular-levels, and beyond, it may be possible to re-interpret the supposed expansion-contraction dynamics of cosmology in relation to the mathematical model presented in this book.

First, flipping the left and right sides of the equation (10.2.1) on $Matter_{Complexity}$ to yield Equation 10.2.3:

$$\left| MS_{System_{Pr}} \cup MS_{System_P} \cup MS_{System_K} \cup MS_{System_N} \right|$$
$$\propto Matter_{Complexity}$$

Eq 10.2.3: Flipped Matter-Complexity Equation

This implies that the manifested-set, MS, is growing at a certain threshold level. Assuming this threshold level, $MS\_Growth_{Threshold}$, is a property of space related to dark energy it may be possible to restate the condition of cosmological expansion and contraction. Hence, so long as the MS is increasing at a certain rate that exceeds $MS\_Growth_{Threshold}$ the level of dark energy is such that the acceleration of galaxies, $Acceleration_{Galaxies}$ exceeds the contracting force of gravity at the universal level, $Gravity_{Universe}$.

Therefore, as in Equations 10.2.4 and 10.2.5:

$$MS\_Growth_{Threshold}: Acceleration_{Galaxies}$$
$$> Gravity_{Universe}$$

Eq. 10.2.4: Expansion of Universe Related to Manifest Set Growth Threshold

Conversely:

$$! MS\_Growth_{Threshold}: Acceleration_{Galaxies}$$
$$< Gravity_{Universe}$$

Eq. 10.2.5: Contraction of Universe Related to Manifest Set Growth Threshold

The relation between $MS\_Growth_{Threshold}$ and Dark Energy is suggested by Equation 10.2.6:

$$MS\_Growth_{Threshold} \propto Dark\ Energy$$

Eq. 10.2.6: Relation Between Manifested Set Growth Threshold and Dark Energy

In this interpretation so long as more of the infinite set at M_2 results in the manifested-set, MS_{System}, the universe keeps growing. In other words, so long as matter tends towards super-matter, the universe continues with expansion. It seems therefore that matter is a dynamic state linked to the very life of the universe.

Further if the relative distances between some of the emergences discussed in this book are considered a pattern emerges. The quantum level is dealing in distances of the order of 10^{-35}m. The cellular level is dealing with distances of the order of 10^{-6}m. The astronomical level is dealing with distances as large as 10^{26}m – the estimated size of the universe. It can be observed therefore that the cell is approximately in the center with roughly 30 orders of magnitude on either side.

Such symmetry that we are at now with the human being at the mid-point between two known material extremes, seems to suggest a cosmic balance of scales with human-will and other faculties such as awareness as a potential

418

arbiter. But this we already know is the condition for the activation of meta-levels and also therefore for the creation of super-matter. Lack of will for continued functional-richness will in this view lead to a reversal in the Manifested Set due to a reversal in the quantization trend and consequently to a return of this universe to a state of collapse or unfulfilled absorption into the initiating Light. Matter hence collapses into naught. In that case it will be another possible Big Bang initiated through the interplay of multiple layers of Light, and a birth following such death as depicted by the ancient Ouroboros symbol of a dragon or snake eating its own tail. The easting of its own tail can be mathematically depicted by a condition of x_T reverting to x_U hence resulting in collapse as captured by Equation 10.2.7, Light-Space-Time Collapse:

$$Light - Space - Time\ Collapse =$$

$$\begin{Vmatrix} \begin{bmatrix} c_\infty : [Pr, Po, K, H] \\ (\downarrow R_{C_K} = f(R_{C_\infty})) \\ c_K : [S_{Pr}, S_{Po}, S_K, S_H] \\ (\downarrow R_{C_N} = f(R_{C_K})) \\ c_N : f(S_{Pr} \times S_{Po} \times S_K \times S_H) \\ (\downarrow R_{C_U} = f(R_{C_N})) \\ c_U : [P, V, M, C] \\ \Uparrow \\ c_{0:[D,W,I,C]} \end{bmatrix}_{Light} \quad \begin{bmatrix} M_3 \rightarrow System_X \\ (\uparrow F \rightarrow I) \\ M_2 \rightarrow S_{System_X} \\ (\uparrow Sig \rightarrow F) \\ M_1 \rightarrow Sig_x \\ (\uparrow > P_x) \\ U \rightarrow x_U \end{bmatrix}_{Space} \\[2em] U \rightarrow \begin{bmatrix} M_3 : -\infty \leq t \leq \infty \\ \downarrow \\ M_2 : 0 \geq t > \infty \\ \downarrow \\ M_1 : 0 > t > \infty \\ \downarrow \\ t \sim E_{Cell}; TC: M_3 \rightarrow U \\ t \sim E_{\text{▯▯}uman}; TC: U \rightarrow M_3 \end{bmatrix}_{Time} \quad TC \rightarrow x_U \end{Vmatrix}_{\langle x_U | x_U \rangle}$$

Eq. 10.2.7: *Light-Space-Time Collapse*

On the other hand the will for continued functional-richness may lead to continuous quantization, space expansion, super-matter, and a new Ouroboros of eternal return.

A New Ouroboros

Perhaps it is that if the expansion continues the apparent gaps between galaxies filled with extraordinarily high dark-energy becomes ripe for a series of inter-galactic-bangs. And perhaps such a series of inter-galactic-bangs in such a functionally-rich universe precipitates cycles of development based on other speeds of light, that in turn sets up a more conscious fusion of the multiple layers of existence so created, into one incredible Universe where an entirely new type of super-matter can form.

The new Ouroboros will signify a new type of endless creation where richness builds on richness and new types of super-matter on previous super-matter. In Martin Nowak's book, Super Cooperators, he proposes a third principle of evolution: that of cooperation. A first principle is that of mutation, responsible for generating genetic diversity. A second principle is that of selection, that focuses on individuals best suited a certain environments. He calls cooperation the master architect of evolution since from it emerges the constructive side of evolution 'from genes to organisms to language to complex social behavior' (Nowak, 2011). But as suggested here, evolution will occur driven by Light itself until emergent constructs see as Light sees, and become as Light is, even materially.

The human has in a certain sense been the result of 30 orders of microcosmic magnitude implicit to it. Fields, quanta, atoms, cells, the incredible four-fold emergences of properties codified in Light have all contributed to the emergence of the human. Now in a possible reversal of causality the human appears to be positioned to similarly influence and even structure the further development of the macrocosm by mastery of the microcosm.

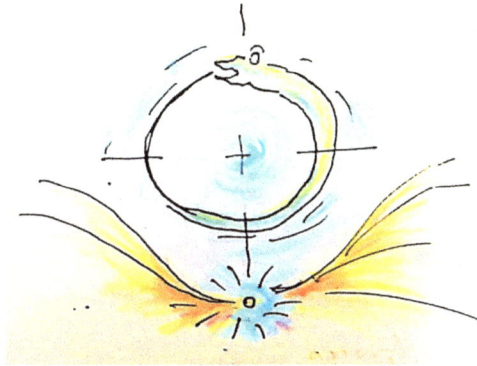

Through will and other human faculties continued states of quantum-certainty resulting in four-fold quantization can occur. Such quantization changes the very nature of space, time, energy, and gravity and can alter the development of the 30 orders of macrocosmic magnitude.

Such audacity is only possible because the human is the latest emergence in a universe of Light where all is Light and all is of Light. Light in material form turning on itself can alter the very nature of what it is by what it is. Such is the future tale of Ouroboros changing from swallowing its tail to entwining itself through a Cosmos in a never-ending spiral toward unforeseen flowerings.

Matter in this view is synonymous with the emergence of the universe and therefore with the Light-Space-Time Emergence Equation (3.1.3). This also implies that in its intent, super-matter is likely nothing other than materialized light in all its infinite possibility. Hence (3.1.3) can be restated as Equation 10.2.8, Super-Matter:

$Super - Matter$

$$= \begin{Vmatrix} \begin{bmatrix} \begin{bmatrix} c_\infty : [Pr, Po, K, H] \\ (\downarrow R_{C_K} = f(R_{C_\infty})) \\ c_K : [S_{Pr}, S_{Po}, S_K, S_H] \\ (\downarrow R_{C_N} = f(R_{C_K})) \\ c_N : f(S_{Pr} \times S_{Po} \times S_K \times S_H) \\ (\downarrow R_{C_U} = f(R_{C_N})) \\ c_U : [P, V, M, C] \\ \Uparrow \\ c_{0:[D,W,I,C]} \end{bmatrix}_{Light} & \begin{bmatrix} M_3 \to System_X \\ (\uparrow F \to I) \\ M_2 \to S_{System_X} \\ (\uparrow Sig \to F) \\ M_1 \to Sig_X \\ (\uparrow > P_{x)} \\ U \to x_U \end{bmatrix}_{Space} \end{bmatrix} \\ \begin{bmatrix} U \to \begin{bmatrix} M_3 : -\infty \le t \le \infty \\ \downarrow \\ M_2 : 0 \ge t > \infty \\ \downarrow \\ M_1 : 0 > t > \infty \\ \downarrow \\ t \sim E_{Cell}; TC: M_3 \to U \\ t \sim E_{Human}; TC: U \to M_3 \end{bmatrix}_{Time} & TC \to x_T \end{bmatrix} \end{Vmatrix}_{\langle x_U | x_T \rangle}$$

Eq. 10.2.8: Super-Matter

Eq. 10.2.8, though, implies that genetic-type code should be depicted by some revised summation form of (3.1.3). Building off (8.2.6) for the genetic code for sustainable global civilization Equation 10.2.9, Future Code, may be summarized as:

Future Code =

$$\left[\begin{array}{c} \left[\begin{array}{c} c_\infty: [Pr, Po, K, H] \\ \left(\downarrow R_{C_K} = f(R_{C_\infty})\right) \\ c_K: [S_{Pr}, S_{Po}, S_K, S_H] \\ \left(\downarrow R_{C_N} = f(R_{C_K})\right) \\ c_N: f(S_{Pr} \times S_{Po} \times S_K \times S_H) \\ \left(\downarrow R_{C_U} = f(R_{C_N})\right) \\ c_U: [P, V, M, C] \\ \Uparrow \\ c_{0:[D,W,I,C]} \end{array}\right]_{Light} \left[\begin{array}{c} M_3 \rightarrow System_X \\ (\uparrow F \rightarrow I) \\ M_2 \rightarrow S_{System_X} \\ (\uparrow Sig \rightarrow F) \\ M_1 \rightarrow Sig_x \\ (\uparrow > P_x) \\ U \rightarrow x_{Sust.Global\ Civilization} \end{array}\right]_{Space} \\ \left[\begin{array}{c} M_3 : -\infty \leq t \leq \infty \\ \downarrow \\ M_2 : 0 \geq t > \infty \\ \downarrow \\ M_1 : 0 > t > \infty \\ \downarrow \\ U \rightarrow \begin{array}{l} t \leq E_{Cell}; TC: M_3 \rightarrow U \\ t \sim E_{Human}; TC: U \rightarrow M_3 \end{array} \end{array}\right]_{Time} \quad TC \rightarrow x_{Future} \end{array}\right| \langle x_U | x_T \rangle$$

$\Rightarrow$

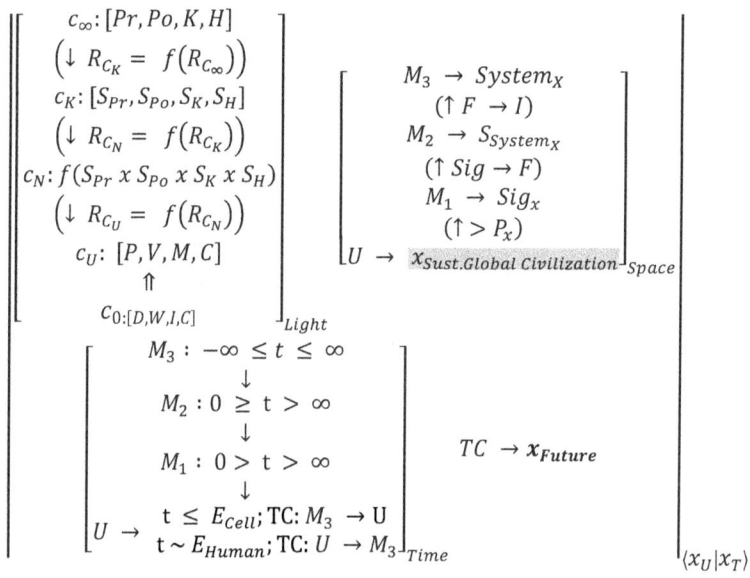

$\langle Space - Time - Energy - Gravity\ Material - Fabric\ Pre - Genetic\ Code \rangle +$

$\langle Electromagnetic\ Spectrum\ Material - Fabric\ Pre - Genetic\ Code \rangle +$

$\langle Quantum - Particle\ Material - Fabric\ Pre - Genetic\ Code \rangle +$

$\langle Atoms\ Pre - Genetic\ Code \rangle$
$\quad + \langle Molecules\ Material - Fabric\ Pre - Genetic\ Code \rangle +$

$LSTE \langle ... \rangle + PGGT < \cdots > + \langle Cells\ Genetic\ Code \rangle + LSTE \langle ... \rangle +$

$(Fundamental\ Capacities\ of\ Self\ Genetic\ Code) + FBLEEE \langle ... \rangle +$

$(Humans\ Genetic\ Code) + LSTE \langle ... \rangle + FBLEEE \langle ... \rangle +$

$(Stable\ Mega - Organization\ Genetic\ Code) + LSTE < \cdots > +$

$$\sum_{n}^{\infty} \begin{pmatrix} \sum Sig_I = Xa + \overline{Yb_{0-n}} \\ where \begin{bmatrix} X \in [S_{System_N}] \\ Y \in [S_{System_{Pr}}, S_{System_P}, S_{System_K}, S_{System_N}] \\ a, b \text{ are integers}; a > b \end{bmatrix} \\ \sum Sig_M = Xa + \overline{Yb_{0-n}} \\ where \begin{bmatrix} X \in [S_{System_K}] \\ Y \in [S_{System_{Pr}}, S_{System_P}, S_{System_K}, S_{System_N}] \\ a, b \text{ are integers}; a > b \end{bmatrix} \\ \sum Sig_V = Xa + \overline{Yb_{0-n}} \\ where \begin{bmatrix} X \in [S_{System_P}] \\ Y \in [S_{System_{Pr}}, S_{System_P}, S_{System_K}, S_{System_N}] \\ a, b \text{ are integers}; a > b \end{bmatrix} \\ \sum Sig_P = Xa + \overline{Yb_{0-n}} \\ where \begin{bmatrix} X \in [S_{System_{Pr}}] \\ Y \in [S_{System_{Pr}}, S_{System_P}, S_{System_K}, S_{System_N}] \\ a, b \text{ are integers}; a > b \end{bmatrix} \end{pmatrix}$$

Eq. 10.2.9 Future Code

Eq. 10.2.9 implies a constant stream of materialized functional-richness, an evolving super-matter, and an evolving post-genetic structure for encapsulating such future code.

In such a view genetics today is only an intermediate information storage and processing mechanism in a journey which began with pre Big-Bang information precipitating into ∞-entanglement, K-entanglement, and N-entanglement pre material-fabric pre-genetic libraries, further precipitating to FBLEE and then the pre-genetic material-fabric, before it transitions now, into many forms of unimaginable post-genetic structures.

Illustrator's Note

This note elaborates the illustrators' rationale for each of the 147 figures, by section.

Introductory Section Figures

Fig 1: The Pyramid of Life and Light

"The Cosmology of Light"- is illustrated as a pyramid of Life and Light. At the base is soil, vegetation, animals, and as it reaches the apex there is vast blue sky. The Sun Light is in the middle.

Section I Figures

Fig 2: The Origins and Possibilities of Genetics

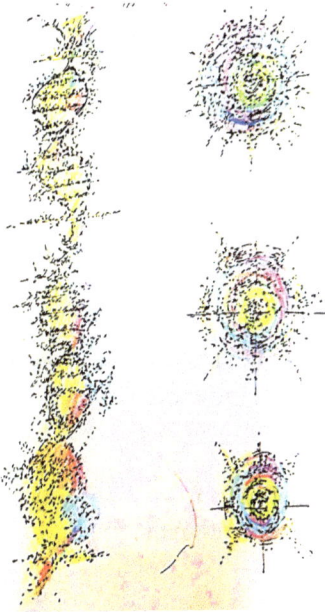

The origins and possibilities of genetics is shown with upward and downward DNA strands. Colors adorning it signify quadrality. The illustration on the right shows a number of spirals enlarging from centers of their own and progressively increasing in size. (The same diagram gets further completed and is added at the end of the book.)

Fig 3: The 8 Minutes & Past, Present, Future

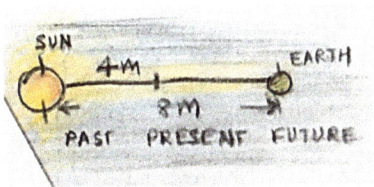

"Imagine traveling on a ray of light from the sun to the earth. Imagine that you are in minute 4 of the approximately 8-minute journey. As you look back you will see that 4 minutes in the past you were at the sun. 4 minutes in the future you will be at the earth." Here it is shown in a diagram.

Fig 4: Thought Experiment on Night Sky

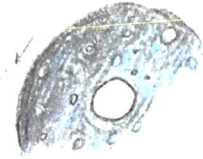

"A further thought experiment may give insight into such omnipresence by considering the night sky." Shown here in a small portion as seen through an opening.

Fig 5: O-O-O-O Omnipresent-Omnipotent-Omniscient-Omninurturing- Light Connects All

O-O-O-O: Omnipresent-Omnipotent-Omniscient-Omninurturing - is shown here. It is also emphasized that it connects at various levels and with all aspects of Light.

Fig 6: This Rearrangement or Elaboration of Implicit Information

"There is implicit in this transformation also a high degree of 'entanglement' as it were". In this figure it is shown as turbulent water waves in a tube, with parallel waves mixing and altering each other at a threshold

430

point. This captures a phenomenon known or resembling one of the laws in fluid mechanics.

Fig 7: Seed Equation, Subtle Structures and Entanglement

"A boson or lepton, or any subsequent emergence of matter is already deeply entangled by virtue of having emanated from the same single starting-point conceptualized as the Big Bang". It is shown here with a seed point with a number of curves being emitted from it - apart from centripetal and divergent also getting related, mingled and entangled.

Fig 8: Light Selectively Slowing Down to Different Levels

"When light is made to slow down in experiments, the range or band that it slows down to deviates only slightly from its relative-to-infinity slower speed in vacuum". This is shown here as ranges and bands through a few strokes of four colored (quadrality) rays.

Fig 9: Islands of Matter

"…then the result of that is the material universe as we see and experience it, also with its division… there are properties in the reality that emerge as a result of this". This is shown here as an island of matter and sudden emergence of matter symbolized by the curves

Fig 10: All Realities Exist Simultaneously

From the womb of infinity diverse curves are emerging showing that 'All realities exist simultaneously'

432

Fig 11: Light, Quanta, Relationship to Genetics

Quadrality here is shown with four point sources and as they emit rays, these intersect, and redefine each to emphasize a passage between light, quanta & genetics.

Fig 12: Quadrality of PVMH-OOOO-PPFM-PVMH

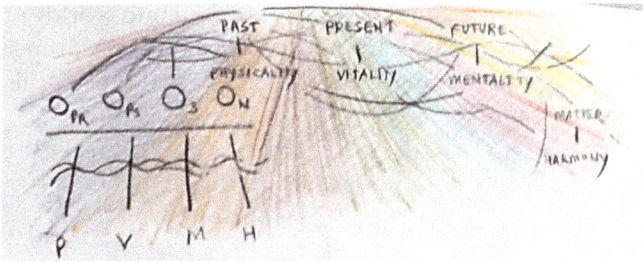

"So whether light is traveling at an infinite speed or the speed we know as c, there is something about the properties it projects, that in essence is the same." This Quadrality of PVMH-OOOO-PPFM-PVMH is represented here through line diagrams.

Fig 13: Quanta are Doorways or an Interface

Shown here is an opening in matrix form of windows in a cave architecture of India (inspired by a real such openings seen in Kanheri caves near Mumbai, India).
"So quanta are a doorway or an interface into worlds of Light, and a doorway by which possibilities in deeper worlds of Light can express themselves here in this material realm."

Fig 14: Light Beam Through an Interface Quanta

"This has to be since quanta are the interface between different layers of light." Shown here in a traditional *Gavaksha* type of opening in a house / palace wall. This is an ancient architecture which is used to usher in the early morning light rays.

Fig 15: The Emergence of Light & Life

Shown here as rising flames from an altar and a significant variant here is the altar has a heart shape emerging with the fire…a Quadrality of Light -- Power, Wisdom, Harmony and Perfection to a luminous ascent.

Fig 16: Some Concepts of Genetics in a Cosmology of Light

'The infinite information codified in Light is the origin of genetics. ..Light in its native state, possessing characteristics of omnipresence, omnipotence, omniscience, and omninurturance, contains all-possibility within it. This all-possibility can be thought of as an infinite amount of information, and its structure related to the four characteristics, as the basis of genetics.' This is shown here by a spiral emanating from single point like creation, also reflective of origin of the world from the big bang and resulting in infinite possibilities in innumerable directions.

435

Section II Figures

"..the mathematical structure of subtle-DNA. In this structure there is a downward-strand and an upward-strand. These strands form the backbone of all possibility.' Here it is symbolized by a pipal tree leaf, half of which is buried in

the soil. The downward movement also gives rise to a new upward strand / movement which signifies the raising upwards of new leaf and new Life.

Fig 18: Ubiquitous Subtle DNA

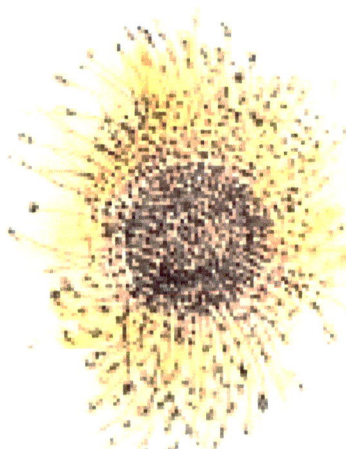

The Light-Matrix is thought of as a seed-equation that provides insight into dynamics of the universe. The Light-Matrix codifies dynamics that may be experienced at the quantum borders of realities created through light traveling at different speeds, and further in the worlds that the quantum veil or window provide access to." This is visualized here as innumerable rays of light emanating from seed clusters

436

from a center. The whole figure then starts resembling sunflower petals and its unique center of seeds.

Fig 19: Hence all Universes Exist in that Created...

"...such superposition also creates a "strand" as it were, that intimately connects together these universes and the realities represented by them. Entanglement creates the dynamic of influence. So it is the combined action of superposition and entanglement that also creates the notion of ubiquity of the structure of "subtle-DNA." Further, 'The Cosmic Serpent: DNA and the Origins of Knowledge' (Narby 1998) notes that the shamans of ancient cultures enter into the molecular realm and actually perceive DNA and accompanying structures at the cellular levels. This is visualized here as spherical shapes symbolizing universes. This has two helical curves creating the shape of the sphere. The helical pair resembles DNA and also kundalini shakti of mystic traditions of the world such as Yoga and Tantra.

Fig 20: Light Matrix First Transformation

"The function itself is a quantization-function that allows some essential characteristics in the point-nature of light to express itself more "materially", relatively speaking, by light traveling at a relatively slower speed. This quantization-function can be thought of as a node in which a higher-level quantum-computation occurs by which infinite possibility begins to disaggregate itself." Here the equation of Light Matrix First Transformation is visualized as an hourglass with smooth curves which coincide with the curves of a dancing lady. At the node is the transformation from lower to higher and she is sporting a rotating ring around her body which moves across the height of the upper and lower hemisphere/ strand of DNA.

438

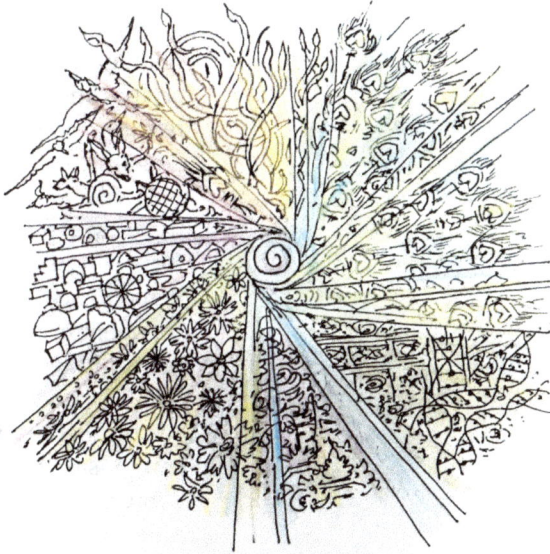

"The ubiquitous-point-instant is a function of the dynamics of that reality where light travels at ∞ speed and is envisioned as being infinitely entangled by virtue of light being omnipresent-omnipotent-omniscient-omninurturing in that realm." This is shown here as a continuous spiral at the center and numerous rays and channels emanating from it. In each section there is one form of existence or manifestation, from atoms, molecues, objects, assemblies, machinery and structures to DNA, cells, to flora, fauna, and species in different stages of evolution. The final section is filled with peacock feathers...symbolizing wisdom-power-beauty and perfection as aspects of Light /of divinity.

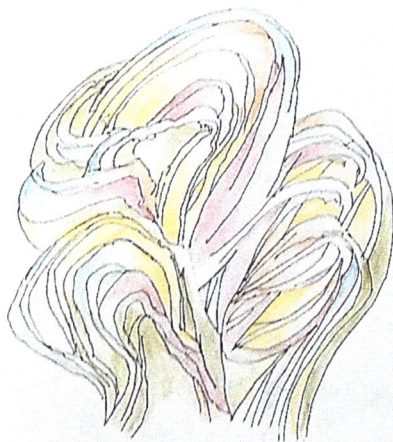

"While the uniqueness of organizations as represented by the Signature is a seed, like any seed there is a process for its emergence…This uniqueness is often hidden until certain conditions are fulfilled So we can say that …mutation may either tap into the subtle-libraries created through the structures already set up by the ∞-entanglement, K-entanglement, and N-entanglement levels, or by a fresh interaction with the layer represented by these nodes…." This is visualized here by the emergence of several bands of various colors / attributes which are entangled and yet result in a unique form finally.

440

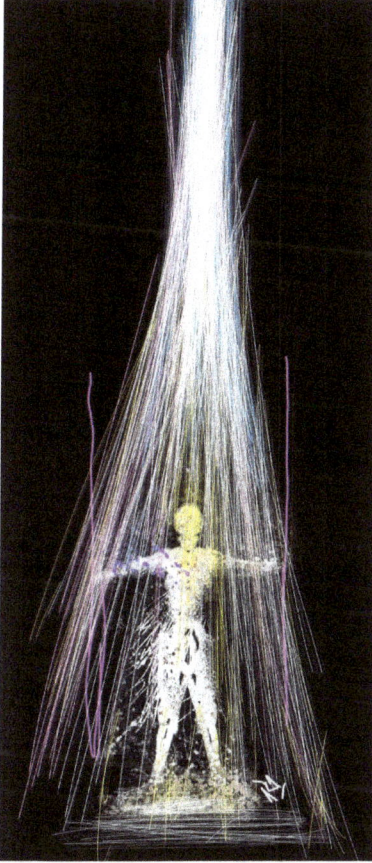

Fig 23: Uniqueness Emerges When Certain Conditions are Fulfilled

"The integral form is a threshold phase, and allows the uniqueness suggested by the Signature to emerge in fuller force or in its 'force' form. The final phase is the 'contextual form' that allows the signature to act with impunity within a considered context." This is visualized here as a human figure standing under a spotlight as if in a theatre (of this world) - asserting its own individuality with spread hands and raised head...the figure is shown here not as a frozen concrete form but rather an emerging form through many lines and particles. Light descends and the uniqueness/ individuality emerges and ascends.

441

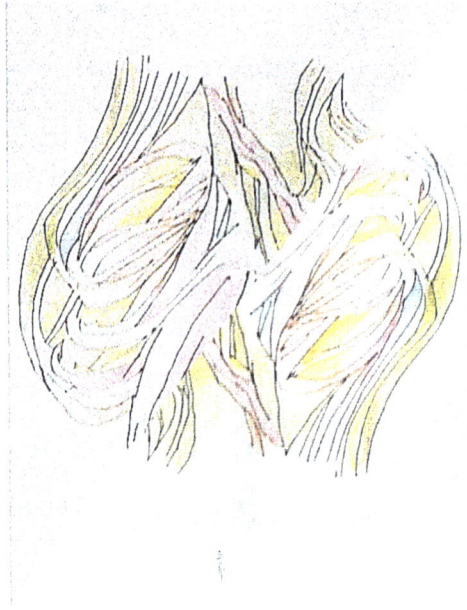

"The physical, the vital, and the mental levels are orientations in which patterns of perceiving, being, behaving are set in their ways. Each pattern has its purpose and its limitation, and it can be argued that being able to learn from each orientation and yet being able to move beyond that, is the next logical step in any developmental model." This is visualized here by mirroring Fig. 22 which is now rearranged in two halves wherein each is an upside-down mirror-reflection of the other. The textures of both are also kept subtly different. The result is an altogether different pattern.

442

Fig 25: The Inherent Creativity

"The inherent creativity in Light is summarized by four overarching properties in the nature of a point and its attendant '∞-entanglement' and thereafter this deep fount of creativity is seen everywhere as a range of creative forces, the architectural forces which define the possibility inherent in any system." This is visualized here as a pleasant creative combination of colours and patterns ... which resembles a beatific cluster of trees with thin lines of branches and points signifying flowers / spring.

Fig 26: Signature Becomes Active at Meta Level

The signature or uniqueness of the physical (Sig_P) becomes active at meta-level 1 (M_1). This acts as a force and thus leads to a breakthrough to the next level - meta-level gets dynamic and begins to act at the once 'untransformed' level (U) further modifying it. This transformation represents the inherent creativity - and the dynamic driving of any mutation sequence

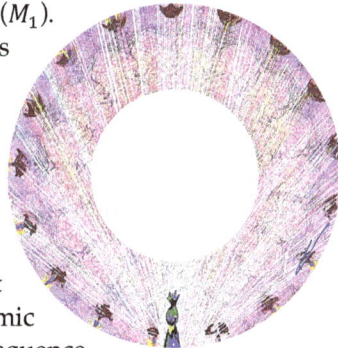

within physical-type systems. This is visualized here as a ring with light on the periphery and a peacock head emerging from beneath.

443

Section III Figures

"The mathematics relevant to genetic mutation is like a structure of the downward and upward strands of subtle-DNA. The downward-strand caused by light slowing down in quantized-decelerations, concretizes more of the information in light. The upward-strand is a time-variable sequence wholly determined by the nature of the layers in the downward-strand. The time-variability is a result of interplay of the active influences emanating from the downward-strand…." This is visualized here by superimposing The Vitruvian Man (Italian: L'uomo vitruviano; [ˈlwɔːmo vitruˈvjaːno]) by the Italian Renaissance artist and scientist Leonardo da Vinci but adapted here to fit it in a combination of circle and square (the two trigonometric figures which are also the beginning of Vaastu shastra, mandala, and yantra traditions. Here the corners are shown with roots or branches emerging signifying life. The central portion has a double helix symbolizing DNA and the combination of upward and downward strands.

444

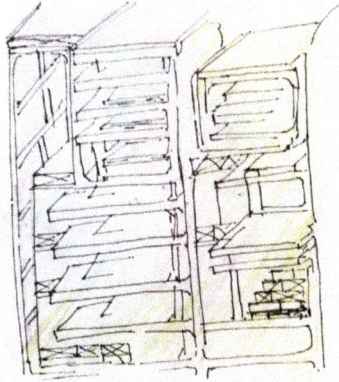

The logic of cosmic fundamentals and of the very basis of matter is deeply ingrained in every part, every layer and every space, time. It is shown in this book in several matrix forms of equations. Here it is visualized as a physical structure of a box and sections with drawers or trays at different layers. It also connects with the idea of subtle libraries or plethora of information stored somewhere.

From Schrodinger's Equation & Multiple Layers of Light, Uncertainty, to Leonardo da Vinci's famous Vitruvian Man (which was mentioned in earlier illustration) and in an attempt to model matter as a wave rather than as a particle. All is now sort of harmonized here and significant additions or alterations in this include lotus petals imposed on plain trigonometry; the female figure in traditional attire with an ascending serpent or helix in her foundation, also showing creative energy; she is carrying a series of pots on her head - a view common in the country side - but also depicting the containerization of information in equations and in genetics. Block structures containing wave / water as well.

446

(For figures 30,31,32)   The uncertainty relation may suggest the phenomenon of function-precipitation. As mentioned in Eq 3.4.2, there is Integration of Levels (Leveraging Heisenberg's Uncertainty Relation). This is depicted here by collage of several images used in the preceding figures and also adding new images and visualizations to it.

Fig 33: The Process of Quantization

"The Light-Space-Time Emergence equation (3.1.3) derived in Chapter 3.1, and reproduced here for convenience, offers some insight into the process of quantization." Here while visualizing all such ideas by strokes of intersecting lines

449

and curves, what emerged was a shape resembling the Shankha or conch shell. A great symbol of cultural significance, wherein cosmic powers like air, sound, light, water are part of its making. Matter and life are merged in one entity. The shankha is adorned here with several other symbols including Sri Yantra, mandala and others which have spiritual and mystic meanings.

Fig 34-35-36-37: Space, Time, Matter, and Gravity Mandala as Light Slows Down

These figures are trying to visualize space, time, matter, and gravity mandalas as light slows down. "Hence space, time, matter, and gravity, that arise when Light slows down, can be thought of as emergent phenomena. The emergence itself can be thought of as a function of the Light-Matrix." "This quantization of Space, which holds the seeds of knowledge of all that will emerge, hence, is modeled in the following way as specified by Equation 3.5.2...The quantization of Time, which holds the inevitability of the seeds or knowledge emerging in a phased maturity, Energy, which through a process of

450

containment can result in Matter...." These are some of the points which helped in this depiction as mandala drawings.

Fig 38: Energy & Matter

While the energy and matter dwaita- adwaita (duality & ultimate unity) and the associated science, mathematics, cosmology and philosophy is perpetually captivating, here it is visualized in a five-branch star with whirls of light rays within. At the center is the symbol of a creative womb, or a zero / infinity center. 'The Effect of Levels of Light on Genetic Mutation' is seen in this section through several more details.

"The Light-Space-Time Emergence equation (3.1.3) being iterative can be used to model emergence from simpler to more complex four-fold manifestations…In other words (3.1.3) suggests a computational approach to the development of the universe. Pre-genetic and genetic information may be thought of as the output of such computation. This computational approach involves multiple layers of reality as suggested by (3.1.3) and is driven more by a process of qualified determinism…'Effect of Levels of Light on Mutation Using Simplified Version of Light-Space-Time-Emergence.'" This has prompted the mandala design in many layers in this figure. It is kept here in just dark lines and white light emanating from it and no particular colors are added to the mandala.

Fig 40: Emergence of Unique Form from Formless as Light-Space-Time Emergence Transforms

"...the systematization of the mutational-sequence, while relatively transformed organizations have opened to the active influence of meta-levels, relatively untransformed organizations remain under the influence of untransformed physical, vital, mental, and integral dynamics at U. This modeling can also be applied to mutation." These passages have prompted to look at this constructive mutation as if a form is emerging from the formless, as when a skilled craftsman chisels out the unwanted part of the raw stone. The image shown here is of a side facing deity with a benign smile, and a calm, composed face and sparkling eyes...signifying all four aspects of the Quadrality. Notably the image is not shown completely black as in a stone sculpture. It has golden shades to show light emanating from its form.

Fig 41: Activation and Potential Change of FBLEE & the Flame Rising Above

"Traversing the buffer implies that meta-levels are active, moreover they have to become active in a special way for four-base logic-encoding ecosystems (FBLEE) to be affected." This potential 'Effect of Levels of Light on Pre-Genetic or Genetic Information' and activation and potential change of FBLEE is visualized here as a flame rising up from a torch. Its blue beginning…yellow, orange and red parts in subtle soothing shades, and the almost invisible tip in vast infinite sky …while the bottom portion of it, apparent like a torch is shown in very faint horizontal layers, again it culminates in a point in the vast infinite in conscience above.

"Line 1 from the top in the matrix is simply a static form of (3.6.1) the Simplified Light-Space-Time Emergence equation...." These passages were visualized as a serene seashore wherein blue waves, sand, sky and the sun are seen as different levels and yet in an almost intermingling and unifying whole.

Fig 43: "The call from below, as it were, may invoke some function that already exists in the subtle-libraries 'above'"

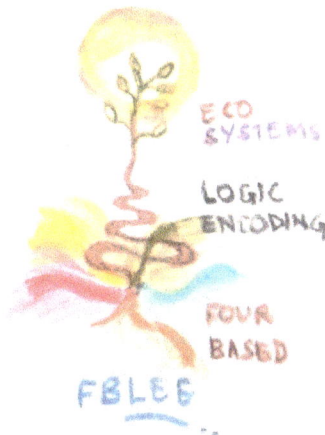

The essential dynamics leading to possible mutations at the genetic level can more clearly be highlighted further. "The call from below, as it were, may invoke some function that already exists in the subtle-libraries 'above', so that some already existing function may influence FBLEE through ∞-entanglement, K- entanglement, or N-entanglement. This may be thought of as a key-and-lock mechanism, where a deep enough visceral urge from below acting as the key, opens an entangled lock to alter FBLEE as per the visceral urge." This is seen here as four sections, four based roots, rising up in spiral movement and emerging as a plant ascending to the sun.

Fig 44: May be Thought of as Some Lock & Key Mechanism

The 'lock & key' part of the quote referred to in the previous illustration is visualized here as a mechanical lock and key.

"But the nature of the operations is designated by Q: Quantization; F: Fragmentation; R: Random'. Z_Q, then, implies that the full quantization originating from updated four-base logic-encoding ecosystems (FBLEE) can take place, and will result in lasting material change at the pre-genetic or genetic level." This is literally deciphered here as symbols for each stage drawn sequentially. Quantization by a measuring scale, 'fragmentation' by divisions in the whole, and 'random' by further scattering of the mass in random places and heights.

Fig 46-47: "Infinite Potential Superposed States that then Collapse into a Material Possibility"

"This wavefunction equation deals with dynamics behind the surface, and particularly of as many as infinite potential superposed states that then collapse into a material possibility in time and space at U. Such superposition is another way of saying that there is wholeness behind the surface…such wholeness is represented by L, S, T, the Light, Space, and Time matrices respectively." This has inspired these two illustrations. The one above shows a channel with particles / entities flowing in, and they collapse at a junction to emerge as some material entity on other side - the image resembles a twisted leaf stem or similar tubular structure which has sap of a tree flowing within.

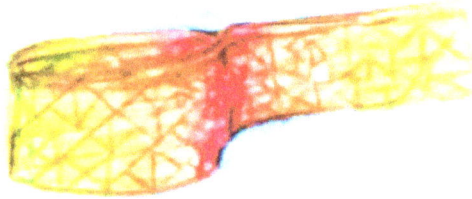

The second figure resembles a mechanical forging, such as takes place on a connecting rod required for an automobile. Its changing shape at a juncture is due to force and temperature variables applied to it there. The typical diagram resembles computer aided engineering (CAE) wherein finite element analysis is done with grid formation, and mechanical as well as thermal stresses are measured, to result in the emergent design of the right dimensions and shape.

"The application of the vertical component of the new function being proposed, DI_V, to this core matrix will yield the nature or 'strength' of the state (x) or orientation under consideration. If the untransformed or U layer is strongest, implying that the habitual patterns that keep an organization locked into its untransformed way of operation are still very active, then the nature of the output of DI_V, notated by x-state, will be x_U. If the habitual patterns have been overcome then the strength of the x-state increases since it is the dynamics of M_1 or Sig_x that are now active...." This is shown here as a palm of a minor prince, barely covering the top part of a Rajdanda (symbolic shaft to be held to discharge duties of the ruler and govern as per righteousness). Usually associated with one who is forced to sit on the throne as successor after the demise of an able king. One can see that this habituated pattern of choosing a successor by heredity principle is 'Habitual patterns that keep an organization locked into its untransformed way of operation.' The Rajdanda design is also significant with a layered spherical top and bottom and a line dividing the upper and lower hemispheres - showing a design resembling the different layers and levels of Light.

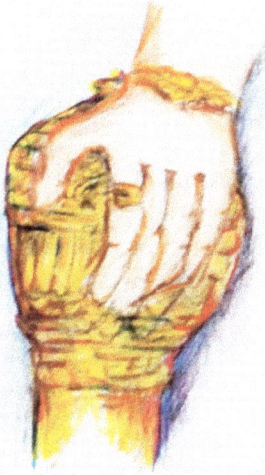

459

"Hence, this mathematical model is suggesting that any situation, including micro and nano-level situations, rather than having a random outcome, has a 'qualified deterministic' outcome."…"It is not the principle of causation itself which has broken down in modern physics, but rather the traditional formulation of it." This is shown here by a series of links in a linear chain of cause and effect.

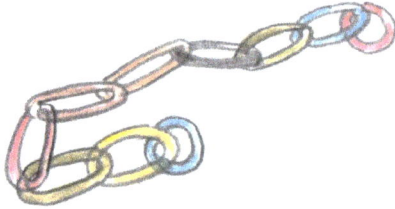

Fig 50: Nature Could Follow Only One Road

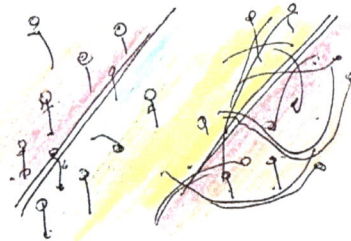

"State A was inevitably succeeded by state B. So far the new science has only been able to say that state A may be followed by state B or C or D or by innumerable other states. It can, it is true, say that B is more likely than C, C than D, and so on; it can even specify the relative probabilities of B, C, and D." The old science of strict linearity and new science of nonlinearity but not indeterminism is illustrated here in two adjoining sketches.

460

"But just because it has to speak in terms of probabilities, it cannot speak with certainty which state will follow which; this is a matter which lies on the knees of the gods – whatever gods there may be." This is shown here by ancient Greek or Gandhar style architecture of a seating image of Buddha or any other deity in a stone sculpture of this style. The anatomy is perfect and the legs released down from the seat / throne have a thin vastra / apparel covering them, but is shown almost visible through the folds and thin veil of the apparel. At the feet is seen a small devotee.

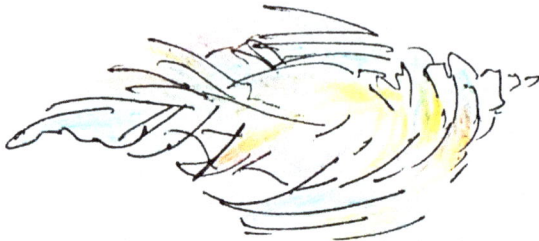

"...at the micro or nano level there may appear to be randomness and a dethroning of the principle of causation. The notion of a multiplicity of levels, each having its impact on the strength of an orientation or base and further on the consequent direction from a multiplicity of possible orientations, is being suggested..." The conch / shankha symbol explained in an earlier illustration is revisualized here with more curves and interrelations.

Section IV Figures

"Light-Space-Time Matrix based computations generates a range of code from an era pre-dating the Big Bang to modern day Global Civilization, to whatever else may emerge"' That means everything that exists has a code and the way the code is housed changes in forms recognized as living, and thereby also implies that the code within all living cells has to do with space, time, energy, gravity. This is also true of the functioning of the electromagnetic spectrum, and even all the stages of material elaboration that precede the emergence of a living cell. This is shown here as a rising creeper which has a vertical axis as support and marches up to the sky and to the light.

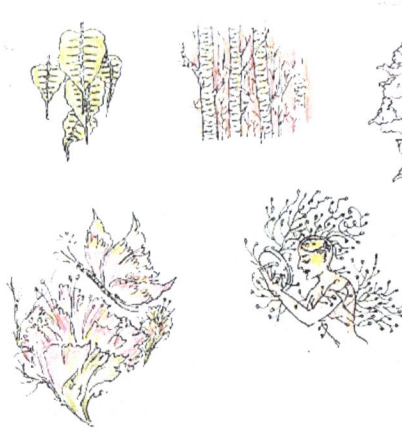

The vast amount of pre-genetic information that has already been generated suggests the means by which this information can be materialized. "The Big Bang puts into play a potentially endless evolution materializing more and more possibility that exists in Light." Even Post-Big Bang there is 'Generation of Material-Fabric, Genetic Code, and Post-Genetic Code.' The emergence through matter, life, and mega-organization, and the occurrence of space-time-energy-gravity quantization shows that the pre-cellular structures are the mechanism for affecting the fabric of material existence. The various fourfold code projections accelerate material evolution. There is tremendous "potential in Light working from the basis of an initial electromagnetic spectrum to the current basis of a possible sustainable global civilization, and beyond ... emergence itself may equally be dictated by the different kinds of entanglement and the vast variety of information in pre-material-fabric subtle-libraries." These paragraphs have inspired these series of illustrations wherein:

464

a) Pipal leaves hanging down, overlapping and connecting to each other, and there symmetry around each one's central axis is also seen in totality in one figure.

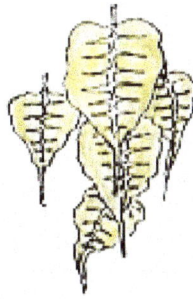

b) while a multiple layered and petaled rose like flower in another.

c) Also seen is a forest with parallel and similar, and yet separate and distinct trees. The variation here, in contrast to the white trunks of trees, are the orange and red flower blossoms without a trace of green leaves in it.

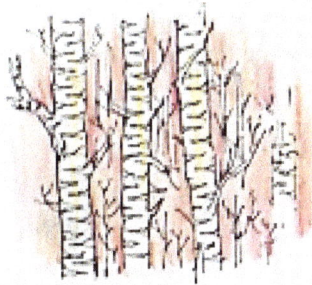

d) In another significant illustration a flower and butterfly are seen as a pair in deep interaction. This is evident by the colors, lines and form of either being inspired by, and even reflecting the other.

e) In another figure a human being is seen looking in mirror and reflecting or introspecting…the real turning point of evolution of a species…and while doing so several creepers of tiny leaves are emerging from the mirror, symbolizing the life and its infinite expressions being fathomed by the thinking power of the person.

"Further, somewhere between the cell and the human, the emergence of vision is said to have sparked an explosion of evolution ... that instrumentation to perceive light, such as vision, would then generate an explosion in evolution." Here any genetic code will be "the materialization of pre-material-fabric or more accurately four-base logic-encoding ecosystems (FBLEE) as a result of the Light-Matrix computation, there will also be a degree of entanglement to such ecosystems. Such entanglement will be different from ∞-entanglement, K-entanglement, and N-entanglement," ... referred to as FBLEE-entanglement. This is visualized here as four armed images of gods especially Vishnu - four arms signify Quadrality: Shankha / Conch, Chakra / Wheel, Gada / Mace, Padma / Lotus are shown here in four corners of a pair of concentric squares (resemble mandala or yantra and also vaastu rachana) symbolizing Wisdom, Perfection, Power, Beauty - O-O-O-O, P-V-M-C and so on. In the inner squares a supplement of each is shown: like a sword and bow and arrow as weapon supplements to the mace, and so on for every corner. In another view is seen a stepped structure like a pyramid where on each point of projection from the top, again is a symbol / entity. The four-armed deities are very common in Indian culture and are called Chaturbhuj. However, this interpretation from cosmology of light as done here might be very novel. The stepped structure in elevation is

467

like a platform in northeast India's traditional Vishnu worship units - a 'Satra' (place of worship). It also seems like an altar, a Yajna kunda / chiti or part of temple architecture.

Fig 61: Shri Vishnu, Shiva Shakti, Duality, Quadrality

".. space, time, energy, and gravity quantization to create an holistic 'ecosystem' with its own 'electromagnetic spectrum logic' as it were." The concepts in the previous illustrations are further visualized to add the complementarity of opposites to it. Shiva-Shakti, Sri Vishnu are seen here in the Ardhanarishwar forms depicted in cave architecture like Elephanta caves or in Linga purana apart from several other texts of India. The same appears as Yin-Yang in Chinese tradition. Shown here in four armed Quadrality and also with their vehicles, the whole is placed on a tree leaf.

Section V Figures

Fig 62: Plant Chlorophyll & Human Hemoglobin

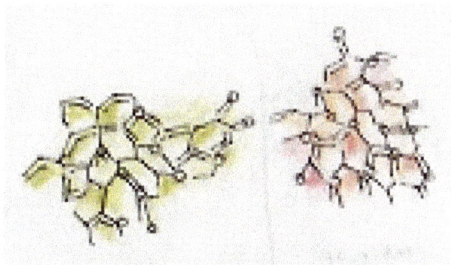

Visualized here as sketches which resemble each other and shows connectedness, rootedness, analogous universe; the same consciousness in all living being whether made of plant cells or human blood cells.

In pursuit of generating the pre-genetic code for matter, the four underlying properties of Light that we call Harmony, Knowledge, Power, and Presence are of the

essence of the speed with which the electromagnetic

spectrum moves, the wave-range within the electromagnetic spectrum, the energy-gradient within, and its implicit mass-potential.

Section VI Figures

Fig 63,64, 65 to 71: Generating the Pre-Genetic Code for
Matter and Containerization of Matter Through Sacred
Geometry

Some of the concepts which are being visualized here are
--the electromagnetic spectrum, the mass-possibilities
due to the electromagnetic spectrum, a continuous
process of computation, quantum-realms and quantization to create the realities of quantum particles, including bosons, and atoms, and so to generate the material-fabric, pre-genetic code for matter, containerization of matter, light-space-time emergence generating pre-genetic code, the material-fabric, macro space-time-energy-gravity level, and the electromagnetic spectrum level, universality versus localization of space, time, energy, and gravity and so on. These are visualized as a series of mandalas and sacred geometry designs, each different from every other and yet with some shared patterns.

471

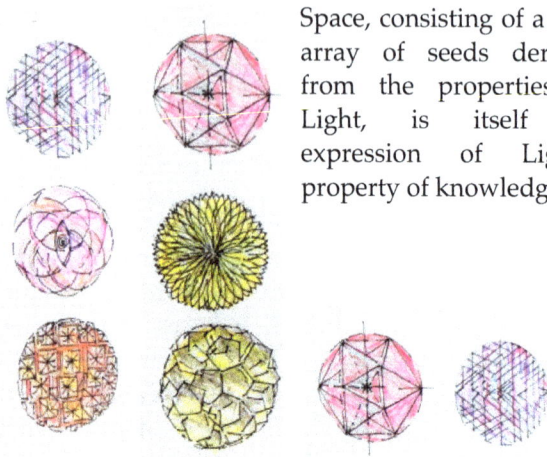

Space, consisting of a vast array of seeds derived from the properties of Light, is itself an expression of Light's property of knowledge.

Fig 72,73, ... Onwards

These are the Artists replicas of the corresponding sketches in text. Use of four colors symbolic of Quadrality and shades add to the illustrations.

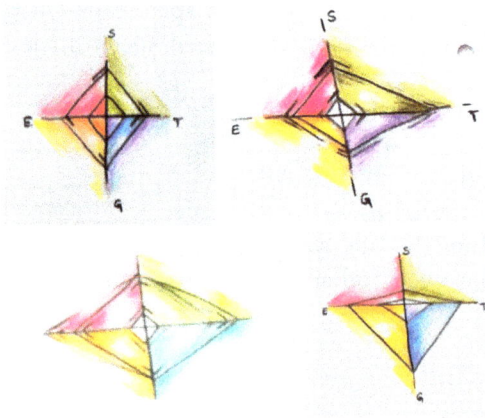

472

Generation of material fabric is shown here in a pair of digital mandalas with two axis symmetry.

Fig 80,81,82: Generation of Material-Fabric, Pre-Genetic Quantum Particle Code

Generation of Material-Fabric, Pre-Genetic Quantum Particle Code includes physical (related to Presence), vital (related to Power), mental (related to Knowledge), and connection. These are factors playing a critical part in allowing the material-fabric to express infinite diversity.

'Typically, it is the process as captured by the Space-Matrix ... patterns at the untransformed layer, U, will need to be overcome ... it is only with the advent of the human-system that the automaticity of the action of meta-levels is reversed" ... Access to this zone is a prerequisite for constructive mutation to occur, as designated by the element" ...

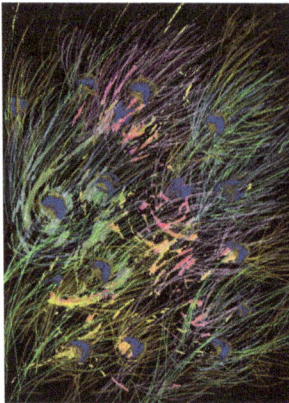

"So, we see that again the underlying properties of light - Knowledge, Power, Harmony, and Presence - emerge as quarks, leptons, bosons, and Higgs-bosons respectively." This part is visualized here by a series of three illustrations. The first two have many nuclei and sharp and forceful strokes of emanation at various levels and possibilities from each one of them. The first two were illustrated using a white and black background respectively. The third one further develops this in a combination of sorts and uses a detailing which resembles peacock feathers.

474

Fig 83: Summarizing at Quantum Particle Stage ...

'Summarizing, after the quantum particle stage iterations of the Light-Space-Time Emergence equation are complete, the following code-segments will have been generated as specified by Equation 6.2.7, Active Material-Fabric Pre-Genetic Code Segments at Quantum Particle Stage." This is illustrated here with a tree emerging from the soil with a water reservoir close by, and the blue sky behind.

Fig 84,85,86 to 89: Generation of Material-Fabric Pre-Genetic Boson, Gluon, Photon Code

These generations of code are illustrated with a series of mandalas, each unique, and yet based on the same principle and symmetry.

475

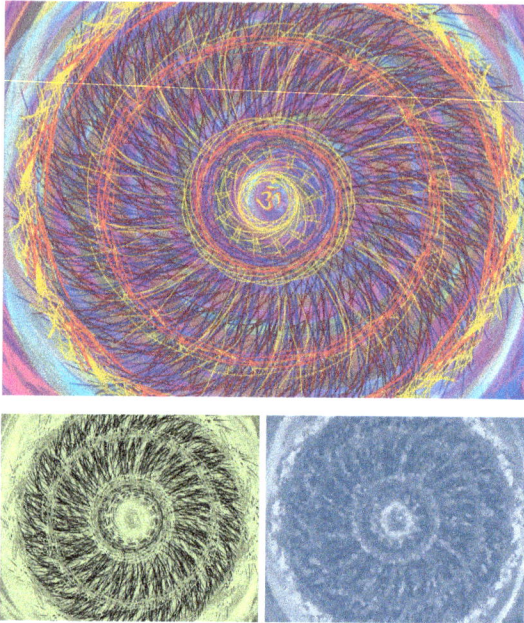

Illustrated with a series of mandalas, each unique and yet based on same principle.

Section VII Figures

Fig 93: Generating the Genetic Code for Life

"This section explores the quantum-level computation and the generation of genetic code that is associated with the emergence of life through the cell, complex human attributes such as thoughts and feelings, and uniqueness of individuality." This is shown here with a rainbow and sun, with life waking up in the dawn, and birds spreading their wings.

Fig 94: PNS -PSNS-SNS

The PNS is shown here with two of its components PSNS (parasympathetic nervous system) and SNS (sympathetic nervous system) - two sides of a weighing balance.

Fig 95: The Emergence of a Quadrumvirate Cell-based Structure

"The emergence of a quadrumvirate cell-based structure specified by four molecular plans is significant in that the genetic code is now being housed in a post material-fabric

structure. It is likely the collective and simultaneous action of nucleic acids, polysaccharides, lipids, and proteins that facilitates this post material-fabric housing ability." Here Light, Space, Time and Emergence are each shown as a separate symbol.

Fig 96: Light's Emergence as Living Cells

"As the Light-Space-Time Emergence equation is iterative, so it can be used to model emergence as it proceeds from simpler four-fold to more complex four-fold manifestations."

This is shown here as several tiny cells / particles of life emerging in different curves. The bright yellow color is used to signify light.

"The generation of the code-segments for space-time-energy-gravity quantization at the time of the Big Bang, the generation of the FBLEE (four-base logic-encoding ecosystem) code-segment and ... the MF (material-fabric) code segment ... are combined together into one equation so that the

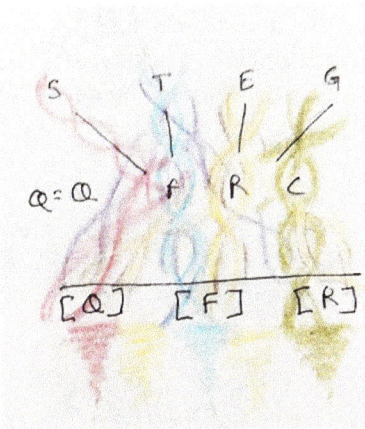

FBLEE process is apparently transparent." Here all the variables - PVMC, STEG, QFR are visualized together in a single illustration.

Fig 98, 99: Active Pregenetic & Genetic Code at the Cellular Stage

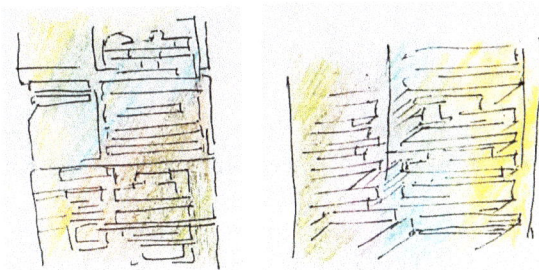

479

The matrices in the section for 'Active Pre-Genetic and Genetic Code at the Cellular Stage' are visualized here literally, as sections / cabinets in a space, storing information packets which may be derived as per a program focused on desired function.

Fig 100, 101, 102, 103: Generation of Cellular Level Genetic Code - Urges, Desires, Wills - As Seen in Common Human Life and Nature

These illustrations depict a slightly different visualization for Urges, Desires, Wills - as seen in common human life and Nature. The urge for shelter, basic necessities, proximity to nature, trees, seashore, sunshine is depicted here in various landscape sketches.

480

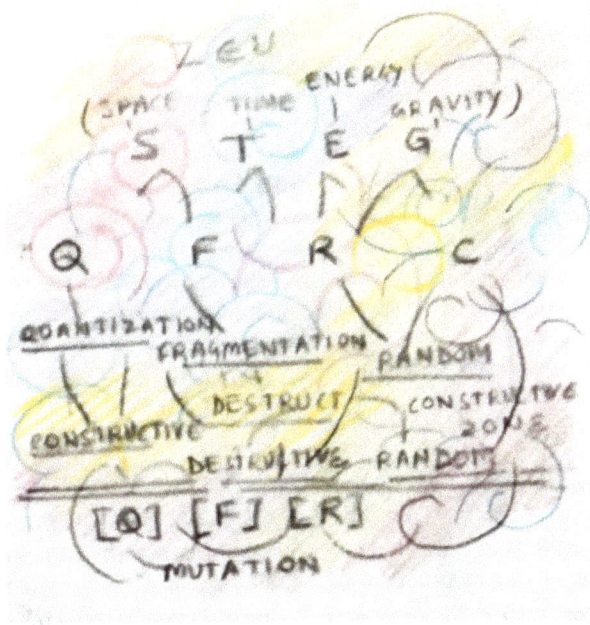

The quadrality, FLBEE, Q F R, S T E G, and all such variables that contribute to the basic genetics' concepts, are seen together and joined with a ray diagram. "At the core individuals can be thought of as projections of seeds formed in a vast continent of Light. So that core must always be there, but it is often covered by surface dynamics of the physical, the vital, and the mental. These too are formed from properties of Light, but since the light has been separated from its source these movements and dynamics are incomplete in their nature."

"…the architecture of truer individuality, elaborates the sets for Presence, Power, Knowledge, and Harmony, each containing multiple elements." Here a different take is attempted for these visualizations. While the basic elements like eyes, nose, mouth, ears remain the same, every animal still looks quite unique and by a subtle change in curve and line is symbolizing a unique type of individuality, depicting knowledge, compassion, fear, inertia, innocence, beauty, brute power and so on. It's amazing to experience this, how Nature in Her evolution creates such variations in the same basic idea - say an idea of creating 'A face of a four legged animal.'

Depicting another variation in the illustration of some formative variables.

Section VIII Figures

Fig 111: Human Welfare, Emergence of Complex Organization and Quadrality

While generating 'Genetic-Code for Complex Organization' - having traced the computations resulting in the emergence of space-time-energy-gravity, electromagnetic spectrum, matter, life, and even human becoming as manifest in sensation, urges, feelings, and thoughts, and truer individuality - the computation resulting in the emergence of complex organizations and possible genetic code related to that is depicted by the continuously enlarging spiral with a square rotation as the variant, to emphasize Quadrality, and as having the human figure illustrated by Leonardo (discussed earlier) at its center.

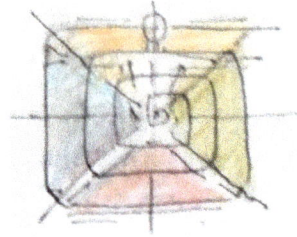

Fig 112: Generation of Cellular-Level Genetic Code for Stable Mega-Organizations

A conventional DNA sketch is drawn here with a variant in which a bold root-and-branch spread is seen in every cell. Further, the sun, moon, and stars are part of the cellular structure depicting the unity in microcosm and macrocosm as revealed in modern science, as well as in ancient Vedanta or its interpretation by Sri Aurobindo. The stable mega organizations will be a logically fitting middle of these two extremes. 'Light's Emergence as Stable Mega-Organization' is seen here in progressive steps – "The emergence of the four properties of Light through the primary fourfold emergence of space-time-energy-gravity, the electromagnetic spectrum, through matter, through life, and even through human becoming as manifest in sensation, urges, feelings, and thoughts, and truer individuality. At each of these levels it seems it is a balanced combination of all four of the properties that creates stability and makes that structure or organism or being, sustainable. And this has to be since the four properties occur simultaneously in light and must be impelled to hold to that relationship even when projected from it."

Fig 113, 114: Sustainability of CAS

The graph of 'Sustainability of Complex Adaptive Systems' is shown here in two variations. The first illustration is connected via green rope in the background of a steeped structure, which has several windows in it like an ancient architecture.

The second illustration takes it even further and projects this graph to resemble a temple, cathedral, mosque, or any place of worship with a pyramid like top - shikhara or gopura or stupa. These remind us also of some deeper connections and "organization, such as is responsible for the architecture and cohesiveness of sustainable mega-organizations, has the possibility of altering FBLEE and post material fabric housing structures so long as the bases involved are driven primarily by a meta-level."

Fig 115: Starting from the Light Matrix

The Light-Matrix specifies the architecture of mega-organization, and this is visualized here as an ascending path with smaller branches emerging like a mighty river giving life and smaller rivulets connected to it.

Fig 116: Generation of Cellular Level Genetic Code

Symbolically shown here as a small spherical entity in which crystallization along the walls has started. These small crystals will soon join each other to enlarge themselves.

Fig 117, 118, 119: Generation of Cellular-Level Genetic Code for Stable Mega-Organizations

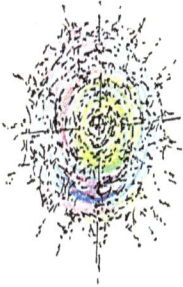

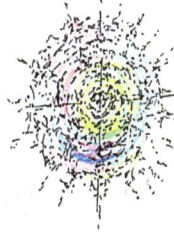

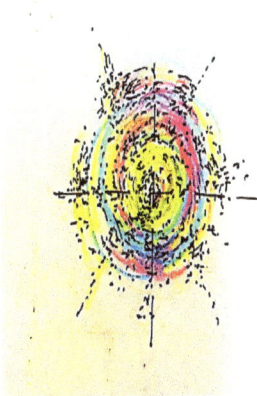

The concepts are shown here by particles moving out in centrifugal forces in a spiral, and ultimately resulting in several possibilities of creating mega level stable organizations. From a galactic level to the aerial view of the city of Auroville, we can see a similar spiral movement in architectural forces in a stable and integral manifestation.

487

Fig 120, 121: Summarizing and Reiterating a Few Points

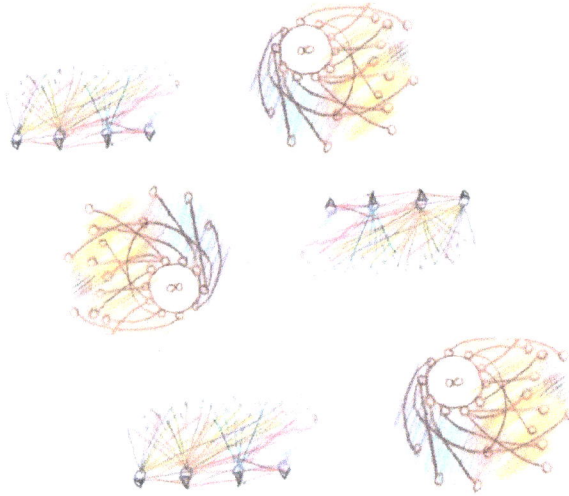

Two of the earlier sketches are rearranged here in three layers of alternate mirroring.

Light's Emergence as Sustainable Global Civilization is shown here as a small cube with light source at one of the apex.

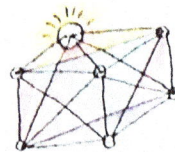

488

The architecture of sustainable global civilization --
"...may awaken from the cumulative Light-constructs already materialized. Such a call may invoke the precipitation of some sustainable-global-civilization meta-function that exists as a mathematical possibility 'above' in some subtle-library. This may be thought of as a key-and-lock mechanism, where a deep enough visceral...." This matrix and layered approach invoking higher powers and their descent, is captured first in its elemental matrix form as if a mechanical system.

Fig 123: Unlocking

The phrase again is depicted here in a literal way of lock and key.

While 'Generating Cellular-Level Genetic Code for Global Sustainable Civilization' the four factors of space, time, energy, and gravity are shown here in four symbols - one in each quadrant.

490

While 'Generating Cellular-Level Genetic Code for Global Sustainable Organization', components of the 'Knowledge-Power- Nurturing- Service Family' are shown separately in separate quadrants. An Indian scholar reading and teaching manuscripts is 'Knowledge', Japanese Samurai taking a stance for attack with a sword in his hand is 'Power', a production of resources and its distribution to all is 'Nurturing', and a service to ensure all facilities kept ready for life of all is 'Service'.

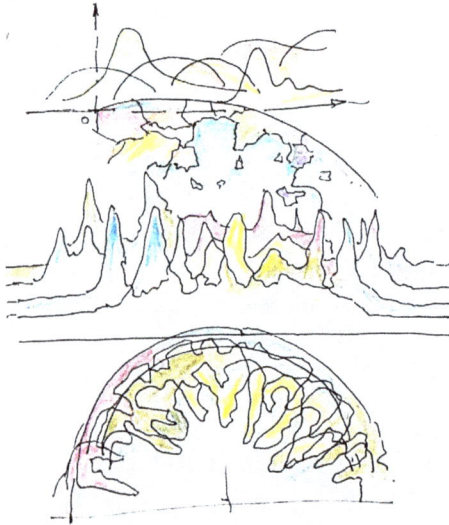

Summarizing, after the cellular stage iterations, including the sustainable global civilization of the Light-Space-Time Emergence equation are complete, and adding the concepts from the psychological model of history by Sri Aurobindo as described in his book 'The Human Cycle' and further adding Fritjof Capra's Turning point model of rise and fall of civilizations, the illustration is brought together in three variants with depiction and dimensions on one axis, two and three axis. The third is shown as on an inner surface of an upper hemisphere.

Section IX Figures

Fig 127, 128: A Practical Look at Genetic Mutation

The two illustrations attempt to capture how genetic mutation may work in a light-based model of reality. We will get its insight by leveraging the discussion on cells to look at how genetic mutation may occur in a case of an adapting insect.

The quantum fluctuation is said to be due to an always buzzing pregnant-infinity, that lies behind everything even when it is seemingly still. This is depicted here as a pregnant female deity / lady / Mother resting on a bed of infinity and surrounded by galaxies while Light in rainbow colors overarches her.

494

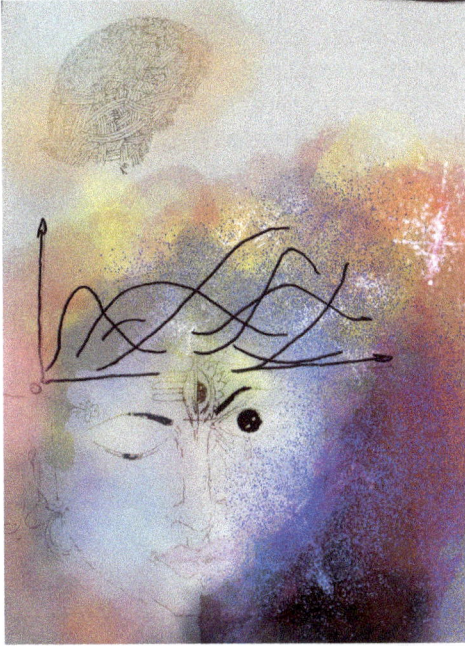

"So while the breaking of patterns as specified by the Space-Matrix was a first step in the process, the subsequent creation of a new meta-function can be thought of as a second step. This new meta-function can be thought of as happening due to some combined functioning of the Light and Space Matrices. ...Recall that space-quantization further structures space so that seeds of knowledge or knowledge-potential to do with the new meta-function are now resident in the material ecosystem associated with the meta-function."

The breaking of patterns and new creation makes one recall 'Rudra' mentioned in ancient Indian texts, and who is also mentioned by Sri Aurobindo while explaining 'The Human Cycle.' This reference was also used in one of the earlier volumes of the Cosmology of Light series, and here it is depicted with more color and digitized effects, to emphasize further the clash of global civilizations, and the contemplation, peace and fury of His three eyes as he balances his dance between creation and destruction.

It was previously mentioned that "...the purpose of a particular new type of specialized cell that exists to protect the insect against predator", and "Recall that energy-quantization allows matter to be formed, and in this case, will have to do with the incremental material changes that the species will go through in forming a hard protective shell...Recall that gravity-quantization has to do with inter-relation between the species and surrounding objects and will change the very nature of gravity to allow a subtle new balance in the species interaction with its surrounding so that the deep urge of the species can more easily be fulfilled."

This very significant idea is explored here in several ways.

1) Here is a sketch wherein a certain flower whose sap is sucked by some insect sitting on its petals, transforms the petal into a tongue and gulps the insect. The transition between vegetable kingdom and animals is part of evolution theory, there are indeed some vegetables who attract and eat insects. Commenting on The Life Divine, The Mother has also mentioned in one of her talks that while evolution of consciousness happens in stages - vegetation, insects, animals, human - there are many in-between species also, which Nature experiments with before completing transformations to the next level.

2) There is also a tale of a species of small fish terrorized by giant fish, that forms a cluster of several small fishes, and together they make a shape resembling a giant fish to counterattack the real giant fish. This call out a transformation - not biological but psychological and on a collective basis.

3) Another extremely significant tale is of the fish incarnation of Vishnu: Matsyavatara. The tale existed from Vedic times to the Puranas and in medieval literature also. Here a tiny fish asks a king for shelter and to save him from the demon in the ocean. The king first saw the fish in his palm when he was offering oblation to the Sun (Light). He takes the fish home, and as he gives him all that is needed, the fish continues to grow. The fish is transferred to a pot, then to a pond, to a lake, to a river, and then finally to the ocean. The fish kills the demon in his half-fish half-Deity form, and then saves the world from

destruction by a giant flood by pulling a boat across the ocean to safe shores where a new creation starts. The same tale appears in the Bible as the story of Noah's arch.

3) In another very significant tale an apparently inanimate pillar when kicked in contempt by a demon, breaks and from it emerges a deity in half lion and half god form:

Narsimha avatar - an incarnation of Vishnu in several scriptures. Here again the silent and helpless oppressed is helped by a new manifestation that counterattacks as the destroyer.

4) In medieval times we have an incident when a place for a new empire's capital was chosen, after seeing a rabbit being chased by a tiger. The rabbit was almost frightened to death and was about to give up. However, as the rabbit crossed a particular line, he got sudden aggressive power, and he turned back to attack the tiger.

5) In another illustration a feeble insect viciously attacked by rising hoods of serpents, turns back to become an eagle and attacks the snakes with vehement and swift movements.

The following illustration is showing the mathematical equations of the Light matrix series.

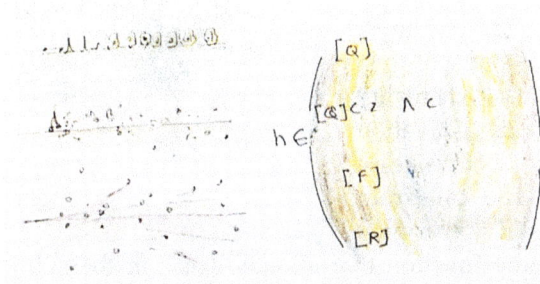

The last illustration in this series is a collage of the preceding examples.

Fig 141: Deeper Nation Essence Case with Sand-Clock or Damaroo of Rudra

"... Assume that there is a deep wanting felt by a threshold number of people. Perhaps this wanting, a step towards a global sustainable civilization, is for each nation to operate more from their true essence, or I-state, as opposed to fundamentalist form. In other words, a meta-function,

$Sig_{Nation-essence}$, must be created that will then organize its own reality ... This implies activation and potential change of FBLEE. The call from below, as it were, may invoke some function that already exists in the subtle-libraries...." This is depicted here in a form of meeting of two hemispheres, or sand-clock, which also resembles a damaroo - a percussion instrument in the hands of the Rudra aspect of Shiva. In the other view it looks like the continuous enlarging spiral from a nucleus manifesting in innumerable ways and directions.

500

Section X Figures

Fig 142: Super-Matter and a Possible Future of Genetics

While summarizing the previous sections and drawing insights into the nature of matter and genetics itself, there is a culmination in a general equation for future genetic-type code, and a reflection on genetics itself. Super-matter can be thought of as the result of a conscious will and cohesive wanting that causes a deliberate process of quantization by which the very fabric of matter is changed. Quantization occurs in any case even when Nature's automatic working drives change. This is visualized as free-flowing, luminous, and graceful lines forming a figure of a feminine deity with spread hands, ready to take flight from the springboards of matter to the heights of Super-matter.

Fig 143: To Super-Matter & Post Genetic Structure

The transformation to super-matter can be accelerated, broadened, and heightened, so that a wider and higher possibility of function may consciously materialize. The nature of Matter, dark matter and future of genetics is explored now while acknowledging the Complexity of Matter, Dark Matter, & Dark Energy. "The will for continued functional-richness may lead to continuous quantization, space expansion, super-matter, and a new Ouroboros of eternal return."

This is visualized with the rise from the bottom layer, through the conical ascent to the middle zone, and again goes in expansion in the upper layer. Such a figure resembles many ideas and images such as a sand-clock, pipal tree, and several other symbols reminding us of some insights of new physics, of Cosmology of Light and also of ancient mystic traditions especially those of India. Apart from a route which is straight up, there is also a route from the side, through a curve which meets the upper layer. This signifies that there are nonlinear solutions, more than one path to the same destination, of going beyond stereotypes, of circumventing a problem artistically / creatively when logic stumbles. It is a passage to super-matter, by taking higher flight from levels of matter and mind. The grid structure resembles meshes in finite element analysis, also showing the importance of elements and relations in the whole.

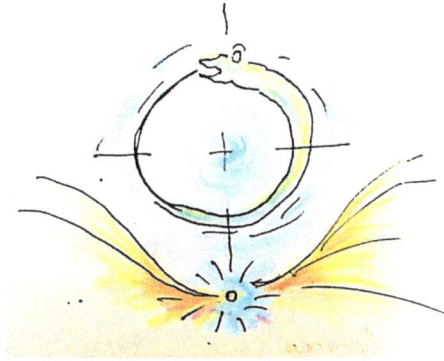

Fig 144: Such is the Future Tale of Ouroboros

A New Ouroboros is discussed here, wherein a snake is eating its own tail. It is preceded and succeeded by the curves of transformation whose junction point coincides with the central axis of the ouroboros.

Fig 145:   Can Alter the Development of 30 Orders of Magnitude

"Through will and other human faculties continued states of quantum-certainty resulting in four-fold quantization can occur.   Such quantization changes the very nature of space, time, energy, and gravity and can alter the development of the 30 orders of macrocosmic magnitude."

Visualized here as a new ouroboros – a snake eating its
own tail. A pair is seen rising up in a helix formation -
which on closer examination appears as a double-helix -
ascending from the dark bottom of the inconscient to a
luminous future.

Fig 147: The Origins and Possibilities of Genetics

Multiple layers of a double helix are seen in gradual growth as depicted on the left, with more sophisticated stages illustrated on the right.

Relevant Background and Follow-up Information

The Author's Early Books

1. The Flowering of Management
2. India's Contribution to Management

The Fractal Series

1. Connecting Inner Power with Global Change: The Fractal Ladder
2. Redesigning the Stock Market: A Fractal Approach
3. The Flower Chronicles: A Radical Approach to Systems and Organizational Development
4. The Fractal Organization: Creating Enterprises of Tomorrow

The Cosmology of Light Series

1. A Story of Light: A Simple Exploration of the Creation and Dynamics of this Universe and Others
2. Oceans of Innovation: The Mathematical Heart of Complex Systems
3. Emergence: A Mathematical Journey from the Big Bang to Sustainable Global Civilization
4. Quantum Certainty: A Mathematics of Natural and Sustainable Human History
5. Super-Matter: Functional Richness in an Expanding Universe
6. Cosmology of Light: A Mathematical Integration of Matter, Life, History & Civilization, Universe, and Self

The Application of Cosmology of Light Series

1. The Emperor's Quantum Computer: An Alternative Light-Centered Interpretation of Quanta, Superposition, Entanglement and the Computing that Arises from it
2. The Origins and Possibilities of Genetics: A Mathematical Exploration in a Cosmology of Light
3. The Second Singularity: A Mathematical Exploration of AI-Based and Other Singularities in a Cosmology of Light
4. Triumph of Love: A Mathematical Exploration of Being, Becoming, Life, and Transhumanism in a Cosmology of Light

The Artistic Interpretation of Cosmology of Light Series

1. The Mandala Illustrated Story of Light
2. Musings on Light: A Meditative, Non-Mathematical Summary of a Cosmology of Light
3. The Illustrated Oceans of Innovation: The Mathematical Heart of Complex Systems Depicted in Indian Arts
4. Emergence Illustrated: A Mathematical Journey from the Big Bang to Sustainable Global Civilization Depicted with Indian Mythological Arts
5. The Dawn of Flame-Beings: Mythological Musings Based on a Cosmology of Light
6. Quantum Certainty Illustrated: A Mathematical Journey Through Natural and Sustainable Human History Depicted with Art
7. Super-Matter Illustrated: Functional Richness in an Expanding Universe Depicted with Drawings

8. Cosmology of Light Illustrated: A Mathematical Integration of Matter, Life, History Depicted with Art
9. The Emperor's Quantum Computer Illustrated: An Alternative Light-Centered Interpretation of Quanta, Superposition, and Entanglement and the Computing that Arises from it Illustrated with Art
10. The Origin and Possibilities of Genetics Illustrated: A Mathematical Exploration in a Cosmology of Light

Note on Genesis of Books

In the earlier stage I wrote 'The Flowering of Management' and 'India's Contribution to Management'. The impetus for both these books was similar in that they were reactions to the environment that I was placed in at the time. When I first began working in the corporate world the reality of the environment struck me as decidedly anachronistic. I had a different sense of what life could offer and wrote 'The Flowering of Management' to capture aspects of a vision I thought corporations and money existed for. Similarly, when I wrote 'India's Contribution to Management' it was the result of the dissatisfaction I experienced when confronted with the prevalent interpretation of India. This was precipitated by my working with a US-based company, A.T. Kearney, in India. I sought to reverse that interpretation with 'India's Contribution to Management' which aimed to capture my understanding of the essence and deeper capacity of synthesis of India.

The next phase was marked by a strong interest in fractals that primarily stemmed from my beginning to see similar patterns in seemingly distinct areas of life. I wrote

'Connecting Inner Power with Global Change: The Fractal Ladder' as a theoretical framework of fractals. The fractals that I envisioned included the added complexities of emotional and thought components. This was followed by 'Redesigning the Stock Market: A Fractal Approach' which was an application of the theoretical fractal framework to the then recent global financial crises of 2008. 'The Flower Chronicles' sought to make the gist of the ubiquitous fractal I had described in the previous two books easily graspable at the visceral level primarily through many practical examples spanning diverse walks of life. 'The Fractal Organization: Creating Enterprises of Tomorrow' was a comprehensive summary of the fractal framework that included the basic theory, the applications, and a practical field guide that had been developed while I was working at the Organizational Development group at Stanford University Medical Center.

The most recent phase has focused on creating a mathematical framework to take the previously developed fractal framework further. The development of such a mathematical framework that seeks to frame innovation in complex adaptive systems was also the focus of my doctoral work. This gave birth to an exciting period and will result in multiple series of books.

The first series, comprising of six books, extended my inquiry into mathematics and complex adaptive systems to an interesting limit culminating in the nature of Light and the Cosmos. The fractal mathematics I propose is at the heart of this series: Cosmology of Light.

The first book, 'A Story of Light: A Simple Exploration of the Creation and Dynamics of this Universe and Others' contains the main ideas in the mathematics, in non-mathematical terms, that are further explored mathematically in the remaining books in this series. The second book 'Oceans of Innovation: The Mathematical Heart of Complex Systems' describes my interpretation of the mathematical foundation of complex systems. The third book, 'Emergence: A Mathematical Journey from the Big Bang to Sustainable Global Civilization' applies the mathematics to several layers of matter and life. The fourth book, 'Quantum Certainty: A Mathematics of Natural and Sustainable Human History' describes a process culminating in space, time, energy, and gravity quantization by which history is made. The fifth book, 'Super-Matter: Functional Richness in an Expanding Universe' describes a process for the creation of super-matter-based on a need for continued functional-richness. A link is made between the resulting quantization of space and an expanding universe. The final book, 'Cosmology of Light: A Mathematical Integration of Matter, Life, History & Civilization, Universe, and Self' proposes an integrated mathematical framework that flows through all things, hence unifying matter, light, civilization, history, universe, and self.

The second series further explores the implications of "one mathematics flowing through all things". The first book in this series 'The Emperor's Quantum Computer: An Alternative Light-Centered Interpretation of Quanta, Superposition, Entanglement and the Computing that Arises from it' describes an alternative narrative for quantum computing to the one commonly expressed today. The second book in the series, 'The Origin and Possibilities of Genetics: A Mathematical Exploration in a Cosmology of Light' explores pre-genetic, genetic, and post-genetic possibilities in a cosmology of light. This book, 'The Second Singularity: A Mathematical Exploration of AI-Based and Other Singularities in a Cosmology of Light' explores the limits of AI-based singularities with respect to light-based singularities. This book, the final in this series explores transhumanism in a cosmology of light.

The third series, Artistic Interpretation of Cosmology of Light, is intended to make Cosmology of Light more accessible by interpreting it artistically. The first book in the series is 'The Mandala Illustrated Story of Light'. The objective is to lead the reader through the story of light using mandalas as an aid in the journey. The second book 'Musings on Light' is a meditative book set to graphical illustrations. The illustrations focus on 50 key concepts derived from the ten-book joint Cosmology of Light series. The third book 'The Illustrated Oceans of Innovation: The Mathematical Heart of Complex Systems Depicted in Indian Arts' uses Indian Arts to illustrate the mathematical heart of complex systems. The fourth book 'Emergence Illustrated: A Mathematical Journey from the Big Bang to Sustainable Global Civilization Depicted with Indian Mythological Arts' uses Indian mythological arts to express the concepts of Emergence. The fifth book 'The Dawn of Flame-Beings: Mythological Musings Based on a Cosmology of Light' uses illustrations to help focus on the mythological aspects of a cosmology of light. The sixth book is 'Quantum Certainty Illustrated: A Mathematical Journey of Natural and Sustainable Human History Depicted with Art' that leverages pencil drawings and other art to shed insight into the mathematical development of history. The seventh book, 'Super-Matter Illustrated: Functional Richness in an Expanding Universe Depicted with Drawings,' suggests the destiny of matter and its implicit relationship with an expanding universe. The pencil drawings provide visceral insight into the concepts and mathematics in the book. The eighth book is 'Cosmology of Light Illustrated: A Mathematical Integration of Matter, Life, History Depicted with Art'. The ninth book is 'The Emperor's Quantum Computer Illustrated: An Alternative Light-Centered Interpretation of Quanta, Superposition, and Entanglement and the Computing that Arises from it Illustrated with Art'. The tenth and current book is 'The

Origins and Possibilities of Genetics Illustrated: A
Mathematical Exploration in a Cosmology of Light.

About the Author

Dr. Pravir Malik is the founder and chief technologist at
QIQuantum and the Forbes Technology Council group
leader for Quantum Computing.

He is a systems thinker and approaches computation at
the quantum level by seeing the layers of matter and life
as a single system. This perception has resulted in the
articulation of a fractal model connecting patterns at the
cellular, atomic, and quantum particle levels to root
patterns at the quantum level and has been elaborated
through a ten-volume mathematical treatise on a
cosmology of light, and numerous IEEE technical articles.

Dr. Malik has a patent pending on a fourfold fractal-based
quantum computer. Dr. Malik's quantum computational
theory derives from a unified theory and mathematics of
organization developed over a couple of decades. In all,
he has penned over 25 books related to this. In his
doctoral dissertation on the mathematics of innovation in
complex adaptive systems, he created a top-down in
contrast to the already existing bottom-up mathematics
for complex systems. Beyond its application in quantum
systems, this mathematics has been practically applied to
the development of pricing systems, emotional
intelligence systems, and AI, and theoretically applied to
the areas of genetics, pharmaceutical technology,
nanotechnology, space sciences, and computational
theory.

Dr. Malik has funded activities at QIQuantum through
modeling behavioral, organizational, and other complex
systems to help different stakeholders practically
navigate and understand possible futures. In his capacity

as the Forbes Technology Council group leader for Quantum Computing, he has led numerous quantum computing events focused on alternative trajectories for quantum computation. The group that he helped found earlier in 2023 has now grown to over 140 executives belonging to Forbes Council member companies.

During the pandemic, he led a multi-part Organizational Sciences Certification program for Forbes, which was attended by executives from 250 companies. This certification was based on his views of complex systems, cosmology of light, and the importance of maturing teams through emotional intelligence. Dr. Malik believes that each individual is endowed with a sophisticated EQ radar-system that can give profound insight into how complex systems are changing. He developed a practical EQ-based App that has been used by the Indian Armed Forces, Zappos, Stanford University Medical Center, and Nucor, amongst other organizations, to help teams reach higher levels of self-directed operational maturity.

Previously, Dr. Malik has held a number of diverse leadership positions. These include being Head of Organizational Sciences and Head of Pricing Operations at Zappos.com - an eCommerce company, Managing Director of BSR - a global sustainability and environmental consulting company, and a member of the Founding Team of A.T. Kearney India - a global operations-focused management consulting company.

He has a Ph.D. in Technology Management with a focus on Mathematics of Innovation in Complex Adaptive Systems from the University of Pretoria, an MBA from Northwestern University's J.L. Kellogg Graduate School of Management, an MS in Computer Science from the University of Florida with a focus on AI, and a BSE in Computer Engineering from the Case Western Reserve University.

Pravir is a global citizen who has worked, lived, and been educated in many parts of the world.

Dr. Narendra has three decades of extensive experience in academics, industry, projects, grassroot work on social and cultural issues, editing magazines and journal & his recent passion has been in multidisciplinary research and its artistic expressions.

He has completed his graduation and post-graduation in Production engineering from University of Mumbai and ranked first in his Master's program. He completed several technology, management and entrepreneurship certifications from reputed institutes like IIT Bombay, IIM Ahmedabad, ISB Hyderabad, NEN-STVP and IIM B. He completed his Ph.D. from Hindu University of America. His thesis was on an unusual subject: 'A study of philosophy and Futurology of Artificial Intelligence in the light of Sri Aurobindo's integral thought.'

Inspired by the message of Swami Vivekananda and Sri Aurobindo for future of India and humanity and intrigued by the shift in foundations of Physics and the future shock, he switched over from his industrial career and worked as full-time dedicated worker for a leading NGO, Vivekananda Kendra, Kanyakumari and was posted in Northeast India in various remote areas of Assam and Arunachal Pradesh after extensive training in Kanyakumari for one year. Perplexed by the complex socioeconomic developmental especially cultural and human problems there, he started studying works of Sri Aurobindo more deeply. He was given a research project which was later published as a book named 'Ashwattha' by Vivekananda Kendra. This book has hundreds of illustrations by him and since then he has written many papers for conferences, journals, and magazines on diverse topics from technology, whole brain thinking, creativity and innovation, management, Indian arts, consciousness studies, history, evolution of consciousness, appropriate technology and sustainability, science and technology in Ancient India, Indian ethos in management, Indian philosophy, psychology and Sri Aurobindo studies. His paper on Indian ethos in management won the award from BMA for best paper for the year. He has served in industry and in academics holding various key positions including principal, project head, academic consultant, etc. He was benefitted in this journey by association and guidance from many experts.

Dr. Narendra has worked as Principal for eight years and faculty for nine years in engineering fields and he is proactive in assessing training & development needs and effectively aligning programs / interventions with business /social objectives. He is deft in designing innovative programs for industry as well as academics while catering to emerging industry needs. He has a distinguished inclination to social responsibility exemplified in interdisciplinary research projects, field work and development of innovative projects dedicated to integral and futuristic studies and also for discovering the underlying currents of culture and psychology in it. At present he is working as Project Director for Vivekananda Prabodhini, Mumbai. He also provides Academic consultancy especially for distant, digital and open university models wherein students from underprivileged and remote areas have urge to learn further.

However more than all these things, he is an artist by heart and has developed passion for sketching, painting in traditional as well as digital art over the years. His sketches are now part of several magazines, blogs, books and forms. He has found that illustrating a text helps in deeper thought expression and provides further insights especially for path breaking and transdisciplinary subjects such as this book.

Selected Author Online Presence

- Amazon Author Page: https://www.amazon.com/Pravir-Malik/e/B002JVAEZE
- LinkedIn Profile: https://www.linkedin.com/in/pravirmalik/
- Forbes Profile: https://councils.forbes.com/profile/Pravir-Malik-Chief-Technologist-QIQuantum/2258eb58-6350-495f-8d26-16581994fcc4
- Google Scholar Page: https://scholar.google.com/citations?user=7DWWWZ8AAAAJ&hl=en
- IEEE Profile: https://ieeexplore.ieee.org/author/37086022058
- YouTube Page: https://www.youtube.com/user/Aurosoorya
- Twitter: https://twitter.com/PravirMalik
- Research Gate Profile: https://www.researchgate.net/profile/Pravir_Malik
- Eventbrite: https://www.eventbrite.com/o/pravir-malik-30159112262
- Medium: https://medium.com/@PravirMalik
- Company website: http://www.deepordertechnologies.com/

REFERENCES

1. Arabatzis, T. 2006. Representing Electrons: A Biological Approach to Theoretical Entities. University of Chicago. Chicago
2. Beinhocker, E. 2006. The Origin of Wealth. Boston: Harvard Business School Press.
3. Chown, M. 1990. Can Photons Travel Faster Than Light? New Scientist 126(1711)
4. Cottingham, W & Greenwood, D. 2007. An Introduction to the Standard Model of Particle Physics. Cambridge University Press. Cambridge
5. Deppe, A. 2013. Therapy with Light, a Practitioner's Guide. Strategic Book Publishing.
6. Diamond, J. 2005. Collapse: How Societies Choose to Fail or Succeed. Viking Books: New York
7. Young, E. 2016. I Contain Multitudes: The Microbes Within Us and a Grander View of Life. New York: HarperCollins.
8. Einstein, A. 1995. Relativity: The Special and General Theory. New York: Broadway Books.
9. Feynman, RP. 1985. QED The Strange Theory of Light and Matter. New Jersey: Princeton University Press
10. Fuller, B. 1982. Synergetics: Explorations in the Geometry of Thinking. MacMillan Publishing Co.: New York
11. Goodsell, David. 2010. The Machinery of Life. New York: Springer
12. Gottlieb, M. 2013. The Feynman Lectures on Physics: III. California Institute of Technology. http://www.feynmanlectures.caltech.edu/III_1 6.html
13. Gray, T. 2009. The Elements: A Visual Exploration of Every Known Atom in the Universe. Black Dog & Levental Publishers. New York.

14. Gubser, S. 2010. The Little Book of String Theory. Princeton University Press

15. Harvard-Edu. 2014.
    http://news.harvard.edu/gazette/story/2014/04/harvard-to-sign-on-to-united-nations-supported-principles-for-responsible-investment/

16. Harvard-Smithsonian Center for Astrophysics. 2004.
    https://www.cfa.harvard.edu/seuforum/de_whatmight.htm#

17. Hawking, Stephen. 1988. A Brief History of Time. New York: Bantam Books

18. Heiserman, D. 1991. Exploring Chemical Elements and their Compounds. McGraw-Hill. New York.

19. Holland, P. 1995. The Quantum Theory of Motion: An Account of the de Broglie-Bohm Causal Interpretation of Quantum Mechanics. Cambridge: Cambridge University Press.

20. Hope, Chris; Fowler, Stephen J. (2007). "A Critical Review of Sustainable Business Indices and Their Impact". Journal of Business Ethics. Vol. 76. S. 243–252. Springer: New York

21. Hyperphysics. 2016. Department of Physics and Astronomy. Georgia State University.
    http://hyperphysics.phy-astr.gsu.edu/hbase/mod3.html#c1

22. Isaacson, W. 2008. Einstein: His Life and Universe. Simon and Schuster. New York.

23. Jeans, J. 1932. The Mysterious Universe. Cambridge University Press.

24. Kaufmann, S. 1995. At Home in the Universe. New York: Oxford University Press.

25. Kaufmann, S. 2003. The Adjacent Possible, Edge.
    https://www.edge.org/conversation/the-adjacent-possible

26. Lloyd, S. 2007. Programming the Universe: A Quantum Computer Scientist Takes On the Cosmos. New York: Vintage

27. Logue, A. 2008. Socially Responsible Investing for Dummies. Wiley Publishing: New Jersey

28. Lorentz, H.A. 1925. The Science of Nature. Vol. 25, p 1008. Springer

29. Malik, P. 2009. Connecting Inner Power with Global Change: The Fractal Ladder. New Delhi: Sage Publications

30. Malik, P. 2015. The Fractal Organization. New Delhi: Sage

31. Malik, P. 2017a. Doctoral Thesis. University of Pretoria Graduate School of Technology Management. https://repository.up.ac.za/handle/2263/62779?show=full

32. Malik, P. 2017b. A Story of Light. Amazon Kindle.

33. Malik, P. 2017c. Oceans of Innovation. Amazon Kindle.

34. Malik, P. 2017d. Emergence. Amazon Kindle.

35. Malik, P. 2017e. Quantum Certainty. Amazon Kindle.

36. Malik, P, Pretorius, L, Winzker, D. 2017. Qualified Determinism in Emergent-Technology Complex Adaptive Systems. IEEE TEMSCON.

37. Malik, P. 2018a. Super-Matter. Amazon Kindle.

38. Malik, P. 2018b. Cosmology of Light. Amazon Kindle.

39. Malik, P. 2018c. The Emperor's Quantum Computer. Amazon Kindle.

40. Malik, P, Pretorius, L. 2018. Symmetries of Light and Emergence of Matter. Indian Journal of Science & Technology. http://www.indjst.org/index.php/indjst/article/view/110789.

41. Martel, A. 2018. Light Therapies: A Complete Guide to the Healing Power of Light. Rochester, Vermont: Healing Arts Press

42. Mora, C, Tittensor, D, Adl, S, Simpson, A, Worm, B. 2011. How Many Species Are There on Earth and in the Ocean? Plos.org. http://journals.plos.org/plosbiology/article?id=10.1371/journal.pbio.1001127#abstract0

43. Narby, J. 1998. The Cosmic Serpent: DNA and the Origins of Knowledge. New York: Penguin Putnam.

44. NASA-darkmatter. 2016. http://science.nasa.gov/astrophysics/focus-areas/what-is-dark-energy/

45. NASA-supernova. 2001. News Release Number: STScI-2001-09: "Blast from the Past: Farthest Supernova Ever Seen Sheds Light on Dark Universe". http://hubblesite.org/newscenter/archive/releases/2001/09/

46. NASA-WMAP, 2014. https://map.gsfc.nasa.gov/media/121238/index.html

47. Nowak, M., Highfield, R. 2011. Super Cooperators: Altruism, Evolution, and Why We Need Each Other to Succeed. New York: Free Press.

48. Olive, K.A et al. 2014. Particle Data Group. Chin. Phys. C, 38, 090001.

49. Openshaw, J. 2015. http://www.marketwatch.com/story/socially-responsible-investing-has-beaten-the-sp-500-for-decades-2015-05-21

50. Panek,R. 2010. Dark Energy: The Biggest Mystery in the Universe. Smithsonian Magazine. http://www.smithsonianmag.com/science-

nature/dark-energy-the-biggest-mystery-in-the-universe-9482130/?no-ist

51. Parker, A. 2003. In the Blink of an Eye: How Vision Sparked the Big Bang of Evolution. New York: Basic Books.

52. Particle Data Group. 2015. Lawrence Berkeley National Laboratory. http://www.cpepphysics.org/main universe/universe.html

53. Pauli, W. 1964. Nobel Lectures, Physics 1942 – 1962. Elsevier Publishing Company. Amsterdam.

54. Perkowitz, S. 2011. Slow Light. London: Imperial College Press

55. Planck, M. 1933. Where is Science Going? Ox Bow Press. Connecticut.

56. Portugali, J., 2012. *Self-organization and the city*. New York: Springer Science & Business Media.

57. Prigogine, I. 1977. Time, Structure, and Fluctuations. *Nobelprize.org.* Nobel Media AB 2014. Web. 5 Mar 2016. <http://www.nobelprize.org/nobel_prizes/chemistry/laureates/1977/prigogine-lecture.html>

58. Ridley, M. 1999. Genome: The Autobiography of a Species in 23 Chapters. New York: Harper Perennial.

59. Rovelli, C. Reality Is Not What It Seems. New York: Riverhead Books

60. Salkind, N. 2007. Encyclopedia of Measurement and Statistics. Thousand Oaks: Sage Publications.

61. Snyder, M. 2010. Stanford Medicine. http://med.stanford.edu/news/all-news/2010/03/what-makes-you-unique-not-genes-so-much-as-surrounding-sequences-says-stanford-study.html#.html

62. Spitler, H. 2011. The Syntonic Principle: In Relation to Health and Ocular Problems. Eugene, Or: Resource Publications.

63. Sri Aurobindo. 1971. Social and Political Thought. Sri Aurobindo Ashram Press: Pondicherry

64. Stewart, Ian. 2012. In Pursuit of the Unknown. Basic Books. New York.

65. Smith, J, Szathmary, E. 1995. The Major Transitions in Evolution. Oxford: Oxford University Press

66. Toynbee, A. 1961. A Study of History, Volumes I – XII. Oxford University Press: Oxford

67. Turok, N. 2012. The Universe Within. Anasi Press: Ontario

68. Tweed, M. 2003. Essential Elements: Atoms, Quarks, and the Periodic Table. Walker & Copmany. New York.

69. Van Obbergen, P. 2014. Traite de Couleur Therapie Pratique. Paris: Guy Tredaniel Editeur.

70. Wheeler, J. Ford, K. 2000. Geons, Black Holes, and Quantum Foam: A Life in Physics. New York: W. W. Norton & Co.

71. Whitaker, A._2006. Einstein, Bohr and the Quantum Dilemma: From Quantum Theory to Quantum Information. Cambridge: Cambridge University Press

72. Wilczek, F. 2016. A Beautiful Question: Finding Nature's Deep Design. New York: Penguin Books

73. Willis, A. 2003. The Role of the Global Reporting Initiative's Sustainability Reporting Guidelines in the Social Screening of Investment. Journal of Business Ethics. Volume 43. Springer: New York

74. Wimmel, H. 1992. *Quantum Physics & Observed Reality: A Critical Interpretation of Quantum Mechanics*. World Scientific. p. 2. Bibcode:1992qpor.book.....W. ISBN 978-981-02-1010-6.

75. Wolchover, N. 2013. A Jewel at the Heart of

Quantum Physics. Quanta Magazine .
https://www.quantamagazine.org/20130917-a-jewel-at-the-heart-of-quantum-physics/

76. Wright, R. 2009. Evolution of Compassion.
https://www.ted.com/talks/robert_wright_the_evolution_of_compassion/transcript?language=en. TED 2009.

77. Yates, F.E. 2012. *Self-organizing systems: The emergence of order*. New York: Springer Science & Business Media.